# Mechanization of Horticulture

**Suresh Chander Mandhar** passed his B.Sc. in Agricultural Engineering from Punjab Agricultural University, Ludhiana in 1972 and M.Tech in Design Engineering with Distinction from Indian Institute of Technology, Delhi in 1975. He started his carrier during 1974 at Small Industry Extension Training Institute in Hyderabad and served as Assistant Engineer in Farm Machinery Testing Centre, Department of Farm Power and Machinery, Punjab Agricultural University, Ludhiana from 1975-1980. He tested seed cum fertilizer drills, potato planters, sprayers, potato harvesters and threshers for wheat and paddy. He also surveyed the manufacturing technologies of farm machinery in Punjab during this period. He was also teaching Farm Power and Machinery and Mechanical Engineering Courses.

He was appointed Scientist (S-2) Farm Machinery and Power at Indian Institute of Horticultural Research, Bangalore during 1980 and he served Indian Institute of Horticultural Research (ICAR), Banglore for about 33 years and retired as Principal Scientist and Head, Section of Agricultural Engineering. He believed in end to end mechanization of horticultural crops. To attain these he had developed vegetable nursery raising machinery and transplanter for raised bed cultivation of vegetable crops. He also developed seed drill for sowing onion under rainfed cultivation and weeder for raised bed cultivated crops.

The harvesters for mango, sapota, guava and lime have been developed. For mechanisation of mango pickle industry the peeler, slicer and cube cutter has been developed. For mechanization of garlic pickle industry the bulb breaker and garlic peeler has been developed. He also planned and developed the complete campus of Indian Institute Horticultural Research, Bangalore. The rain water harvesting system, irrigation system and drainage systems were developed.

S.C. Mandhar is life member of Indian Society of Agricultural Engineers, Society for Promotion of Horticulture and Institute of Agriculture Technologists, Bangalore. He has received "Dr. J.S. Pruthi Award 1999 and Kejriwal Award 2005"from Association of Food Scientists and Technologists, India for his innovative work on development of machinery for mechanisation of mango and garlic pickle industry. He also received Team Award from Indian Society of Agricultural Engineers for his work on development of machinery for vegetable nurseries. He had been member of Bureau of Indian Standards, Sectional Committee, FAD 20 for Standardization of Agricultural Produce Processing and Milling Machinery and Surface Covered Cultivation Structures. He was also member of National Horticultural Board delegation to visit "EIMA International Agriculture and Gardening Machinery Exhibition" from November 7-11, 2012 at Bologna, Italy. His present interest is end to end mechanization of cultivation of horticultural crops.

# Mechanization of Horticulture

**S.C. Mandhar**
Formerly Principal Scientist & Head
Section of Agricultural Engineering
Indian Institute of Horticultural Research
Bangalore

**2014**
**Daya Publishing House®**
*A Division of*
**Astral International Pvt. Ltd.**
**New Delhi - 110 002**

ISBN 9789351302650



*Published by* : **Daya Publishing House®**
A Division of
**Astral International Pvt. Ltd.**
– ISO 9001:2008 Certified Company –
4760-61/23, Ansari Road, Darya Ganj
New Delhi-110 002
Ph. 011-43549197, 23278134
E-mail: info@astralint.com
Website: www.astralint.com

*Laser Typesetting* : **Classic Computer Services**, Delhi - 110 035

*Printed at* : **Replika Press Pvt. Ltd.**

PRINTED IN INDIA

*This book is dedicated to.............*

## Dr. K.L. Chadha

Ph.D (Hort.), D.Sc.(Honoris Causa), FNAAS, FHSI
(Padma Shri Awardee)
Father of Modern Horticulture in India
Former Deputy Director General, ICAR, New Delhi
Horticulture Commissioner, Government of India
Executive Director, National Horticulture Board, India &
Director, Indian Institute of Horticultural Research, Bangalore

*For his 50 years of invaluable service and extra-ordinary leadership to horticultural renaissance of India and the inspiration & guidance in bringing out this volume on Mechanization in Horticulture.*

**Prof. S.R. Verma**
Ph.D.(Ag. Engg.), FNAAS, FIE, FISAE
Ex-Dean, College of Agricultural Engineering
Punjab Agricultural University, Ludhiana

14, Goods Friend Colony
Barewal Road,
Ludhiana-141012 (Punjab)
R: R:0161-2551096, M:09872092597
Email: srverma10@yahoo.com

# Foreword

Our Country has made impressive strides in boosting agriculture production and productivity through her well-orchestrated Green Revolution beginning mid-60's. The country achieved food production of 255 M-tonnes during 2011-12. The 3-4 fold increase in agricultural production has been realized through introduction of high yielding seed varieties, use of optimum water, energy and chemicals in higher and balanced doses apart from appropriate Govt. polices and incentives. Despite many positive Indian agriculture is presently faced with many problem and challenges such as decline in factor productivity, higher cost of production low quality of produce as well as degradation and depletion of natural resources. The climate change has further aggravated the situation. Diversification of agriculture is being recommended to alleviate some of these problems. Horticulture is among the key areas for effecting diversification of Indian agriculture. A National Horticultural Mission has been launched in the country. Presently the status of horticultural mechanization for critical operations such as planting/transplanting, plant protection, harvesting as well as post-harvest handling, processing and value addition for different fruits, vegetables, floriculture, spices etc. is quite low.

The book entitled *"Mechanization of Horticulture"* authored by Er. S.C. Mandhar is a pioneering effort to collate and provide the information in one compilation. It deals with different operations for production and post-harvest processing and management of horticultural crops and commodities. This book is essentially a Monograph which has been divided into 16 chapters. Chapters I and II summarise the current horticultural scenario as well as status of agricultural mechanization in India. Very useful information has been provided in the text to enable the reader to understand the agronomic practices and soil conditions for growing of horticultural crops. Chapter III through XII deal with various operations, *viz.* seed bed preparation, planting, nursery raising, soil working tools, transplanting of seedlings weeding and mulching, pruning, spraying and harvesting. Information regarding different types of equipment available in India and abroad has been included in these chapters. The text amply illustrated.

Thus, it will serve as a useful reference book for teachers, researchers, extension workers, farmers, students as well as other state holder. The last three chapters *viz* chapters XIII through XVI deal with packaging of horticultural produce post harvest management and storage, green house design and construction as well as irrigation systems.

Er. Mandhar has taken pains to collect relevant information regarding different types of gadgets and equipment for mechanization of horticultural crops. It is hoped that the information provided in this Monograph will accelerate the pace of horticultural mechanization through speedy R&D, technology transfer as well as commercial production of different types of equipment in India. It would be desirable if in the subsequent edition of this Monograph, recommendations regarding each horticultural crop and related farming systems could be provided in a package form.

The efforts of Er. Suresh Chander Mandhar for compiling this Monograph deserve to be commended. I extend my best wishes for his praiseworthy contribution.

**Prof. S. R. Verma**

# Preface

India was importing wheat and milk powder during 1960's as sufficient production of food was not there. The author has seen the green revolution happening through adoption of improved and dwarf varieties of wheat and paddy in Punjab and making of India self sufficient in food. He has also seen the adoption of animal drawn disc harrows and seed cum fertilizer drills for cultivation of wheat. The manufacture and adoption of tractors and tractor operated disc harrows, cultivators and seed cum fertilizer drills increased area under cultivation in Punjab. The mechanization of wheat was almost completed by use of threshers and combine harvesters. The assured irrigation through the electric or diesel operated pump sets increased the yield of crops. Similar trend is being followed for other crops in Punjab and other states also.

However, the work of mechanization of agricultural and horticultural crops is not yet complete. We have yet to develop the cultivation systems for end to end mechanization of agricultural and horticultural crops. The vegetable crops are labour intensive, short duration and more profitable and need mechanization through development of raised bed cultivation systems, nursery raising machinery, seed planters, seedling transplanters, weeders and harvesters for saving the labour and to reduce the cost of cultivation. Similarly for fruit crop cultivation pit hole digger, weeder, basin cleaner, sprayer, pruner and harvesters need development. Similar action is required for mechanization of plantation, spices and ornamental crops.

This book is a small step in educating our students, farmers and industrialists in mechanization of horticultural crops. A large number of websites and videos of machinery available throughout the world has been gone through and large number of photographs of internationally available machines has been referred in this book. The sources of availability of the machines have been referred in photographs and in the references given in the book. The author has seen the mechanization of horticultural crops in USA during his training at Michigan State University during 1989. He has also seen the machineries available throughout the world during his visit to "EIMA International Agriculture and Gardening Machinery Exhibition" at Bologna, Italy. Efforts have been made to compile the information on all type of horticulture machines available in the world. He hopes that this book will be a small step toward mechanization of horticulture in India.

**S.C. Mandhar**

Formerly, Principal Scientist & Head
Section of Agricultural Engineering
Indian Institute of Horticultural Research
Bangalore

# Contents

# ABBREVIATIONS

The following abbreviations are used for names of organizations and scientific and technical terms.

## Organizations

| | |
|---|---|
| AAIDU | Sam Higginbottom Institute of Agriculture, Technology and Sciences Allahabad Agricultural Institute (Deemed University) |
| AAU | Assam Agricultural University |
| AIAMMA | All India Agriculture Machinery Manufacturer's Association |
| ANGRAU | Acharya NG Ranga Agricultural University |
| ASABE | American Society of Agricultural and Biological Engineers |
| BAU | Birsa Agricultural University |
| CARI | Central Agricultural Research Institute |
| CAU | Central Agricultural University |
| CCSHAU | CCS Haryana Agricultural University |
| CIAE | Central Institute of Agricultural Engineering |
| CISH | Central Institute of Sub-tropical Horticulture |
| CIPHET | Central Institute of Post Harvest Engineering and Technology |
| CFMTTI | Central Farm Machinery Training and Testing Institute |
| CMERI | Central Mechanical Engineering Research Institute |
| CPCRI | Central Plantation Crop Research Institute |
| CPRI | Central Potato Research Institute |
| CSRI | Cyber Security Research Insttitute |
| Dr. PDKV | Dr. Panjabrao Deshmukh Krishi Vidyapeeth |
| DBSKKV | Dr. Balasaheb Sawant Konkan Krishi VidhyaPeeth |
| DOPR | Directorate of Oil Palm Research |
| DOGR | Directorate of Onion and Garlic Research |
| FAO | Food and Agriculture Organization |
| FICCI | Federation of Indian Chambers of Commerce and Industries |
| GBPUAT | G B Pant University of Agriculture & Technology |
| IARI | Indian Agricultural Research Institute |

| | |
|---|---|
| IASRI | Indian Agricultural Statistics Research Institute |
| ICAR | Indian Council of Agricultural Research |
| ICRISAT | International Crops Research Institute for the Semi-Arid Tropics |
| IGAU | Indira Gandhi Agricultural University |
| IGFRI | Indian Grassland and Fodder Research Institute |
| IIHR | Indian Institute of Horticultural Research |
| IISR | Indian Institute of Sugarcane Research |
| IIT | Indian Institute of Technology |
| IPI | International Potash Institute |
| ISAE | Indian Society of Agricultural Engineers |
| IISR | Indian Institute of Spices Research |
| HPKV | CSK Himachal P Krishi Vishvavidyalaya |
| JNKVV | Jawahar Lal Nehru Krishi Vishwa Vidyalaya |
| KAU | Kerala Agricultural University |
| NDUAT | Narendra Dev University of Agriculture & Technology |
| NEH | ICAR Research Complex for NEH Region |
| NCPAH | National Committee on Plasticulture Applications in Horticulture |
| NID | National Innovation Foundation |
| NRCG | National Research Centre for Grapes |
| NRCSS | National Research Centre for Seed Spices |
| MPKV | Mahatma Phule Krishi Vidyapeeth |
| MPUAT | Maharana Pratap University of Agriculture & Technology |
| OUAT | Orissa University of Agriculture & Technology |
| PAU | Punjab Agricultural University |
| TAMMM | Tractor and Agricultural Machinery Manufacturer's Meet |
| RAU | Rajendra Agricultural University |
| TERI | Tata Energy Research Institute |
| TMA | Tractors Manufacturers Association |
| UAS | University of Agricultural Sciences |

## Technical Terms and Scientific Units

| | |
|---|---|
| AMD | Arithmetic mean diameter |
| Cm | Centimeter |
| EC | Electric Conductivity |
| ET | Evapo transpiration |
| GIS | Geographic Information System |
| GMD | Geometric mean diameter |
| HB | Brinell hardness |
| HDPE | High density polyethylene |

| | |
|---|---|
| H | Hour |
| Ha | Hectare |
| HP | Horse power |
| HV | High volume |
| kg | kilogram |
| kg/cm$^2$ | kilogram per square centimeter |
| kg/ha | kilogram per hectare |
| km | kilometer |
| km/h | kilometer per hour |
| kpa | kilo pascal |
| kv | kilo volt |
| kw/ha | kilo watt per hectare |
| KW | kilo watt per tone |
| l | Litre |
| l/h | Litre per hour |
| l/ha | Litre per hectare |
| LDPE | Low Density polyethylene |
| lph | Litre per hour |
| lpm | Litre per minute |
| min | Minute |
| M H | Million hectare |
| MT | Metric Tonne |
| MT/ha | Metric tone per hectare |
| m$^3$/h | Cubic meter per hour |
| rpm | Revolution per minute |
| ppm | Parts per million |
| PVC | Poly vinyl chloride |
| sec | Second |
| SMD | Sauter Mean Diameter |
| t/ha | Tonnes per hectare |
| ULV | Ultra Low Volume |
| V | Volt |
| VMD | Volume mean diameter |
| W | Watt |

# *Chapter 1*
# Indian Horticulture Scenario

## 1.1 Status of Horticulture in India

India grows wide group of horticultural crops namely fruits; vegetables, tropical tuber crops and mushroom; ornamental crops; medicinal and aromatic plants, spices and plantation crops. Major emphasis was laid on horticulture from mid eighties onwards as a mean of diversification for making agriculture more profitable through efficient land use, optimum utilization of natural resources (soil, water and environment) and creating skilled employment for rural masses especially women folk. The past efforts have been rewarding and India has now emerged as the largest producer of coconut, arecanut, cashew, ginger, turmeric, black pepper and the second largest producer of fruits (Table 1.1), vegetables (Table 1.2) and tea. Among the new crops, kiwi, olive, gherkins, kinnow, oil palm and noni have been successfully introduced for commercial cultivation.

Production base of horticultural crops has been expanding since independence. From eighth plan (1992) onwards this sector has witnessed tremendous growth in area, production and productivity. The area and production under horticultural crops was 20.876 million ha and 223.089 million tonnes respectively during 2009-10 (Table 1.3) . India produced fruits in 6.329 million ha, vegetables in 7.985 million ha, spice in 2.464 million ha and plantation crops in 3.265 million ha during 2009-10.

During 2001-02 to 2009-10 the area under vegetable crops and production increased by 3.3% and 5.3% respectively per year. The area under fruit crops and production increased by 5.9 % and 6.6 % respectively per year during the same period. The area under cultivation of horticultural crops & production has been increasing; the percentage increase in area & production is given in Table 1.4A.

The area under horticultural crops has been increasing in various states. The area under various horticultural crops was about 13.0 % area of the total cultivated area Chadha and Choudhary (2007). The States/ UTs namely Chhatishgarh, Gujarat, Haryana, Jharkhand, Madhya Pradesh, Nagaland, Punjab, Uttar Pradesh, Dadar & Nagar Haveli have less than 10 per cent coverage of area under horticultural crops. The States namely Andhra Pradesh, Assam, Bihar, Karnataka, Maharashtra, Orissa, Rajasthan, Tamil Nadu and Uttarakhand covers about 10-25 per cent area under horticultural crops. The six states namely Arunachal Pradesh, Himachal Pradesh, Jammu & Kashmir, Meghalaya, Sikkim, Pondicherry covers about 25-50% of their area under horticulture while the states Goa, Kerala, Manipur, Andman & Nicobar, Delhi, Lakshdweep have more than 50 per cent of their area and are saturated for further area expansion in horticulture.

India is at number two in fruit production and vegetable production with the production of about 71.52 million tonnes and 133.74 million tonnes per year respectively (Table 1.1 and Table 1.2). It can be noticed that the productivity of the fruit and vegetables is more in countries having mechanized farming. China is having more production of fruit and vegetables due to more area under cultivation, however the productivity is at par with India.

**Table 1.1:** Fruit Production in Major Countries of the World (2009-10)

| Country | Area (Million ha) | Production (Million MT) | Productivity (MT/ha) |
|---|---|---|---|
| China | 11.05 | 109.62 | 9.9 |
| India | 6.33 | 71.52 | 11.3 |
| Brazil | 2.45 | 38.68 | 15.8 |
| USA | 1.16 | 27.76 | 23.9 |
| Italy | 1.27 | 17.65 | 13.9 |
| Spain | 1.74 | 16.28 | 9.3 |
| Mexico | 1.20 | 16.12 | 13.4 |
| Indonesia | 0.72 | 16.03 | 22.2 |
| Philippines | 1.12 | 15.62 | 13.9 |
| Iran | 1.26 | 13.18 | 10.4 |
| Turkey | 1.04 | 12.83 | 12.4 |
| Others | 26.24 | 228.95 | 8.7 |
| World | 55.01 | 579.92 | 10.5 |

**Table 1.2:** Vegetable Production in Major Countries of the World (2009-10)

| Country | Area (Million ha) | Production (Million MT) | Productivity (MT/ha) |
|---|---|---|---|
| China | 24.08 | 457.83 | 19.0 |
| India | 7.98 | 133.74 | 16.7 |
| USA | 1.16 | 37.25 | 32.1 |
| Turkey | 1.11 | 27.16 | 24.6 |
| Egypt | 0.72 | 19.17 | 26.5 |
| Iran | 0.59 | 14.97 | 25.3 |
| Russian Fed | 0.77 | 14.35 | 18.7 |
| Italy | 0.52 | 13.70 | 26.1 |
| Spain | 0.35 | 12.62 | 35.8 |
| Maxico | 0.67 | 12.10 | 18.1 |
| Rep of Korea | 0.31 | 11.28 | 36.7 |
| Others | 16.88 | 221.29 | 13.1 |
| World | 53.71 | 931.85 | 17.3 |

**Table 1.3:** Area and Production of Horticultural Crops in India (2009-10)

| Crop | Area (000'ha) | Production (000'MT) | Productivity (MT/ha) |
|---|---|---|---|
| Vegetables | 7985 | 133738 | 16.75 |
| Fruits | 6329 | 71516 | 11.30 |
| Plantation crops | 3265 | 11928 | 3.65 |
| Spices | 2464 | 4016 | 1.63 |
| Flowers | 183 | 1021 | 5.58 |
| Medicinal & Aromatic Plants | 509 | 573 | 1.13 |
| Almond & Walnut | 142 | 193 | 1.36 |
| Honey | - | 65 | - |
| Mushroom | - | 41 | - |
| Total | 20876 | 223089 | 10.69 |

**Table 1.4.A:** Increase of Area and Production of Horticultural Crops in India 1990-91 to 2009-10

| | **Annual increase in area (% per year)** | | **Annual increase in production (% per year)** | |
|---|---|---|---|---|
| **Crop** | **1991-92 to 2001-02** | **2001-02 to 2009-10** | **1991-92 to 2001-02** | **2001-02 to 2009-10** |
| Vegetables | 1.0 | 3.3 | 4.2 | 5.3 |
| Fruits | 3.4 | 5.9 | 4.1 | 6.6 |
| Plantation crops | 2.6 | 1.1 | 2.6 | 2.6 |
| Spices | 4.9 | -3.0 | 7.1 | 1.0 |
| Flowers | NA | 7.1 | NA | 8.4 |
| Medicinal & Aromatic Plants | NA | NA | NA | NA |
| Almond & Walnut | NA | 2.5 | NA | 6.8 |
| Total | 2.7 | 2.9 | 4.2 | 5.5 |

**Table 1.4**: State wise Area Under Horticulture Crops in Relation to Total Area Under Cultivation (2007-08)

| **State/UTs** | **% Area under Horticulture** | **State/UTs** | **% Area under Horticulture** |
|---|---|---|---|
| Andhra Pradesh | 11.41 | Nagaland | 7.20 |
| Arunachal Pradesh | 40.15 | Orissa | 10.91 |
| Assam | 14.34 | Punjab | 4.85 |
| Bihar | 19.63 | Rajasthan | 14.80 |
| Chhattisgarh | 2.75 | Sikkim | 58.80 |
| Goa | 70.85 | Tamil Nadu | 16.86 |
| Gujarat | 8.57 | Tripura | 28.47 |
| Haryana | 6.50 | Uttarakhand | 15.20 |
| Himachal Pradesh | 44.85 | Uttar Pradesh | 6.41 |
| Jammu & Kashmir | 34.13 | West Bengal | 25.89 |
| Jharkhand | 4.73 | Andaman & Nicobar | 100.00 |
| Karnataka | 12.73 | Chandigarh | 10.00 |

*Contd...*

*Table 1.4: contd...*

| | | | |
|---|---|---|---|
| Kerala | 73.33 | Dadar & Nagar Haveli | 9.17 |
| Madhya Pradesh | 4.61 | Daman & Diu | 12.50 |
| Maharashtra | 10.23 | Delhi | 83.32 |
| Manipura | 57.21 | Lakshdweep | 100.00 |
| Meghalaya | 31.25 | Pondicherry | 31.24 |
| Mizoram | 18.00 | National | 13.08 |

## 1.2 Status of Fruit Crops

The global fruit production during 2009-10 was 579.92 million tonnes. During the past 15 years, production of fruits in China has gone upto 3 times, which helped China to become the largest producer of fruits (109.62 million tonnes) in the world. It contributes 18.9% share in global production and has largest area under fruit production. India is the second largest producer of fruits by producing 71.52 million tonnes and contributes 12.3% share in global fruit production. Brazil occupies the third position, with 38.68 million tonnes and contributes 6.7% in global production. The other countries among top ten producers are USA, Italy, Spain, Mexico, Turkey, Iran, and Indonesia (Table 1.1).

India occupies first place in production of mango, banana, litchi, papaya, pomegranate, sapota and aonla, second place in limes and lemons and fifth place in pineapple production of the world. The productivity of grapes in India is highest in the world. India has higher national average productivity in banana and sapota compared to world average productivity, while in citrus, mango, apple, guava, pineapple, papaya, India has substantially low productivity in comparison to the world average, Chadha and Choudhary (2007).

The major fruit crops of economic importance based on area under cultivation in India, which need mechanization in order of preference are given in Table 1.5.

The area under cultivation of fruit crops in different states during 1991-92, 2001-02 and 2009-10 and growth rate is given in Table 1.6.

India accounts for an area of 6.33 million ha under fruit crops with a production of 71.52 million tonnes. During the period 1991-92 to 2001-02 and 2001-02 to 2009-10 the yearly growth in area was observed to be 3.4% and 5.9% respectively. Maharashtra had highest area under fruit cultivation 1.541 million ha and the yearly growth rates for last two decades was 8.6% and 12.9% respectively. Andhra Pradesh had second position with an area of 921 thousand ha and 6.3% and 6.0% growth respectively. Uttar Pradesh (356.7 thousand ha), Gujarat (352.9 thousand ha), Karnataka (351.0 thousand ha), Orissa (302.1 thousand ha), Bihar (293.6 thousand ha), Kerala (296.7 thousand ha) followed by Tamil Nadu (291.6 thousand ha) in area under cultivation of fruit crops.

**Table 1.5:** Area, Production and Productivity of Major Fruit Crops (2009-10)

| Crop | Area ('000 ha) | Production ('000 MT) | Productivity (MT/ha) |
|---|---|---|---|
| Mango | 2312 | 15027 | 6.5 |
| Citrus | 987 | 9638 | 9.8 |
| Banana | 770 | 26470 | 34.4 |
| Apple | 283 | 1777 | 6.3 |
| Guava | 220 | 2572 | 11.7 |
| Sapota | 159 | 1347 | 8.5 |
| Pomegranate | 125 | 820 | 6.6 |
| Grape | 106 | 881 | 8.3 |
| Papaya | 96 | 3913 | 40.9 |
| Pineapple | 92 | 1387 | 15.1 |
| Litchi | 74 | 483 | 6.5 |

The state wise production and productivity of major fruit crops is given in Table 1.7.

**Mango**

India produces approximately 40.0% of the world's mango and occupies top position among mango growing countries. China and Thailand stand second and third among mango producing countries in the world. Brazil, China, Thailand and Mexico have higher average national productivity than India. India has an average productivity of 6.5 tonnes/ha which is about half of the productivity compared to Brazil. The area under mango in India is 2.31 million ha with a total production of 15.0 million tonnes (Table 1.5). Mango is grown in about 38 per cent area and contributes 21.0 per cent in total fruit production in India. It is grown in almost all the states of India. Andhra Pradesh tops the list of mango producing states, with highest acreage as well as production. Other major mango producing states are Uttar Pradesh, Karnataka, Bihar, Gujarat, Tamil Nadu, Maharashtra, West Bengal, Orissa and Kerala. Uttarakhand, Madhya Pradesh and Punjab have very low production.

**Citrus**

India occupies sixth position among citrus growing countries in the world. Maximum productivity is obtained in USA (35.11 tonne/ha) followed by Brazil, Spain and Italy. In India citrus ranks second in area with 0.987 million ha and third in production with 9.64 million tonnes and productivity of 9.8 tonnes/ha (Table 1.5). Limes, lemons, sweet orange and mandarin cover bulk of the area under citrus fruits.

**Table 1.6:** Area Under Cultivation of Fruit Crops and Growth Rate for Major States in India (2009-10)

| State/ UTs | Area (000 ha) | | | Yearly growth rate (%) | |
|---|---|---|---|---|---|
| | 1991-92 | 2001-02 | 2009-10 | 1991-92 to 2001-02 | 2001-02 to 2009-10 |
| **India** | **2875** | **4010** | **6329** | **3.4** | **5.9** |
| Maharashtra | 256.1 | 582.8 | 1540.6 | 8.6 | 12.9 |
| Andhra Pradesh | 313.1 | 575.8 | 921.1 | 6.3 | 6.0 |
| Uttar Pradesh | 453.7 | 288.3 | 356.7 | -4.5 | 2.7 |
| Gujarat | 84.6 . | 149.0 | 352.9 | 5.8 | 11.4 |
| Karnataka | 209.2 | 257.1 | 351.0 | 2.1 | 4.0 |
| Orissa | 136.3 | 225.0 | 302.1 | 5.1 | 3.8 |
| Bihar | 266.9 | 272.3 | 293.6 | 0.2 | 0.9 |
| Kerala | 236.3 | 234.5 | 296.7 | 0.0 | 3.0 |
| Tamil Nadu | 136.2 | 227.5 | 291.6 | 5.3 | 3.2 |
| J&K | 119.1 | 142.2 | 209.8 | 1.8 | 5.0 |
| Himachal Pradesh | 157.2 | 223.0 | 208.4 | 3.6 | -0.8 |
| West Bengal | 111.2 | 147.6 | 208.3 | 2.9 | 4.4 |
| Uttarakhand | 150.5 | 197.5 | 193.8 | 2.8 | -0.3 |
| Chhatisgarh | - | 14.4 | 125.3 | - | 31.1 |
| Assam | 72.3 | 110.8 | 117.3 | 4.4 | 0.7 |
| Madhya Pradesh | 64.7 | 46.6 | 113.1 | -3.3 | 11.7 |
| Arunachal Pradesh | 20.2 | 41.6 | 72.0 | 7.5 | 7.1 |
| Punjab | 72.7 | 37.5 | 67.6 | -6.4 | 7.6 |
| Haryana | 13.9 | 31.3 | 41.5 | 8.5 | 3.6 |
| Manipur | 19.8 | 26.1 | 38.4 | 2.8 | 4.9 |
| Jharkhand | - | 31.5 | 37.7 | - | 2.3 |
| Tripura | 44.9 | 28.3 | 36.9 | -4.5 | 3.4 |
| Meghalaya | 24.2 | 24.0 | 32.9 | 0.0 | 4.0 |
| Rajasthan | 22.8 | 22.1 | 32.1 | -0.3 | 4.8 |
| Mizoram | 9.3 | 19.0 | 27.1 | 7.4 | 4.5 |
| Goa | 11.0 | 10.7 | 11.0 | -0.3 | 0.3 |

Mandarins are produced in Maharashtra, Madhya Pradesh, Assam, West Bengal, Mizoram, Nagaland, Tripura, Arunachal Pradesh, Tamil Nadu, Rajasthan, Sikkim, Meghalaya and Manipur. Kinnow is cultivated exclusively in Punjab, Haryana, Rajasthan.

Maximum area and production of sweet orange is reported from Andhra Pradesh followed by Maharashtra, while maximum productivity is recorded from Madhya Pradesh. Limes are mostly grown in Andhra Pradesh, Karnataka, Gujarat, Maharashtra, Rajasthan, Madhya Pradesh and Tamil Nadu, whereas lemon is mostly grown in hilly areas of Assam, Nagaland, Tripura, Manipur, Mizoram, Uttarakhand and Punjab. Maximum production of lime takes place in Andhra Pradesh, followed by Karnataka, Gujarat, Bihar and Maharashtra. India is only next to Mexico in production of limes and lemons. Highest productivity of limes is in Karnataka and medium in Andhra Pradesh, Gujarat and Maharashtra. The lowest productivity of lime is in Tamil Nadu.

**Banana**

India is the largest producer of banana in the world. Other major banana producing countries are Brazil, China, Equador, Philippines, Indonesia, Costa Rica, Mexico, Thailand, Burundi, Columbia, and Vietnam. Ecuador, Philippines, Costa Rica and Colombia, produce much less banana than India, but they are specialized in producing first grade table bananas and maximum export of banana is done by these countries. Among fruits in India, banana ranks third in area with 0.77 million ha and first in production with 26.47 million tonnes covering 12 % of area and contributing 37.0% of fruit production. The state of Tamil Nadu ranks first in area while production and productivity is highest in Maharashtra (Table 1.7). Lower yields in Assam (15.1MT/ ha), Kerala (7.9 MT/ ha), Andhra Pradesh (35.0 MT/ha) and Karnataka (20.4 MT/ha) are mainly due to varying cultivation practices adopted and extent of inputs applied for cultivation. Lower yields in Kerala are mainly due to cultivation of low yielding cultivars like Nendran. The phenomenal increase in production and productivity in certain pockets has been due to adoption of high density planting, use of tissue-cultured seedlings and drip irrigation. The overall increase is productivity in Maharashtra and Gujarat is inexplicable.

**Apple**

India has large area under apple cultivation next to China and Russian Federation, it ranks ninth in production among apple-growing countries of the world, because, of low productivity. Apple is the fourth major fruit crop of India and occupies a total area of 0.283 million ha with production of 1.78 million tonnes. Leading apple producing states are Jammu and Kashmir, Himachal Pradesh, Uttarakhand and Arunachal Pradesh. Maximum area, production and productivity under apple is reported from Jammu and Kashmir. Himachal Pradesh has the second largest area under apple but production and productivity are very low compared to Jammu and Kashmir. In Uttarakhand and Arunachal Pradesh, productivity is still lower than Himachal Pradesh.

**Guava**

Guava is cultivated in sizeable area in Brazil, Mexico, Thailand, USA (Hawaii), New Zealand, Philippines, China, Indonesia, Venezuela and Australia, etc. The best quality guavas are cultivated and produced in Uttar Pradesh particularly in the Allahabad region. In India it is cultivated in an area of 0.220 million ha with a total production of 1.78 million tonnes (Table 1.5). The maximum area under guava is found in Uttar Pradesh, followed by Maharashtra and Bihar. The highest productivity of guava is observed in Punjab (21.2 tonnes/ha ) followed by Karnataka (19.3 tonnes/ha), Lowest productivity being in Maharashtra (7.7 tonnes/ha) and Orissa (7.1 tonnes/ha).

**Sapota**

Sapota is one of the important tropical fruit crops of India. It is cultivated in 1,59,000 ha with a total production of 13,47,000 tonnes. Major sapota producing states are Karnataka, Maharashtra, Gujarat, Tamil Nadu and Andhra Pradesh. It is also produced to some extent in West Bengal and Orissa. The productivity is more or less the same in all the states, except that it is very high in Tamil Nadu and very low in Maharashtra (Table 1.7).

**Pomegranate**

Pomegranate is an important fruit crop of India. Presently it is being cultivated in 0.125 million ha area with a total production of 0.820 million tonnes. India occupies top position among pomegranate growing countries in the world, producing about 50% of the total world production. Maharashtra is the leading pomegranate producing state having an area of around 98,900 ha followed by Karnataka (13,200 ha). Pomegranate is also grown as wild in forests of Jammu and Kashmir and Himachal Pradesh and 400 and 100 tonnes of *Anardana* is marketed from Jammu and Kashmir and Himachal Pradesh, respectively.

**Pineapple**

Major pineapple growing countries in the world are Thailand, Philippines, Brazil, China and India. India ranks second in area and fifth in production of pineapples in the world. In India, pineapple is cultivated in an area of 92,000 ha with a total production of 1.387 million tonnes. It is abundantly grown in West Bengal, Assam, NE region, Kerala, Bihar and Karnataka, with maximum area in Assam. Productivity of pineapple is the highest in Karnataka followed by West Bengal and Bihar. Productivity of pineapples in most of the NE states is less than the average national pineapple productivity, which needs to be improved through suitable management practices (Table 1.7).

**Papaya**

The major papaya producing countries are Brazil, Nigeria, Mexico, India, Indonesia, Ethiopia, Congo Democratic Republic, Peru, China, Columbia, Thailand and Venezuela. India is the fourth important producer of papaya in the world. Papaya ranks nineth in area with 96,000 ha and fourth in production with 3.91 million tonnes of the total production (Table 1.5). Major production of papaya takes place in Andhra Pradesh, followed by Gujarat, Karnataka and West Bengal. The highest fruit productivity was observed in Andhra Pradesh followed by Karnataka (Table 1.7).

## Grapes

The major grape growing countries are Italy, France, USA, Spain, China, Turkey, Iran, Argentina, Australia, Chile, Germany, South Africa and India. India's high productivity in grape has made it to reach 12th position in the World as far as production is concerned, but from acreage point of view, India's position is 33rd among grape-growing countries in the World. In India grapes are cultivated in 1,06,000 ha with a total production of 8,81,000 tonnes and productivity of 8.3 tonnes /ha. Maximum area and production of grapes is in Maharashtra, followed by Karnataka and Tamil Nadu. In Andhra Pradesh, Punjab and Haryana, grapes are cultivated only to a limited extent. However, there is again a recent surge in grape production in Andhra Pradesh (Table 1.7).

## Litchi

India is the largest producer of litchi in the world. Other major litchi producing countries are China, Taiwan, Thailand, South Africa, Madagascar and Australia. Although authentic data on litchi-production is not available, total world litchi production is estimated to be 6-6.5 lakh tonnes. In India, the total area under litchi is 74,000 ha with production of 4,83,000 tonnes (Table 1.5). Litchi is largely grown in Bihar, West Bengal , Jharkhand and Assam (Table 1.7). It is also grown in small pockets in Punjab, Chhatisgarh, Uttarakhand, Haryana, Orissa and N.E region.

**Table 1.7:** State-wise Production and Productivity of Fruit Crops (2009-10)

| Fruit | State | Area (000'ha) | Production (000'MT) | Productivity (MT/ha)) |
|---|---|---|---|---|
| **Mango** | Andhra Pradesh | 480.4 | 4058.3 | 8.4 |
| | Uttar Pradesh | 276.4 | 3588.0 | 13.0 |
| | Karnataka | 153.8 | 1694.0 | 11.0 |
| | Bihar | 146.0 | 995.9 | 6.8 |
| | Gujarat | 121.5 | 856.7 | 7.0 |
| | Tamil Nadu | 132.7 | 636.3 | 4.8 |
| | Maharashtra | 474.5 | 597.0 | 1.3 |
| | West Bengal | 88.1 | 578.0 | 6.6 |
| | Orissa | 177.6 | 577.5 | 3.3 |
| | Kerala | 63.8 | 373.2 | 5.9 |

*Contd...*

*Table 1.7: contd..*

| | | | | |
|---|---|---|---|---|
| **Citrus** | Andhra Pradesh | 281.3 | 3883.7 | 13.8 |
| | Maharashtra | 287.6 | 1725.1 | 6.0 |
| | Punjab | 42.4 | 905.1 | 21.3 |
| | Madhya Pradesh | 38.3 | 677.8 | 17.7 |
| | Gujarat | 37.1 | 385.6 | 10.4 |
| | Rajasthan | 18.6 | 312.7 | 16.8 |
| | Karnataka | 15.1 | 312.5 | 20.7 |
| | Orissa | 27.4 | 258.8 | 9.4 |
| | Assam | 24.2 | 225.9 | 9.3 |
| | Uttrakhand | 26.7 | 126.6 | 4.7 |
| **Banana** | Maharashtra | 85.0 | 5200.0 | 61.2 |
| | Tamil Nadu | 113.7 | 4980.9 | 43.8 |
| | Gujarat | 61.9 | 3779.8 | 61.0 |
| | Andhra Pradesh | 80.6 | 2819.6 | 35.0 |
| | Karnataka | 104.4 | 2132.3 | 20.4 |
| | Madhya Pradesh | 33.0 | 1459.8 | 44.2 |
| | Bihar | 31.5 | 1435.3 | 45.6 |
| | West Bengal | 41.0 | 982.2 | 24.0 |
| | Assam | 53.4 | 805.2 | 15.1 |
| | Kerala | 51.3 | 406.2 | 7.9 |
| **Apple** | Jammu & Kashmir | 133.7 | 1332.8 | 10.0 |
| | Himachal Pradesh | 97.2 | 510.2 | 5.2 |
| | Uttrakhand | 32.7 | 132.3 | 4.1 |
| | Arunachal Pradesh | 12.8 | 10.0 | 0.8 |
| **Guava** | Uttar Pradesh | 39.9 | 486.7 | 12.2 |
| | Maharashtra | 33.5 | 258.0 | 7.7 |
| | Bihar | 29.2 | 231.5 | 7.9 |
| | West Bengal | 13.4 | 175.7 | 13.1 |
| | Punjab | 8.0 | 169.3 | 21.2 |
| | Gujarat | 9.8 | 156.6 | 15.9 |
| | Andhra Pradesh | 10.1 | 150.8 | 15.0 |
| | Karnataka | 7.2 | 138.8 | 19.3 |
| | Orissa | 14.1 | 100.0 | 7.1 |
| | Tamil Nadu | 7.0 | 92.5 | 13.2 |

*Contd...*

*Table 1.7: Contd...*

| | | | | |
|---|---|---|---|---|
| **Sapota** | Karnataka | 29.3 | 360.0 | 12.3 |
| | Maharashtra | 69.1 | 298.0 | 4.3 |
| | Gujarat | 27.4 | 272.6 | 10.0 |
| | Tamil Nadu | 6.9 | 173.5 | 25.0 |
| | Andhra Pradesh | 17.1 | 171.4 | 10.0 |
| | West Bengal | 4.0 | 43.4 | 10.9 |
| | Orissa | 3.3 | 16.6 | 5.0 |
| **Pomegranate** | Maharashtra | 98.9 | 555.5 | 5.6 |
| | Karnataka | 13.2 | 138.5 | 10.5 |
| | Andhra Pradesh | 5.6 | 56.4 | 10.0 |
| | Gujarat | 4.4 | 45.6 | 10.4 |
| **Pineapple** | West Bengal | 9.6 | 293.8 | 30.5 |
| | Assam | 14.2 | 223.0 | 15.7 |
| | Karnataka | 2.8 | 177.2 | 63.3 |
| | Bihar | 4.7 | 125.0 | 26.4 |
| | Tripura | 6.8 | 117.5 | 17.3 |
| | Meghalaya | 10.8 | 106.8 | 9.9 |
| | Manipur | 12.1 | 103.5 | 8.6 |
| | Kerala | 9.8 | 80.8 | 8.2 |
| | Arunachal Pradesh | 10.9 | 34.4 | 3.2 |
| **Papaya** | Andhra Pradesh | 18.8 | 1500.7 | 80.0 |
| | Gujarat | 15.3 | 832.9 | 54.3 |
| | Karnataka | 5.8 | 419.3 | 72.3 |
| | West Bengal | 11.1 | 321.8 | 29.1 |
| | Chhatisgarh | 9.3 | 211.7 | 22.7 |
| | Madhya Pradesh | 1.7 | 192.9 | 115.6 |
| | Assam | 7.2 | 119.1 | 16.5 |
| | Kerala | 17.7 | 80.7 | 4.6 |
| | Tamil Nadu | 0.4 | 73.9 | 180.2 |
| **Grapes** | Maharashtra | 82.0 | 440.0 | 5.4 |
| | Karnataka | 17.4 | 317.6 | 18.3 |
| | Tamil Nadu | 2.6 | 44.1 | 16.8 |
| | Andhra Pradesh | 1.4 | 29.8 | 21.0 |
| | Punjab | 0.5 | 15.5 | 28.6 |

*Contd...*

*Table 1.7: contd...*

| | | | | |
|---|---|---|---|---|
| **Litchi** | Bihar | 30.6 | 215.1 | 7.0 |
| | West Bengal | 8.4 | 82.4 | 9.8 |
| | Jharkhand | 4.3 | 51.1 | 12.0 |
| | Assam | 5.1 | 39.2 | 7.7 |
| | Punjab | 1.6 | 22.4 | 14.1 |
| | Chhatisgarh | 3.0 | 17.8 | 6.0 |
| | Uttarakhand | 9.1 | 15.7 | 1.7 |
| | Tripura | 2.8 | 17.0 | 6.1 |
| | Orissa | 4.3 | 17.1 | 4.0 |

## 1.3 Status of Vegetable Crops

China is the leading producer of vegetables in the world. India is the second largest vegetable producer with 13.3% production share in the world and USA is the third. India is also the second largest producer of most of the important vegetable crops at global level, namely brinjal, cabbage, cauliflower, pea, onion and tomato except for potato and okra where its ranking is number 3 and 1, Chadha and Choudhary (2007). However, it falls behind in productivity in most of the crops except for tomato where India is ranking at number 1. In okra, it is at par with world productivity and in cauliflower it is quite close to the world productivity.

Total vegetable production in India before independence was 15 million tonnes and since independence for decades the growth rate was stabilized around 0.5%. The impetus on vegetable research and policy intervention to promote vegetable crops witnessed a sudden spurt in growth rate of area of cultivation to 3.3% per year during the last decade. The area under cultivation with production and productivity is given in Table 1.8. The perusal of Table 1.9 indicates that rate of increase of area under cultivation increased from 1.0 % during last decade to 3.3 % due to technological adoption. The potential technological interventions with improved gene pool and precise management can very well take growth rate at national level to nearby 6% per annum. This may lead to increased production of vegetable crops close to 140 million tonnes by 2012, which may go in a long way to achieve the much-awaited target of making available 300 g vegetables per capita per day. As the labour is becoming scare and costly mechanization will play a major roll in cultivation. The state-wise production and productivity of major vegetable crops is given in Table 1.10.

**Potato**

There has been a steady increase in the area and production in potato over different plan periods. The increase was substantial between 1997-2007 (IX and X plan periods). The productivity was stagnant between 1992-2002 (VIII and IX plan periods) Chadha and Choudhary (2007). Uttar Pradesh is the leading state in area, production and productivity of potato followed by West Bengal and Bihar (Table 1.10).

**Table 1.8:** Area, production and productivity of major vegetable crops in India (2009-10)

| Crop | Area (000'ha) | Production (000'MT) | Productivity (MT/ha) |
|---|---|---|---|
| Potato | 1835 | 36577 | 19.9 |
| Onion | 756 | 12159 | 16.1 |
| Tomato | 634 | 12433 | 19.6 |
| Brinjal | 612 | 10563 | 17.2 |
| Okra | 452 | 4803 | 10.6 |
| Peas | 365 | 3029 | 8.3 |
| Cauliflower | 348 | 6569 | 18.9 |
| Cabbage | 331 | 7281 | 22.0 |
| Tapioca | 232 | 8060 | 34.8 |
| Sweet Potato | 119 | 1095 | 9.2 |

**Table 1.9:** Area Under Cultivation of Vegetable Crops and Growth Rate for Major States (2009-10)

| State/ UTs | Area (000 ha) | | | Yearly growth rate (%) | |
|---|---|---|---|---|---|
| | 1991-92 | 2001-02 | 2009-10 | 1991-92 to 2001-02 | 2001-02 to 2009-10 |
| India | 5593 | 6156 | 7985 | 1.0 | 3.3 |
| West Bengal | 456.0 | 1139.0 | 1302.7 | 9.6 | 1.7 |
| Uttar Pradesh | 576.7 | 777.9 | 1020.1 | 3.0 | 3.4 |
| Bihar | 843.3 | 578.9 | 836.0 | -3.7 | 4.7 |
| Orissa | 710.3 | 643.4 | 694.2 | -1.0 | 1.0 |
| Maharashtra | 241.1 | 402.4 | 451.8 | 5.3 | 1.5 |
| Karnataka | 351.1 | 358.1 | 441.2 | 0.2 | 2.6 |
| Gujarat | 114.6 | 232.2 | 406.8 | 7.3 | 7.2 |
| Andhra Pradesh | 155.2 | 222.5 | 331.3 | 3.7 | 5.1 |
| Chhatisgarh | - | 104.1 | 315.4 | - | 14.9 |
| Haryana | 60.8 | 150.4 | 300.9 | 9.5 | 9.1 |
| Tamil Nadu | 889.3 | 213.8 | 263.7 | -13.3 | 2.7 |
| Assam | 222.4 | 237.4 | 255.2 | 0.7 | 0.9 |
| Madhya Pradesh | 177.4 | 136.4 | 250.7 | -2.6 | 7.9 |
| Jharkhand | - | 158.5 | 212.1 | - | 3.7 |
| Punjab | 84.5 | 135.0 | 183.3 | 4.9 | 3.9 |
| Kerala | 202.1 | 114.3 | 151.6 | -5.6 | 3.6 |
| Rajasthan | 62.9 | 99.3 | 131.9 | 4.7 | 3.6 |

*Contd...*

*Table 1.9: contd...*

| | | | | | |
|---|---|---|---|---|---|
| Uattarakhand | 57.1 | 93.8 | 82.6 | 5.1 | -1.6 |
| Himachal Pradesh | 38.7 | 34.6 | 79.8 | -1.1 | 11.0 |
| J&K | 180.3 | 50.8 | 69.8 | -11.9 | 4.1 |
| Meghalaya | 25.9 | 35.7 | 44.3 | 3.3 | 2.7 |
| Delhi | 55.0 | 111.0 | 36.1 | 7.3 | -13.1 |
| Tripura | 30.3 | 31.3 | 32.5 | 0.3 | 0.5 |
| Sikkim | 7.6 | 14.2 | 28.7 | 6.5 | 9.2 |
| Manipur | 11.8 | 10.6 | 19.9 | -1.0 | 8.2 |
| Goa | - | 7.6 | 5.7 | - | -10.5 |
| Arunachal Pradesh | 17.1 | 20.8 | 4.2 | 2.0 | -18.1 |

**Onion**

There has been a constant increase in area under onion during 1992-2007 (VIII, IX and X plan periods). The percentage of increase in production and productivity was remarkable during 1992-2007 ( VIII, IX and X plan periods) Chadha and Choudhary ( 2007). Kharif onion technology with varieties, especially N 53, Agrifound Dark Red and Arka Kalyan have contributed significantly to area expansion, increase in production and productivity. In spite this, there have been period of stringent shortages of onion in the country. Maharashtra has maximum area and production in onion followed by Karnataka in area and Gujarat in production. However, Gujarat leads in productivity followed by Bihar (Table 1.10).

**Tomato**

Highest area and production increase in tomato was witnessed during 1992-1997 (VIII plan), however maximum productivity increase was during 1997-2002 (IX plan) which is being closely followed by 2002-2007 (X plan) Chadha and Choudhary ( 2007). . Large-scale adoption of protected cultivation in tomato has significantly added to the production and productivity. Orissa has highest area under production in tomato; however Andhra Pradesh is leading in production followed by Karnataka (Table 1.10).

**Brinjal**

There has been increase in area and production of brinjal during 1992-1997 (VIII plan), but productivity declined. During 1997-2002 ( IX plan) although increase in area and production had been marginal, a big jump in productivity was evident, Chadha and Choudhary ( 2007). Development of F1 hybrid varieties and their adoption by the farmers led to increase in productivity. West Bengal has highest area under cultivation while the highest productivity is in Karnataka (Table 1.10).

**Tapioca**

The area under tapioca increased during 1987-1997 (VII & VIII plan) periods, which declined during 1997-2002 (IX plan period). The reduction declined marginally during 1992-1997 (VIII plan) but improved during 1997-2007 ( IX and

X) plan periods Chadha and Choudhary (2007). The state of Tamil Nadu leads in area and production of tapioca followed by Kerala (Table 1.10).

**Cabbage**

Cabbage has shown a big leap forward in area, production and productivity during 1992-02 (VIII and IX plans) owing to especially F1 hybrid cultivars, which can be used as a model to be followed for other vegetable crops. However, the percent increase in area and production declined when compared to the previous plan periods Chadha and Choudhary ( 2007). The state of West Bengal leads in area and production of cabbage followed by Orissa (Table 1.10).

**Cauliflower**

The area and production in cauliflower increased during 1992-02 (VIII and IX plan periods), however, productivity increased significantly during 1997-02 (IX plan). The state of West Bengal leads in area and production of cauliflower followed by Bihar (Table 1.10).

**Okra**

There was increase in area and production of okra during 1992-97 (VIII plan), however, productivity increase was not commensurate. There has been consistent increase in area and production of okra during successive plan periods Chadha and Choudhary (2007). The productivity has also been increasing. Development of yellow vein mosaic resistant varieties has also contributed significantly to the increase in production and productivity of this crop West Bengal has highest production of okra (Table 1.10), whereas Andhra Pradesh leads in productivity followed by Jharkhand.

**Pea**

There has been highest increase in area, production and productivity in garden pea during the 1992-97 (VIII plan) due to development and adoption of improved varieties by the farmers, but production and productivity fell drastically during 1997-02 (IX plan). The productivity, however, again attained a positive trend during 2002-07 (X plan) period due to availability of powdery mildew resistant varieties Chadha and Choudhary ( 2007). Uttar Pradesh occupies maximum area under pea and has highest production, followed by Jharkhand and Himachal Pradesh (Table 1.10) and Jharkhand leads in productivity.

**Sweet potato**

The increase in area under sweet potato was steady under different plan periods. However, the production was constant during 1987-02 (VII, VIII and IX plan periods). Whereas, the increase during 2002-07 (X plan period) was significant Chadha and Choudhary (2007). The state of Orissa leads in area and production of sweet potato, but Andhra Pradesh has the highest productivity. West Bengal is the second largest producer of sweet potato (Table 1.10).

**Table 1.10:** State-wise Production and Productivity of Vegetable Crops (2009-10)

| Vegetable | State | Area (000'ha) | Production (000'MT) | Productivity (MT/ha) |
|---|---|---|---|---|
| Potato | Uttar Pradesh | 540.8 | 13447.3 | 24.9 |
| | West Bengal | 370.0 | 8880.0 | 24.0 |
| | Bihar | 313.6 | 5387.2 | 17.2 |
| | Punjab | 83.1 | 2116.5 | 25.5 |
| | Gujarat | 60.1 | 1657.0 | 27.6 |
| | Madhya Pradesh | 60.8 | 608.4 | 10.0 |
| | Assam | 86.6 | 589.7 | 6.8 |
| | Jharkhand | 38.2 | 573.1 | 15.0 |
| | Haryana | 23.0 | 494.8 | 21.5 |
| | Karnataka | 81.1 | 460.3 | 5.7 |
| Onion | Maharashtra | 200.0 | 3146.0 | 15.7 |
| | Karnataka | 141.3 | 2266.2 | 16.0 |
| | Gujarat | 43.4 | 1078.6 | 24.9 |
| | Bihar | 53.0 | 972.0 | 18.3 |
| | Madhya Pradesh | 57.3 | 952.3 | 16.6 |
| | Rajasthan | 45.0 | 742.5 | 16.5 |
| | Andhra Pradesh | 39.0 | 662.6 | 17.0 |
| | Tamil Nadu | 35.3 | 339.7 | 9.6 |
| | Haryana | 18.4 | 330.3 | 17.9 |
| | Uttar Pradesh | 24.3 | 320.3 | 13.2 |
| Tomato | Andhra Pradesh | 87.0 | 1652.1 | 19.0 |
| | Karnataka | 48.3 | 1580.0 | 32.7 |
| | Orissa | 102.9 | 1394.7 | 13.6 |
| | Maharashtra | 50.0 | 1112.5 | 22.3 |
| | West Bengal | 53.5 | 1050.0 | 19.6 |
| | Bihar | 46.5 | 1043.7 | 22.4 |
| | Gujarat | 33.8 | 841.3 | 24.9 |
| | Chhatishgarh | 41.3 | 606. | 14.5 |
| | Tamil Nadu | 26.1 | 519.1 | 19.9 |
| | Jharkhand | 21.8 | 436.2 | 20.0 |
| Brinjal | | | | *Contd...* |
| | West Bengal | 156.8 | 2797.3 | 17.8 |

*Table 1.10 contd...*

| | | | | |
|---|---|---|---|---|
| Brinjal | Orissa | 132.4 | 2135.2 | 16.1 |
| | Bihar | 55.3 | 1198.6 | 21.7 |
| | Gujarat | 65.8 | 1143.6 | 17.4 |
| | Maharashtra | 30.0 | 495.0 | 16.5 |
| | Andhra Pradesh | 24.3 | 485.0 | 20.0 |
| | Jharkhand | 22.7 | 398.3 | 17.5 |
| | Chhatisgarh | 25.2 | 374.5 | 14.9 |
| | Karnataka | 15.7 | 372.7 | 23.7 |
| | Assam | 15.9 | 253.0 | 16.0 |
| | Madhya Pradesh | 20.5 | 246.0 | 12.0 |
| Tapioca | Tamil Nadu | 129.4 | 5113.6 | 39.5 |
| | Kerala | 74.9 | 2525.4 | 33.7 |
| | Andhra Pradesh | 16.5 | 329.5 | 20.0 |
| | Meghalaya | 4.9 | 26.4 | 5.4 |
| | Pondicherry | 0.6 | 25.0 | 40.9 |
| | Assam | 3.6 | 20.8 | 5.7 |
| | Karnataka | 1.0 | 13.1 | 12.8 |
| | Nagaland | 1.0 | 6.0 | 6.3 |
| | Daman & Diu | 0.02 | 0.01 | 0.9 |
| Cabbage | West Bengal | 74.3 | 2059.0 | 27.7 |
| | Orissa | 35.8 | 999.9 | 27.9 |
| | Bihar | 38.7 | 689.9 | 17.8 |
| | Assam | 29.8 | 614.2 | 20.6 |
| | Gujarat | 25.6 | 473.8 | 18.5 |
| | Maharashtra | 17.0 | 442.0 | 26.0 |
| | Jharkhand | 26.8 | 429.3 | 16.0 |
| | Chhatisgarh | 13.7 | 227.8 | 16.7 |
| | Haryana | 12.2 | 218.3 | 17.8 |
| | Karnataka | 8.9 | 176.2 | 19.8 |

*Contd...*

*Table 1.10 contd...*

| | | | | |
|---|---|---|---|---|
| Cauliflower | West Bengal | 69.2 | 1754.1 | 25.4 |
| | Bihar | 62.2 | 1080.1 | 17.4 |
| | Orissa | 46.5 | 675.4 | 14.5 |
| | Haryana | 25.7 | 487.7 | 19.0 |
| | Gujarat | 19.8 | 356.7 | 18.0 |
| | Assam | 20.5 | 351.3 | 17.2 |
| | Jharkhand | 20.6 | 329.8 | 16.0 |
| | Maharashtra | 10.1 | 158.6 | 15.7 |
| | Chhatisgarh | 16.1 | 258.9 | 16.7 |
| Okra | West Bengal | 73.1 | 893.3 | 11.5 |
| | Bihar | 58.3 | 766.6 | 13.2 |
| | Orissa | 73.9 | 651.8 | 8.8 |
| | Andhra Pradesh | 33.8 | 502.1 | 15.0 |
| | Gujarat | 49.5 | 466.2 | 9.4 |
| | Jharkhand | 29.0 | 406.0 | 14.0 |
| | Chhatishgarh | 23.9 | 217.3 | 9.1 |
| | Maharashtra | 27.0 | 175.5 | 6.5 |
| | Assam | 10.9 | 149.7 | 13.8 |
| | Uttar Pradesh | 11.6 | 131.2 | 11.3 |
| Peas | Uttar Pradesh | 159.0 | 1465.8 | 9.2 |
| | Jharkhand | 20.5 | 287.0 | 14.0 |
| | Himachal Pradesh | 21.7 | 237.3 | 10.9 |
| | Madhya Pradesh | 20.0 | 220.0 | 11.0 |
| | West Bengal | 21.1 | 125.6 | 6.0 |
| | Punjab | 19.1 | 115.5 | 6.1 |
| | Chhatisgarh | 10.6 | 96.6 | 9.2 |
| | Uttarakhand | 10.8 | 71.0 | 6.6 |
| | Jammu & Kashmir | 7.6 | 69.9 | 9.2 |
| | Haryana | 10.6 | 64.3 | 6.1 |
| | Bihar | 9.3 | 63.5 | 6.9 |

*Contd...*

*Table 1.10 contd...*

| | | | | |
|---|---|---|---|---|
| Sweet Potato | Orissa | 50.5 | 438.8 | 8.7 |
| | West Bengal | 22.5 | 231.1 | 10.3 |
| | Uttar Pradesh | 16.8 | 227.9 | 13.6 |
| | Assam | 8.9 | 34.3 | 3.9 |
| | Chhatisgarh | 3.6 | 32.4 | 8.9 |
| | Karnataka | 2.4 | 30.9 | 12.9 |
| | Andhra Pradesh | 1.3 | 26.6 | 20.0 |
| | Madhya Pradesh | 4.5 | 26.0 | 5.8 |
| | Meghalaya | 5.5 | 19.2 | 3.5 |
| | Tamil Nadu | 0.8 | 11.5 | 14.9 |

## 1.4 Status of Mushrooms

World produces 61.16 lakh tonnes of cultivated mushrooms annually. The ranking of mushroom production world-wide is button (31%), shiitake (24%), oyster (14%), black ear mushroom (9%), paddy straw mushroom (8%) and milky/ others (the rest). China produces 63% of the world production of mushrooms and ranks first among world's mushroom producing countries.

India produces more than 70,000 MT of mushroom. In India button, oyster, milky and paddy straw mushrooms are grown but button mushroom contributes highest share of production of mushroom. Commercial scale button mushroom cultivation started in India in seventies when some units at Chail and Kasauli (Himachal Pradesh) started growing mushrooms on a large scale. The button mushrooms grown on hills of Himachal were marketed fresh in big cities like Delhi and Mumbai (using insulated ice pack cold stored box transport system). But now button mushroom is grown all over the country, mostly in tropical areas, where raw materials and labour are available at competitive rates as compared to hilly regions. The mushrooms produced in the large commercial units in India are processed and packed in cans/jars for export. This is necessary as button mushroom has short shelf-life of less than a day. The consumption of mushroom is increasing in India.

## 1.5 Status of Spice Crops

Spices, mostly grown under tropical farming systems, have an important position among the horticulture crops with regard to huge domestic consumption, sizeable export earnings and substantial employment generation particularly in the rural sector. India has been a traditional producer and exporter of spices. In fact, in the world, India is associated with spices. Spices are export oriented; low volume and high value crops and are of great commercial importance for India.

The global spices industry is poised for a major leap in the 21st century. Global trade in spices is expected to attain even higher levels due to the anticipated advances in the global food industry. However, with growing opportunities, challenges are also immense. All the concerned implication of the WTO, the growing green sentiment, the Agreement on Sanitary and Phytosanitary measures (ASPS), the Agreement on Technical Barriers to Trade (ATBT) are to be addressed.

Area and production of spices in the country have registered substantial increase over the last fifteen years with average annual growth rate of 2.7% and 6.2% in area and production respectively Chadha and Choudhary (2007). The spices production in India is of the order of 4.01 million tonnes from an area of about 2.46 million ha. Chilli is the major spice crop occupying about 31 percent of area under cultivation and contributing about 30 percent of total spices production in the country. Turmeric accounts for 20% of production and 8% of area, while garlic accounts for 21% of production and 7% of area. Seed spices contribute 17% of production and occupy 41% of area while Pepper contributes 1.2% of production and occupies 8% of area of the total spices in the country.

Area, production and productivity of spices is given in Table 1.11. The major states growing different spices crops is given in Table 1.12.

**Table 1.11:** Area, Production and Productivity of Spices in India (2009-10)

| Spice | Area (000'ha) | Production (000'MT) | Productivity (MT/ha) |
|---|---|---|---|
| Chilli | 767.23 | 1202.94 | 1.6 |
| Garlic | 164.86 | 833.97 | 5.1 |
| Turmeric | 190.96 | 792.98 | 4.4 |
| Ginger | 107.54 | 385.33 | 3.6 |
| Corriander | 360.00 | 236.72 | 0.7 |
| Tamarind | 57.99 | 185.46 | 3.2 |
| Cumin | 377.01 | 156.33 | 0.4 |
| Fennugreek | 43.25 | 57.44 | 1.3 |
| Fennel | 50.67 | 56.55 | 1.1 |
| Pepper | 195.92 | 50.02 | 0.3 |
| Cardamom | 90.20 | 15.72 | 0.2 |
| Other seeds spices | 19.36 | 13.04 | 0.7 |
| Ajowan | 17.06 | 10.27 | 0.6 |
| Nutmeg | 15.05 | 8.00 | 0.5 |
| Tejpat | 3.24 | 7.08 | 2.2 |
| Cinnamon | 1.00 | 1.67 | 1.7 |
| Clove | 2.60 | 1.16 | 0.4 |
| Others | 9.34 | 0.24 | 0.03 |
| Total | 2463.29 | 4015.91 | 1.6 |

**Table 1.12:** Major States Growing the Spices Crop Based on Area Under Cultivation

| Spices | Major States |
|---|---|
| Chillies | Andhra Pradesh, Maharashtra, Orissa, Karnataka, West Bengal, Tamil Nadu, Gujarat |
| Cumin | Gujarat, Uttar Pradesh |
| Coriander | Rajasthan, Uttar Pradesh |
| Pepper | Kerala, Karnataka, Tamil Nadu |
| Turmeric | Tamil Nadu, Andhra Pradesh, Orissa, Karnataka |
| Garlic | Gujarat, Uttar Pradesh, Madhya Pradesh, Rajasthan, Orissa |
| Ginger | Assam, Orissa, Kerala, Karnataka , West Bengal, Sikkim, Mizoram |
| Cardamom | Kerala, Karnataka, Sikkim, Tamil Nadu |
| Fenugreek | Rajasthan, Gujarat, Tamil Nadu |

## 1.6 Status of Ornamental Crops

The floriculture industry in India is characterized by growing traditional flowers (loose flowers) and cut flowers under open field conditions and protected environment conditions respectively. India also has a strong dry flower industry, which provides major contribution to the overall trade. Other segments like fillers, potted plants, seeds and planting material, turf grass industry and value added products also contribute a share in the overall growth of the floriculture sector. The traditional flower cultivation, comprising of growing loose flowers mostly for worship, garland making and decorations, forms the backbone of Indian floriculture, which is mostly in the hands of small and marginal farmers. The domestic industry is growing at an annual rate of 7-10% per annum. During 2003-04 the turnover was Rs. 500 crores. The area under production of flowers is around 1,82,900 hectares concentrated mostly in Tamil Nadu, Andhra Pradesh, Maharashtra, and West Bengal. The industry is characterized by sale of mostly loose flowers (roses, chrysanthemum, jasmine, marigold, crossandra, tuberose etc.) and the export surplus (rose, carnation, gerbera, orchids, anthuriums etc.) from the cut flower industries.

In India nearly 98.5% of flowers are grown under open cultivation and hardly 1.5% flowers are grown under greenhouse cultivation. Hi-Tech floriculture industry in India, is still in its infancy. The corporate sector started its entry into the floriculture sector in the early 1990's. Most of the floriculture projects have adopted technology from either the Netherlands or Israel. Presently the total investment in this sector is Rs. 1,000 crores and is spread over 110 hitech floriculture companies covering an area of 500 ha mostly under rose cultivation. An additional 60 Hi-tech projects have been planned covering around 200 ha. The average investment in each floriculture project ranges between Rs. 3 to 4 crores per hectare. Most units are around 3 to 4 hectares each in size. One of the largest projects is CCL Flowers Ltd., Bangalore, which has more than 25 hectares of area under rose production.

## Traditional Floriculture Sector

India is the second largest producer of flowers after China. About 1,82,900 ha of area is under floriculture producing 10,20,600 MT of loose flowers annually (2009-10). The loose flower sector grew over the Plan periods with the production of flowers almost doubling (1.84 times) between 1992-2007. Chadha and Choudhary (2007). The area expansion during the same period was to a tune of 1.64 times. The traditional flower sector registered an impressive growth during 1992-2002 which grew from 71,000 ha to 1,06,000. During 2002-05 an additional 10,000 ha has been brought under the traditional flowers by the end of 2004-05 to register an overall area of 1,16,000 ha. The production during the corresponding periods was 3,66,000, 5,35,000 and 6,55,000 MT respectively.

Tamil Nadu is the leading producer of loose flowers closely followed by Karnataka both in terms of area and production (Table 1.13). The productivity per hectare was highest in Punjab (48.24 MT/ha) followed by Jharkhand (13.75 MT/ha). On the other hand the productivity of flowers was the least in Madhya Pardesh (0.76 MT/ha) followed by Uttarakhand (0.78 MT/ha) The overall productivity of the country during 2009-10 was recorded at (5.58 MT/ha). Nearly 75% of area under floricultural crops is mainly concentrated in seven states comprising of Tamil Nadu, Karnataka, Andhra Pradesh, West Bengal, Maharashtra, Haryana, Gujarat, Uttar Pradesh and Delhi.

**Table 1.13:** Area, Production and Productivity of Loose Flowers in Different States (2009-10)

| State/Union Territory | Area ('000ha) | Production Loose Flowers (000'MT) | Productivity (MT/ha) |
|---|---|---|---|
| Tamil Nadu | 31.97 | 247.3 | 7.74 |
| Karnataka | 27.01 | 203.9 | 7.55 |
| Andhra Pradesh | 21.40 | 130.3 | 6.09 |
| Maharashtra | 17.51 | 91.1 | 5.20 |
| Punjab | 1.70 | 82.0 | 48.24 |
| Haryana | 6.20 | 60.3 | 9.73 |
| West Bengal | 21.94 | 55.2 | 2.52 |
| Gujarat | 12.53 | 14.95 | 3.95 |
| Orissa | 7.11 | 25.3 | 3.56 |
| Jharkhand | 1.60 | 22.0 | 13.75 |
| Uttar Pradesh | 10.38 | 17.6 | 1.70 |

*Contd...*

*Table 1.13: Contd..*

| | | | |
|---|---|---|---|
| Chhattishgarh | 4.09 | 13.5 | 3.30 |
| Delhi | 5.50 | 5.7 | 1.04 |
| Madhya Pradesh | 6.59 | 5.0 | 0.76 |
| Rajasthan | 3.28 | 4.9 | 1.49 |
| Pondicherry | 0.29 | 2.4 | 8.28 |
| Bihar | 0.20 | 2.3 | 11.5 |
| Uttarakhand | 1.29 | 1.0 | 0.78 |
| Himachal Pradesh | 0.68 | 0.6 | 0.88 |
| Andaman & Nicobar | 0.03 | 0.30 | 10.0 |
| Jammu & Kashmir | 0.13 | 0.2 | 1.54 |
| All India | 182.9 | 1020.6 | 5.58 |

## Cut Flower Sector

Eventhough there has been a sporadic increase in the production of flowers, India's share in global trade is hardly 0.4% compared to the Netherlands (60%), Kenya (10%), Columbia (5%) and Israel (2%). Of the 100 large green houses, 12 green houses are near New Delhi (covering 50 hectares), where the climate is extreme during winter and summer. There are nearly 38 green houses in Bangalore, covering 200 ha. The Maharashtra state has 99 units of greenhouses, of which 15 large greenhouses cover 115 ha and 84 small greenhouses cover 35 ha. The production of cut flowers increased over plan periods to attain a production of 1952 million flowers during X plan period from 615 million cut flowers during the end of 1997 (VIII Plan period) Chadha and Choudhary (2007).

West Bengal leads with a production of 2217.0 million cut flowers followed by Maharashtra with 791.4 million cut flowers. Other major cut flower producing states include Andhra Pradesh, Karnataka, Orissa, Gujarat, Uttarakhand, Uttar Pradesh, Arunachal Pradesh, Jharkhand etc. (Table 1.14).

## 1.7 Status of Medicinal and Aromatic Crops

The medicinal plants based industry is growing at the rate of 7-15% annually. According to a conservative estimate, the value of medicinal plants related trade in India is to the tune of about Rs 5,000 crores per annum while the world trade is about 62 billion US dollars and is expected to grow to the tune of 5 trillion US dollars by the year 2050. The present international situation in the production and trade of essential oils and aroma chemicals is very complex and vibrant. No doubt, the developing countries have a dominant position in world production, but still the competition from developed countries remains very strong. The world production from the developing countries accounts for 55% followed by developed countries (10%). India's contribution in the world production of essential oils has always been significant and showing a steady rise. At present India's position has been rated as third with 16-17% share in terms of production and second with 21 to 22% share in value in the world production. This has been ascribed to the production of menthol mint oil particularly in north India, in huge quantities.

**Table 1.14:** All India Production of Cut Flowers (2009-10)

| State/Union Territory | Cut flowers Nos. (Millions) | State/Union Territory | Cut flowers Nos. (Millions) |
|---|---|---|---|
| West Bengal | 2217.0 | Haryana | 108.4 |
| Maharashtra | 791.4 | Delhi | 103.8 |
| Andhra Pradesh | 620.2 | Himachal Pradesh | 60.5 |
| Karnataka | 586.0 | Sikkim | 20.0 |
| Orissa | 535.6 | Mizoram | 14.2 |
| Gujarat | 506.3 | Jammu & Kashmir | 6.6 |
| Uttarakhand | 341.4 | Nagaland | 1.7 |
| Uttar Pradesh | 295.8 | Bihar | 1.1 |
| Arunachal Pradesh | 286.0 | | |
| Jharkhand | 171.1 | **All India** | **6667.1** |

India is one of the top twelve mega-bio-diversity centres of the world with two hot spots located in the Eastern Himalayas and Western Ghats containing 3,500 and 1,600 endemic species of higher plants, respectively. The Indian biodiversity is immensely rich in medicinal and aromatic plants occurring in diverse eco-systems. The area of 2,72,150 ha presented here for some of the important medicinal plants are on the basis of consolidation made from the fragmented information available in literature and also estimates. (Table 1.15). National Medicinal Plant Board has identified 32 prioritized medicinal plants. The cumulative annual demand of all 32 plants is estimated to be 141398.6 MT during 2004-05.

**Table 1.15:** Area Under Important Medicinal Plants

| Sl. No. | Botanical Name | Vernacular Name(s) | Area (ha) | Cultivating States & Area |
|---|---|---|---|---|
| 1. | *Acorus calamus* | Bhoja, Bauj | 500 | Madhya Pradesh, Bihar, Haryana & Andhra Pradesh |
| 2. | *Adathoda vasica* | Adusa, Ardusa | 1,000 | |
| 3. | *Aegle marmelos* | Bael, Bili | 3,000 | Bihar, West Bengal, Uttar Pradesh and Maharashtra. |

*Contd...*

*Table 1.15: contd...*

| | | | | |
|---|---|---|---|---|
| 4. | *Aloe vera* | Ghritkumari | 1,000 | Rajasthan, Gujarat, Maharashtra, Andhra Pradesh and Tamil Nadu. |
| 5. | *Androraphis Paniculta* | Kalmegh | 2,000 | Many states |
| 6. | *Artemisa annua* | | 100 | Parts of Maharashtra and Gujarat |
| 7. | *Asparagus recimosus* *A. adscendus* | Satavari | 1,000 | Maharashtra, Gujarat, West Bengal and Uttar Pradesh |
| 8. | *Azadirachta indica* | Neem | 2,000 | Maharashtra, Gujarat, Rajasthan, as road side plantation |
| 9. | *Bacopa monnieri* | Jal Brahmi | 100 | West Bengal, Uttar Pradesh and Andhra Pradesh |
| 10. | *Baliospermum montanum* | Danhtimul | 50 | |
| 11. | *Cassia absus* | Chaksu | 50 | - |
| 12. | *Cassia angustifolia* | Senna | 25,000 | Rajasthan, Gujarat, Tamil Nadu, Karnataka, Andhra Pradesh and Maharashtra |
| 13. | *Centella asiatica* | Brahmi | 150 | Assam, West Bengal, Bihar and Uttar Pradesh |
| 14. | *Chlorophytum borivilianum* | Safed musli | 40,000 | Madhya Pradesh, Maharashtra, Gujarat and Rajasthan. At present almost all the states have sizable area. |
| 15. | *Coleus forskohlii* | - | 1,000 | Tamil Nadu, Karnataka and Maharashtra |
| 16. | *Commiphora wightii* | Guggal | 2,000 | Gujarat, Madhya Pradesh, Maharashtra and Rajasthan. |
| 17. | *Curculigo orchioides* | Kali Musli | 100 | Kerala, Tamil Nadu, Madhya Pradesh and Karnataka |
| 18. | *Datura metal* | Dhatura | 200 | - |
| 19. | *Desmodium gangeticum* | Salparni | 1,000 | Tamil Nadu, Kerala, Maharashtra, West Bengal and Uttar Pradesh |
| 20. | *Emblica officinalis* | Amla, Amloki or Anwala | 20,000 | Uttar Pradesh, Gujarat, Rajasthan, Andhra Pradesh, Karnataka and Maharashtra |

*contd...*

*Table 1.15: contd...*

| | | | | |
|---|---|---|---|---|
| 21. | *Garcinia gummigutta* | Bilatti-Amla | 50 | Kerala, Tamil Nadu and Maharashtra |
| 22. | *Gloriosa superba* | Kalihari Vachnag | 1,000 | Tamil Nadu and Himachal Pradesh |
| 23. | *Hemidesmus indicus* | Ananthamul | 100 | - |
| 24. | *Kaemperia galanga* | Chandramula | 300 | Kerala and West Bengal |
| 25. | *Lawsonia inermis* | Mehandi | 35,000 | Rajasthan and to some extend Andhra Pradesh. |
| 26. | *Mucuna pruriens* | Kaunch | 1,000 | Andhra Pradesh, Karnataka, Himachal Pradesh, Gujarat and Maharashtra |
| 27. | *Ocimum sanctum* | Tulsi | 1,000 | Uttar Pradesh, Jammu and Kashmir and Karnataka |
| 28. | *Phyllanthus fraternus* | Jar-amla, Bhuni amla | 100 | Tamil Nadu and Karnataka |
| 29. | *Piper longum* | Pipal, Pipli | 1,500 | Maharashtra, Kerala, Tamil Nadu, Assam and West Bengal |
| 30. | *Plantago ovata* | Isobgul | 1,20,000 | Gujarat, Rajasthan and Madhya Pradesh |
| 31. | *Plumbago zeylanica* | Chitrak Chitawar | 100 | Kerala and Tamil Nadu |
| 32. | *Psoralea corylifolia* | Babchi | 50 | Uttar Pradesh, Madhya Pradesh and Rajasthan |
| 33. | *Rauvolfia serpetina* | Chota Chand, Sarpagandha | 1,000 | West Bengal, Orissa, Madhya Pradesh, Bihar, and Uttar Pradesh |
| 34. | *Solanum surattense* | Kateli, Bhoiringini | 100 | - |
| 35. | *Trachyspermum ammi* | Ajowan | 100 | Rajasthan, Madhya Pradesh and Gujarat |
| 36. | *Vetiveria zizanioides* | Khas-Khas, Khas | 500 | Border crop in Uttar Pradesh, Rajasthan and Bihar |
| 37. | *Withania somnifera* | Ashwagandha | 10,000 | Madhya Pradesh, Uttar Pradesh, Gujarat, Rajasthan, Maharashtra and Andhra Pradesh. |
| Total | | | 2,72,150 | |

## 1.8 Status of Plantation Crops

There are two groups of plantation crops namely–coconut, arecanut , oil palm, cocoa, cashewnut and tea, coffee, rubber. The combined volume of output of tea coffee and rubber is about 2 million tonnes and constitute about 3 per cent share of the total value of agricultural output in India. Coconut is one of the important plantation crop grown in about 2 lakh hectare in India.

The area of production and productivity of major plantation crops is given in Table 1.16.

**Table 1.16:** Area, Production & Productivity of Major Plantation Crops (2009-10).

| Crop | Area ('000 ha) | Production (000'MT) | Productivity (MT/ha) |
|---|---|---|---|
| Coconut | 1895.2 | 10824.3 | 5.7 |
| Cashewnut | 923.0 | 613.0 | 0.7 |
| Arecanut | 400.1 | 478.0 | 1.2 |
| Oil Palm | 210.0 | 976.0 | 20.5 |
| Cocoa | 46.3 | 12.9 | 0.3 |
| Total | 3264.6 | 11928.2 | 3.7 |

**Coconut**

Coconut is grown globally over an area of 12.19 million ha producing 13.68 million MT (copra equivalent) in 93 countries of the world. Asia and Pacific region accounts for a major share of area and production occupying 10.7 million ha (88%) and 9.6 million MT of copra equivalent (70.18%). Indonesia, Philippines and India are the major producers of coconut in the world. The area, production and productivity of coconut have increased at the global level (0.89, 4.36 and 3.79% per year) during the past 15 years (1990-2004). Compared to this, there has been increase in area, production and productivity in the APCC region to the tune of (1.06, 2.09 and 1.07% per year). Among 15 APCC member countries, both positive and negative trends in area and production have been reported Chadha and Choudhary (2007). There was a positive increase in area, production and productivity in India, Indonesia, Kiribati, Marshal Islands and Vanuatu. However, there was significant increase in productivity in Solomon Islands, Thailand and Vietnam despite progressive decrease in area.

Coconut farming and associated activities provide food and livelihood security to more than 10 million people in India. The crop contributes to about Rs. 70,000 million to the country's GDP and about 6 per cent to the edible oil pool. Similarly, coconut industries in the country help to earn foreign exchange to the tune of Rs. 5,000 million per annum. Coconut is also a permanent source of raw material for more than 1,000 oil mills, 15,000 coir industries, 50 desiccated coconut industries and equal number of other industries like activated carbon manufacturing industries, shell powder manufacturing units, solvent extraction units, cocopeat production units etc. which provide continuous employment to nearly 6.5 lakh workers of which 80 per cent are women folk. Consumption of nut for the purpose of Tender Nut Water has increased from 10 to 15%.

Kerala has the highest area under cultivation 7.88 lakh ha, and production 39.92 lakh tonnes however productivity is only 5.07 tonne/ha. Karnataka is second in production with 14.97 lakh tonnes however, productivity is 3.57 tonne/ha. Tamil Nadu is third in production with 3.90 lakh tonnes however productivity is 9.47 tonne/ha. Lakshadweep has only 2,700 ha under cultivation but has highest productivity of 14.81 tonne/ha.

**Table 1.17.** Area, Production and Productivity of Coconut in India (2009-10)

| State/UT | Area ('000 ha) | Production ('000 tonnes) | Productivity (MT/ha) |
|---|---|---|---|
| Kerala | 788.0 | 3992.0 | 5.07 |
| Karnataka | 419.0 | 1497.0 | 3.57 |
| Tamil Nadu | 390.0 | 3692.0 | 9.47 |
| Andhra Pradesh | 104.0 | 667.0 | 6.60 |
| Orissa | 51.0 | 190.0 | 3.73 |
| West Bengal | 28.6 | 245.0 | 8.57 |
| Goa | 25.6 | 88.0 | 3.44 |
| Andaman & Nicobar | 21.7 | 56.0 | 2.58 |
| Maharashtra | 21.0 | 120.0 | 5.71 |
| Assam | 18.8 | 101.0 | 5.37 |
| Gujarat | 16.0 | 108.0 | 6.75 |
| Tripura | 5.8 | 8.0 | 1.38 |
| Lakshadweep | 2.7 | 40.0 | 14.81 |
| Pondicherry | 2.1 | 20.0 | 9.52 |
| Nagaland | 0.9 | 0.3 | 0.33 |
| **All India** | **1895.2** | **10824.3** | **5.71** |

**Cashew**

Cashew is grown in India, Brazil, Vietnam, Tanzania, Mozambique, Indonesia, Sri Lanka and other tropical Asian and African Countries. The area under cashew in the world is 30.62 lakh ha. The world production of cashew is estimated to be around 20.82 lakh MT. India's share in the world raw nut production accounts to about 25%.

In India cashew is grown mainly in Maharashtra, Goa, Karnataka and Kerala along the west coast and Tamil Nadu, Andhra Pradesh, Orissa and West Bengal along the east coast. To a limited extent it is grown in Manipur, Meghalaya, Tripura, Andaman and Nicobar Islands and Chattisgarh and also in plains of Karnataka. At present Andhra Pradesh ranks first in area and Maharashtra in production and productivity in the country. In 1955, cashew in India was grown in an area of 1.1 lakh ha. In the year 2005, cashew was grown over in an area of 8.37 lakh ha. India's raw nut production has increased from 0.79 lakh MT in 1955 to 5.73 lakh MT in 2005-06 Chadha and Choudhary (2007).

The production during 2009-10 was 6.13 lakh tonnes from 9.23 lakh ha (Table 1.18).

**Table 1.18**: Area, Production and Productivity of Cashew in India (2009-10)

| State/UT | Area ('000 ha) | Production ('000 tonnes) | Productivity (MT/ha) |
|---|---|---|---|
| Andhra Pradesh | 183.0 | 99.0 | 0.54 |
| Maharashtra | 175.0 | 198.0 | 1.13 |
| Orissa | 143.0 | 84.0 | 0.59 |
| Tamil Nadu | 133.0 | 60.0 | 0.45 |
| Karnataka | 118.0 | 53.0 | 0.45 |
| Kerala | 72.0 | 66.0 | 0.92 |
| Goa | 55.0 | 26.0 | 0.47 |
| Chhattisgarh | 33.0 | 17.0 | 0.52 |
| West Bengal | 11.0 | 10.0 | 0.91 |
| **All India** | **923.0** | **613.0** | **0.66** |

**Cocoa**

World cocoa production has registered 131.7% increase during 1974-2004 and was estimated at 3.52 million tonnes during 2004-05. The global production was 1.6 million tonnes in 1974, 1.8 million tonnes in 1984 and 2.7 million tonnes in 1994. Africa has the world's largest cocoa producing area. The world's largest cocoa bean producer is Côte d'Ivoire, currently producing more than 1.45 million tonnes beans, which has a growth rate of 2.3 percent/ year. However, yields in Côte d'Ivoire are well below levels of Asia partly because of less use of agricultural inputs. The other major producers are Ghana (0.67 million tonnes), Indonesia (0.49 million tonnes), Nigeria ( 0.37 million tonnes), Brazil (0.17 million tonnes), Cambodia (0.13 million tonnes) and Ecuador (0.09 million tonnes). The variety Forastero comprises 95% of the world production of cocoa, and is the most widely used. Overall, the highest quality of cocoa comes from the variety Criollo which is considered a delicacy. However, Criollo is harder to produce, hence very few countries produce it, with the majority of production coming from Venezuela (Chuao and Porcelana). The Trinitario is a mix between Criollo and Forastero. Netherlands is the leading cocoa processing country, followed by the U.S. Cocoa and its products (including chocolate) are used world-wide. Belgium had the highest per-capita consumption at 5.5 kg, 10 times the world average namely a decade back (1995-96). Chadha and Choudhary (2007).

In India cocoa is cultivated in nearly 46,300 ha with production of nearly 12,900 MT of dry beans. Kerala is the leading producer of cocoa with 11,000 ha of area and 6,300 MT of production with 573 kg/ha in productivity (Table 1.19). Andhra Pardesh has highest area under cultivation. Kerala has highest production (6,300 tonnes) and productivity (573 kg/ha).

**Table 1.19:** Area, Production and Productivity of Cocoa in India (2009-10)

| State | Area ('000 ha) | Production ('000 MT) | Productivity (kg/ha) |
|---|---|---|---|
| Andhra Pradesh | 17.0 | 2.7 | 159 |
| Kerala | 11.0 | 6.3 | 573 |
| Tamil Nadu | 9.3 | 0.9 | 97 |
| Karnataka | 9.0 | 3.0 | 333 |
| **All India** | **46.3** | **12.9** | **279** |

**Arecanut**

Arecanut (*Areca catechu* L.) is one of the important commercial crops of India. Arecanut industry forms economic backbone of nearly 10 million people of India and for many of them it is the sole means of livelihood. The country produces arecanut valued at Rs 3,000 crores annually. India accounts for 56% and 58% of the total area and production of arecanut in the world. The crop is cultivated in India mainly in small and marginal holdings and also as a homestead crop. Its cultivation is concentrated in the states of Karnataka, Kerala, Assam, Maharashtra, Meghalaya, Tamil Nadu, West Bengal, Andhra Pradesh and Andaman & Nicobar Islands. Although the production of arecanut is localized in a few States, the commercial products are widely consumed all over the country by all classes of people. Production of arecanut in the country which stood at 2,56,600 MT from an estimated area of 2,26,600 ha during 1992-93, has reached 4,56,600 MT from an area of 3,74,300 ha during 2004-05. During 2009-10 arecanut production was 4,78,000 tonnes from an area of 4,00,100 ha. Karnataka, Kerala and Assam are the three major states producing arecanut ( Table 1.20).

**Table 1.20:** Area, Production and Productivity of Arecanut in India (2009-10)

| State/UT | Area ('000 ha) | Production ('000 MT) | Productivity (kg/ha) |
|---|---|---|---|
| Karnataka | 184.5 | 224.0 | 1217 |
| Kerala | 97.2 | 112.1 | 1153 |
| Assam | 70.0 | 62.7 | 896 |
| West Bengal | 11.4 | 21.2 | 1860 |
| Meghalaya | 6.6 | 8.2 | 1242 |
| Tamil Nadu | 5.0 | 10.4 | 2200 |
| Tirpura | 4.4 | 8.4 | 1909 |
| Andaman & Nicobar | 4.1 | 6.0 | 1463 |
| Maharashtra | 2.2 | 3.6 | 1636 |
| Goa | 1.9 | 2.8 | 1474 |
| Andhra Pradesh | 0.3 | 0.2 | 667 |
| Nagaland | 0.2 | 1.3 | 6500 |
| Pondicherry | 0.1 | 0.1 | 1000 |
| **All India** | **400.1** | **478** | **1195** |

## Coffee

The unique feature of the global coffee industry is that the production is confined to 60 odd developing countries, whereas, it is largely consumed in the developed countries. The major coffee producers in the world are Brazil (30%), Vietnam (11%), Columbia (10%) and Indonesia (6%). Arabica and Robusta coffee are the two species that are commercially cultivated in the world. Of these, nearly 70% is Arabica and 30% Robusta.

In India, coffee is cultivated in an area of around 4.00 lakh hectares predominantly in Karnataka, Kerala and Tamil Nadu, which contributes about 98 per cent of the total Coffee production. There are about 2.58 lakh coffee holdings in the country, out of which around 2.55 lakh holdings (99%) constitutes the small growers' category (upto 10 hectares) and the balance 1% of the total holdings fall under large grower category having above 10 hectares.

The Indian coffee production in 2010-11 was 3.02 lakh MT comprising 94.14 thousand MT of Arabica and 207.86 thousand MT of Robusta. The current productivity of Indian coffee is around 593 kg/ha. for Arabica and 1037 kg/ha for Robusta. The major coffee producing country like Vietnam (Robusta) and Brazil (Arabica) have much better productivity levels and stronger competitiveness compared to India.

**Table 1.21:** Production of Coffee (2010-11)

| **State** | **Production ('000 MT)** | | |
|---|---|---|---|
| | **Arabica** | **Robusta** | **Total** |
| Karnataka | 74.58 | 139.20 | 213.78 |
| Tamil Nadu | 12.15 | 4.50 | 16.65 |
| Andhra Pradesh | 5.43 | 0.07 | 5.49 |
| Kerala | 1.63 | 64.03 | 65.65 |
| Orissa | 0.26 | 0.00 | 0.26 |
| N.E Region | 0.10 | 0.07 | 0.17 |
| **Total** | **94.14** | **207.86** | **302.00** |

## Tea

India is the 2nd largest producer and consumer of black tea in the world. Tea is grown in 16 States in India, of which Assam, West Bengal, Tamil Nadu and Kerala account for about 96 % of the total tea production. The tea industry provides direct employment to more than a million workers, of which a sizeable number are, women. Additionally, more than two million people derive their livelihood from ancillary activities associated with the industry.

Tea production in India during the year 2009-10 has been estimated at 991.18 million kgs as against 972.77 million kgs achieved in 2008-09. India produced 948.94 million kg tea during (2005-06), 973.07 million kg tea during (2006-07) and 987.02 million kg tea during (2007-08).

India accounts for 1/5th of the global tea area and nearly 1/3rd of the total world production. The area under tea cultivation is 5,78,458 ha with a productivity of 1,705 kg/ha. India consumes about 23.5% of total world consumption of tea, which constitutes about 81% of the tea produced in the country.

**Table 1.22:** Area Under Tea Cultivation (2007)

| **State** | **Area (ha)** | **State** | **Area (ha)** |
|---|---|---|---|
| Assam | 3,21,319 | Kerala | 37,137 |
| West Bengal | 1,15,095 | Karnataka | 2,141 |
| Other Northern States | 22,304 | **All India** | **5,78,458** |
| Tamil Nadu | 80,462 | | |

**Table 1.22A**: Production of Tea in India

| **Year** | **Production (million kg)** |
|---|---|
| 2005-06 | 948.94 |
| 2006-07 | 973.07 |
| 2007-08 | 987.02 |
| 2008-09 | 972.77 |
| 2009-10 | 991.18 |

### Natural Rubber

India is the fourth largest producer (with a share of 8.5% in world production) and consumer among all rubber producing countries, while 2nd in productivity in the world. With regard to rubber area India's position is 5th, the largest area is in Indonesia followed by Thailand while production is more in Thailand than Indonesia. Rubber is grown mainly in Kerala and also in the States of Tamil Nadu, Karnataka, Tripura, Assam, Meghalaya, Nagaland, Mizoram, Manipur, Arunachal Pradesh, Andaman & Nicobar Islands, Goa, Maharashtra, Orissa, West Bengal and Andhra Pradesh. Rubber plantations are spread over 6.87 lakh hectares in the country.

**Table 1.22B:**. Production and Productivity of Natural Rubber in India

| **Year** | **Production ('000Tonnes)** | **Productivity (kg/ha)** |
|---|---|---|
| 2009-10 | 840 | 1,867 |
| 2010-11 | 851 | 1,784 |

## Oil Palm

The cooking oil consumption in India is about 11.85 million MT and about 4.54 million MT is being imported. About 100 thousand hectare has come under Oil Palm cultivation in Andhra Pardesh, Karnataka, Tamil Nadu, Orissa, Chhattisgarh, Kerala, Goa, Tripura, Mizoram and Gujarat. The productivity is around 20 MT/ ha. and oil yield is 17 to 20%.

## Noni

Noni is being cultivated in about 261 ha by 348 farmers. It is cultivated in Maharashtra , Karnataka, Orissa, Andhra Pradesh, Gujarat, Rajasthan, Tamil Nadu, Madhya Pradesh and Andaman & Nicobar Islands. Andaman & Nicobar Islands have maximum number of cultivators.

# 1.9 Exports

India is one of the large producers of horticulture crops in the world. Horticultural crops occupy about 8% of gross cropped area and account for 30% of agricultural GDP. However, the presence of India as an exporter of horticulture products is insignificant. India's share in global horticulture trade was a mere 0.5% in 2004 as per a recent World Bank study ($ 575 million compared with a global trade of $ 108 billion). The quantity and value of export of horticultural products as per APEDA/ NHB are given in Table 1.23.

**Table 1.23:** Quantity and Value of Export of Horticultural Products (APEDA)

| **PRODUCT** | **2007-08** | **2008-09** | **2009-10** | | | |
|---|---|---|---|---|---|---|
| | **Qty ('000 MT)** | **Value (Rs. Crore)** | **Qty ('000 MT)** | **Value (Rs. Crore)** | **Qty ('000 MT)** | **Value (Rs. Crore)** |
| | | | **Fresh Fruits & Vegetables** | | | |
| Fresh onions | 1008.61 | 1036 | 1670.19 | 1828 | 1664.92 | 2319 |
| Other fresh vegetables | 350.24 | 489 | 505.29 | 680 | 419.24 | 732 |
| Dried nuts (walnuts) | 6.72 | 162 | 5.70 | 141 | 9.07 | 198 |
| Fresh mangoes | 54.35 | 127 | 83.70 | 171 | 74.46 | 201 |
| Fresh grapes | 96.96 | 318 | 124.63 | 409 | 131.15 | 545 |
| Other fresh fruits | 207.70 | 305 | 256.77 | 431 | 260.68 | 523 |
| **Total** | **1724.57** | **2437** | **2002.86** | **3659** | **2559.53** | **4518** |
| | | | **Floriculture** | | | |
| Floriculture | 36.24 | 340 | 30.80 | 653 | 26.82 | 294 |
| Fruit and vegetable seeds | 10.16 | 142 | 8.54 | 122 | 8.88 | 145 |
| Total | 46.40 | 482 | 39.33 | 774 | 35.70 | 439 |

*contd...*

*Table 1.23: contd...*

| Processed Fruits & Vegetables | | | | | | |
|---|---|---|---|---|---|---|
| Dried and preserved vegetables | 125.73 | 430 | 147.86 | 496 | 124.61 | 532 |
| Mango pulp | 166.75 | 510 | 173.01 | 753 | 186.20 | 745 |
| Other processed fruits and vegetables | 311.76 | 963 | 387.13 | 1372 | 397.98 | 1436 |
| **Total** | **604.23** | **1902** | **708.00** | **2621** | **708.79** | **2712** |
| **GRAND TOTAL** | **2375.21** | **4822** | **3393.60** | **6769** | **3304.01** | **7670** |

## Major Importing Countries

The major markets for Indian horticulture products are the West Asian countries and UK, Germany, The Netherlands in Europe. The top five destinations for various product groups are listed in Table 1.23a.

Fresh fruits and vegetables comprise almost 35% of the world trade in horticulture, out of which, almost two-third is accounted for by four items, namely, citrus, banana, apple and grape. The other important items are mangoes, papaya and kiwi. Among the vegetables, which account for about 22% in the world trade in horticulture, the major items are tomato, onion, potato, bean, pea, mushroom, asparagus and capsicum.

Processed fruits and vegetables account for about 20% and 17% of the world horticultural trade, respectively. Among the processed fruits, 41% trade is of fruit juices and 12% of dried fruits. Similarly, among processed vegetables, the major items are mushrooms, gherkins and frozen pre-cut vegetables.

**Table 1.23a**: Major Importing Countries

| **Product Group** | **Major Importing Countries** |
|---|---|
| Fresh Onions | Bangladesh, Malaysia, UAE, Sri Lanka, Bahrain |
| Other fresh vegetables | UAE, Nepal, UK, Saudi Arabia, USA |
| Walnuts | Spain, Germany, UK, Egypt, The Netherlands |
| Fresh mangoes | Bangladesh, UAE, Saudi Arabia, UK, Nepal |
| Fresh Grapes | The Netherlands, UK, UAE, Bangladesh, Germany |
| Floriculture | USA, The Netherlands, UK, Germany, Japan |
| Fruits & Vegetable Seeds | USA, Pakistan, Japan, Bangladesh, The Netherlands |
| Other fresh fruits | Bangladesh, UAE, Nepal, Saudi Arabia, The Netherlands |
| Dried & Preserved Veg. | Bangladesh, Sri Lanka, USA, UK, UAE |
| Mango Pulp | Saudi Arabia, The Netherlands, Yemen, UAE, Kuwait |
| Pickles & Chutneys | Russia, USA, France, Belgium, Spain |
| Other processed fruits/Veg. | USA, Israel, Saudi Arabia, UK, The Netherlands |

Floriculture trade is approx. 6.3 billion US Dollars, of which cut flowers are about 86%.

India is exporting fresh fruits, vegetables, processed products of fruits and vegetables, cut & dried flowers, medicinal and aromatic plants, seeds, spices, cashew kernels and their products, tea and coffee. The total value of export of these commodities increased from Rs. 29723 million in 1991-92 to 124175 million in 2005-06. Horticulture produce and products account for about 35 per cent of the total value of exports of agricultural commodities from India during 2005-06. Of these, export of cashew leads all horticultural commodities followed by spices, tea and coffee. During past one decade the maximum increase in exports was observed for floriculture.

## Fresh fruits

India accounts for about 12% of global fruit production and exports a wide variety of horticultural products, only a handful of commodities or products account for the bulk of this trade. India is one of the largest producer of mango, banana, grape, litchi etc. With the vast majority of its output being consumed domestically, India's share of global exports is only 0.3% for fruits. However, fresh fruits make up around 11% of India's horticultural exports. Grapes and mangoes account for close to 60% of India's exports of fresh fruits.

India is the pre-eminent producer of fresh mango in the world. With an average annual production of about 15.0 million MT, India accounts for 40% of world production. India's role in global mango trade is quite modest as it accounts for about 5% of world exports. As a major exporter, it is ranked sixth behind Mexico, Brazil, Pakistan, Peru and the Philippines even though the combined production of these five major exporters is less than half of that of India.

The grape production in India is of the order of 0.88 million MT. and the current production represents about 2% of the global grape production. Grapes are a highly seasonal crop and there is a potential window in the European market between March and April, at the end of the main southern hemisphere production season, in South Africa and Chile and before Egypt and Turkey enter the market. India is one of the few countries that can produce good quality fresh grape at this time of the year. Thompson Seedless is the main variety, which is exported. This variety is becoming more popular in Europe as the market increasingly prefers green and seedless grapes. Most of the Indian grapes are sold in Europe through the network of Supermarkets.

As in the case of mango, India is also the largest producer of Bananas in the world. With annnual production of about 26 million MT, India accounts for about 24% of the world production. Indian banana production is poly-clonal with over 200 cultivars of which about 40 are of commercial value. Only about 0.7% of banana production is exported which is mainly due to high domestic consumption and absence of the protocols for sea transportation.

Pomegranate is another fruit getting export momentum recently. One, which has contributed to this increase is the development of a new variety Bhagva by the Indian researchers which has an attractive colour appealing to the European market. Traditionally most of India's pomegranate exports have been to the Middle East countries but the interest in the European market has increased due to increasing health consciousness and the medicinal properties of this fruit. The current level of export is about 10,000 MT out of a total production of about 82,000 MT, Maharashtra being the main producing state.

## Fresh vegetables

India accounts for 14% of vegetable production in the world and it ranks second after China. The export of fresh vegetables including onions during 2009-10 was of the order of Rs 3051 crores out of which Onions contributed to Rs 2319 crores. Other major export products are Mushrooms, Green Peas, Egg Plants and Okra. Major markets for Indian vegetables are Bangladesh, Malaysia, UAE, Sri Lanka and Nepal.

## Processed fruits and vegetables

Major processed fruit exports include mango pulp which accounts for close to 50% of the value of this category as well as pickles and chutneys from various fruits including mangoes. Gherkins, dehydrated onion, and mixed frozen vegetables account for 60% of processed vegetable exports. In 2009-10 exports of processed fruits and vegetables accounted for Rs.2712 crore. The major markets are Bangladesh, Sri Lanka, USA, UK and UAE besides other countries of the Middle East.

## Flowers cut and dried

The growth in floriculture in India has been phenomenal in the last decade or so. The area under flower cultivation has tripled from 53,000 hectares (1993-94) to 1,60,700 hectares (2007-08). This period also witnessed a new dimension in the floriculture business, when a large number of units set up in clusters around Bangalore, Pune, Delhi and Hyderabad started production of cut flowers primarily for export markets. This export-oriented sector after the initial burst of considerable success faced certain viability problems, particularly in North India where the cost of setting up such units was higher. This industry has now stabilized with about forty units in the 2 largest clusters of Bangalore and Pune having sustained fresh cut flower exports. The total area under organized floriculture (greenhouse based units) is currently estimated at around 500 hectares.

Today, we find a wider variety of flowers cultivated on a commercial basis including carnation, gerbera, lilium, orchids, anthurium and many others being marketed all over the country. The diversity of agro-climatic conditions available in India presents it with a capability to cultivate a large number of flowers.

The value of exports of floriculture products from India has shown very significant growth, from Rs.18.83 crores (1993-94) to Rs.210.00 crores (2004-05) and Rs.294.00 crores during 2009-10. The dried flowers and other plant parts contribute to about 70% of the exports in the floriculture segment. India is still a marginal player in the world floriculture trade indicating the strong potential that can be exploited in the sector. Major markets for Indian floriculture products are USA, The Netherlands, UK, Germany and Japan.

## Plants and seeds

India is also an exporter of seeds of fruits and vegetables. The export during the year 2009- 10 has been of the order of Rs 145 crores. The major markets are USA, Pakistan, Japan, Bangladesh and The Netherlands. The main exported seeds are Sugar Beet, Beet, Clover, Rye Grass, Timothey Grass, Cabmeseeds, Cauliflower, Onion, Pea and Radish.

## Medicinal and aromatic plants

According to the report of the World Health Organisation (WHO), a large population of the world relies on the traditional systems of medicines, largely plant based to meet their primary health care needs. India at present exports herbal materials and medicines to the tune of Rs. 446 crores only while it has been estimated that this can be raised to Rs. 3,000 crores. The Chinese export based on plants including raw drugs, therapeutics and other is estimated to be around Rs. 18,000 to 22,000 crores. In view of the innate Indian strengths which interalia include diverse eco-systems, technical and farming capacity and a strong manufacturing sector, the medicinal plants area can become a huge export opportunity after fulfilling domestic needs. India exports crude drugs mainly to developed countries, *viz.* USA, Germany, France, Switzerland, UK and Japan, who share between them 75 to 80 percent of the total export of crude drugs from India. The principal herbal drugs that have been finding a good market in foreign countries are Aconite Aloe, Belladona, Acorus, Cinchona, Cassia tora, Dioscorea, Digitalis, Ephedra, Plantago (Isabgol), Cassia (Senna) etc.

## Spices

Indian spices accounts for more than 5% of the total agricultural export earnings in the country. More than 0.5 million tonnes (10-15% of the total spices production) of spices are exported annually to more than 150 countries around the world. World trade in spices is in the order of 7.50 lakh tonnes valued at 1650 million US $. India commands a formidable position in the World Spice Trade with 48% share in Volume and 44% in Value. It is estimated that, with a variety of spices in its production list almost one half of the world demand is being met by Indian exports. The share of value added products is estimated at 66% of the total spices export. But most of this was low end value addition. There is both need and potential for increasing the relative share, more particularly in the case of value added spice exports.

Spices exports have registered substantial growth during the last five years, registering an annual average growth rate of 21% in value and 8% in volume. During the year 2010-11, spices export from India has registered an all time high both in terms of quantity and value. In 2010-11 the export of spices from India has been 525,750 tonnes valued Rs. 6840.71 crores (US $ 1502.85 Million) as against 502,750 tonnes valued Rs. 5560.50 crores (US $ 1173.75 Million) in 2009-10, registering an increase of 28% in dollar terms of value and 5% in volume.

India can now boast as the monopoly supplier of spice oils and oleoresins the world over. In the case of curry powders, spice powders, spice mixtures and spices in consumer packs, India is in a formidable position. During the last one-decade India has improved the share of the value added products in the export basket to more than 55%.

**Table 1.24:** Export of Spices from India

| Year | Quantity (tonnes) | Value (Rs in Lakhs) |
|---|---|---|
| 1991-92 | 142,104 | 38096.76 |
| 1996-97 | 225,295 | 123071.77 |
| 2001-02 | 243,203 | 194054.88 |
| 2005-06 | 320,527 | 229525.00 |
| 2007-08 | 444,250 | 443550.00 |
| 2008-09 | 470,520 | 530025.50 |
| 2009-10 | 502,750 | 556050.00 |

The major constraint faced by India in exporting spices is the fall in unit prices of spices in the international markets. This is due to the increased supply from newly emerged competitors like Vietnam, Thailand, China, Guatemala etc. These producers have no domestic market, which made them to push their entire produce to the international market, making the traditional exporters like India to bear the brunt. India though a major producer of spices can only export around 7% of its production due to the high domestic demand. The situation is gradually changing to make India one of the major importer of spices. In some traditional item like pepper, where India was once a major player, the situation is drastically changed to make India a major importer. The major interventions required are production of sufficient surpluses for export, improved quality standards, good manufacturing practices, export driven packaging and promotion and value addition.

## Arecanut

Arecanut is a commodity, which has a very limited export potential. The bulk of production of arecanut is consumed within the country. However, a small quantity of arecanut is exported mainly meant for the Indian settlers abroad. Quantity of export was at a range of 330 MT to 823 MT during the period 1994-95 to 2000-01. But the export increased substantially during the last couple of years. Export from India was 1809 MT valued at Rs. 1174 lakhs during 2003-04 and was 3695 MT valued at Rs 20.66 crores during 2004-05. The main countries to which arecanut is exported are Indonesia, UAE, Maldives, Nepal, Afghanistan, Pakistan etc.

**Table 1.25:** Item-wise Export of Spices from India

| Item | 2007-08 | | 2008-09 | | 2009-10 | |
|---|---|---|---|---|---|---|
| | Qty ('000 MT) | Value (Rs. Crore) | Qty ('000 MT) | Value (Rs. Crore) | Qty ('000 MT) | Value (Rs. Crore) |
| Chilli | 209,000 | 1098 | 1,88,000 | 1081 | 2,04,000 | 1292 |
| Turmeric | 49,250 | 157 | 52,500 | 249 | 50,750 | 381 |
| Cumin | 28,000 | 292 | 52,550 | 544 | 49,750 | 548 |
| Coriander | 26,000 | 110 | 30,200 | 204 | 47,250 | 226 |
| Fenugreek | 11,100 | 33 | 20,750 | 72 | 21,000 | 70 |
| Pepper | 35,000 | 520 | 25,250 | 414 | 19,750 | 314 |
| Mint Products (3) | 21,100 | 1281 | 20,500 | 1420 | 19,000 | 1190 |
| Other Seeds (1) | 8,850 | 31 | 17,500 | 65 | 15,500 | 59 |
| Curry Powder | 11,500 | 111 | 13,250 | 164 | 14,300 | 189 |
| Tamarind | 11,250 | 3,100 | 11,500 | 41 | 12,200 | 47 |
| Garlic | 675 | 4 | 760 | 4 | 10,750 | 30 |
| Other Spices (2) | 7,750 | 50 | 8,500 | 65 | 8,000 | 78 |
| Fennel | 5,250 | 29 | 8,675 | 43 | 6,800 | 56 |
| Oils & Oleoresins | 6,600 | 563 | 6,850 | 721 | 6,750 | 709 |
| Ginger | 6,700 | 28 | 5,000 | 35 | 5,500 | 47 |
| Celery | 2,900 | 13 | 3,650 | 23 | 5,000 | 27 |
| Nutmeg & Mace | 1,300 | 29 | 2,155 | 61 | 3,275 | 92 |
| Cardamom(S) | 500 | 25 | 750 | 47 | 1,975 | 166 |
| Cardamom(L) | 1,325 | 15 | 1,875 | 23 | 1,000 | 18 |
| Vanilla | 200 | 18 | 305 | 27 | 200 | 23 |
| Total | 444,250 | 4436 | 470,520 | 5300 | 502,750 | 5561 |
| Value in Million US$ | | 1101.80 | | 1168.40 | | 1173.75 |

(1) Include Bishops Weed (Ajwanseed), Dill Seed, Poppy Seed, Aniseed, Mustard etc.

(2) Include Asafoetida, Cinnamon, Cassia, Cambodge, Saffron, Spices (Nes) etc.

(3) Include Menthol, Menthol Crystals and Mint Oils.

## Cashew

India has been exporting the cashew kernels since early part of 20th century. The country is at present exporting annually about 1.08 lakh tonne of cashew kernels worth over Rs. 2906/- crores. At present India produces annually about 6.1 lakh tonne of raw cashewnuts. However, India needs about 11 – 12 lakh tonne of raw cashewnuts per annum for feeding over 1700 cashew processing units. In order to meet the requirement of the processing industry, India imports annually about 5-6 lakh tonne of raw cashewnuts from African and other countries.

Over the years, both the export earnings as well as quality of kernels has been increasing. Export earnings has been on the increase since 1955. India has earned an all time high export earnings of Rs. 2906 Crores during 2009-10 (Table 1.26a).

Cashew has been mostly exported to USA, UAE, Netherlands, Japan, UK, Saudi Arabia, France, Spain, Belgium, Greece, Egypt, Australia, Germany, Syria, Norway and Jordan. The exports during 2004-06 and 2007-2010 are given in Table 1.26 and Table 1.26a. respectively.

Cashew industry provides employment to over 5.5 lakh workers annually, mostly women. Between 1980 and 1985, although export earnings increased, quantity of cashew kernels exported decreased. Since 1985, there is a steady growth in the quantity of cashew kernels exported. Inspite of the domestic production of over 5 lakh tonne and import of over 5.5 lakh tonne of raw cashewnuts, the total availability of raw cashewnuts is not able to keep pace of processing. The availability of raw cashewnuts for importing by India may gradually decline or may all together stop. Hence, there is urgent need to increase the domestic raw cashewnut production and become self sufficient in raw cashewnut production.

**Table 1.26:** Trade Analysis of Cashew (2004-05 to 2005-06)

| **Item** | **2004-05** | **2005-06** |
|---|---|---|
| Processing units | 1707 Nos. | 1707 Nos. |
| Import of raw cashewnuts (quantity) | 5.78 lakh tonne | 5.65 lakh tonne |
| Import of raw cashewnuts (value) | Rs. 2191 Crores | Rs. 2163 Crores |
| Export of cashew kernel (quantity) | 1.27 lakh tonne | 1.14 lakh tonne |
| Export of cashew kernel (value) | Rs. 2709 Crores | Rs. 2515 Crores |
| Export of CNSL (quantity) | 7474 tonne | 6405 tonne |
| Export of CNSL (value) | Rs. 8 Crores | Rs. 7 Crores |
| Foreign exchange earning (Kernel + CNSL) | Rs. 2717 Crores | Rs. 2522 Crores |
| Net foreign exchange earning | Rs. 526 Crores | Rs. 359 Crores |

**Table 1.26 a:** Export of cashew (2007-08 to 2009-10)

| Year | Quantity (Lakh tonnes) | Value (Rs. Crores) |
|---|---|---|
| 2007-08 | 1.14 | 2289 |
| 2008-09 | 1.10 | 2988 |
| 2009-10 | 1.08 | 2906 |

## Cocoa

Though India has gained foreign exchange by way of export of cocoa products, cocoa beans are being imported. Cocoa Industry in India has expanded to a considerable extent in the current decade. At present there are more than 15 companies in the field. These industries have the capacity to process 30,000MT of which present availability is only 30%. In addition to utilizing the domestic products, import of dry beans to the level of 4000-6000MT along with other cocoa products are also taking place to cater the needs of the industry. During 2005-06, we have gained a foreign exchange of Rs.24.80 crores by exporting cocoa beans/ cocoa products, whereas we have imported cocoa beans, paste, butter, powder etc from other cocoa growing countries at a cost of Rs.112.865 crores so as to meet the indigenous consumption and export.

## Coconut

The value added products and by-products of coconut which are traded globally are Copra, Coconut oil, Copra meal, Desiccated coconut, Shell charcoal, Activated carbon and Coir & Coir products. The global trade of coconut products is around 1.5 billion US$.

The trend of export and import of coconut products in India is not very encouraging. Even though India's export earnings from coconut and coconut products have been increasing at a compound growth rate of 12.40 per cent per annum, the rate of growth in the import of the products in the same period is 13.50 per cent per annum. Among the commodities in the export basket, coir and coir products account for the bulk of exports from this sector. Coconut oil and other coconut products make a very negligible contribution. This is mainly on account of the fact that the domestic prices of coconut products generally rule above the international price. Further, almost the entire production in the country is consumed in the domestic markets itself. However, since India, is a prominent coconut growing country with a production of 10.8 million tonne from 1.90 million hectares of coconut cultivation, there is a potential for exports, on account of the unique quality, aroma and flavour of Indian coconut. Indian coconut oil as toiletry oil through value addition has made its presence in the international market despite its higher price. Among the imported items, coconut oil, oilcake and desiccated coconut powder are the prominent commodities.

**Table 1.26 b**: Export Data of Coconut and its Products

| Sl. No. | Item | 2008-09 | | 2009-10 | | 2010-11 | |
|---|---|---|---|---|---|---|---|
| | | Quantity | Value | Quantity | Value | Quantity | Value |
| 1. | Coconut (fresh) | 6814 | 938 | 13621 | 1890 | 16423 | 2738 |
| 2. | Coconut (dried) | 1787 | 619 | 2958 | 1119 | 5295 | 2584 |
| 3. | Desiccated coconut | 2173 | 459 | 2050 | 464 | 4190 | 953 |
| 4. | Other coconut excluding fresh/ dried | 8007 | 1883 | 8823 | 2567 | 11217 | 4061 |
| 5. | Coconut oil-crude | 53 | 34 | 14 | 16 | 70 | 40 |
| 6. | Coconut oil-refined | 9801 | 5807 | 5067 | 3987 | 4181 | 3952 |
| 7. | Other residues of coconut | 123 | 19 | 271 | 44 | 11314 | 673 |
| 8. | Oilcake (solvent/ expeller) | 77 | 10 | 346 | 41 | 3 | 1 |
| 9. | Coconut shell–raw | 1165 | 183 | 1987 | 304 | 466 | 174 |
| 10. | Shell charcoal | 17279 | 2449 | 39938 | 2430 | 15862 | 3020 |
| 11. | Copra | 13587 | 5580 | 22997 | 9113 | 16927 | 8821 |
| | **Total** | **17981** | | **21976** | | **27015** | |

**Table 1.26c:** Import Data of Coconut and its Products

| | | | | | | | |
|---|---|---|---|---|---|---|---|
| 1. | Oilcake (solvent/ expeller) | 22135 | 2148 | 70543 | 5393 | 35508 | 3069 |
| 2. | Coconut oil-refined | 12231 | 6753 | 3113 | 1377 | 1063 | 564 |
| 3. | Coconut oil-crude | 2998 | 1389 | 10696 | 3926 | 0 | 0 |
| 4. | Shell charcoal | 186 | 6 | 107 | 11 | 2 | 0.3 |
| 5. | Other products | 142 | 12 | 52 | 9 | 10 | 4 |
| | **Total** | **10308** | | **10717** | 0 | **3638** | |

## Coffee

Coffee is the second largest traded commodity in the world next only to Petroleum. In India, coffee is predominantly an export-oriented commodity and over 70% of the coffee produced in the country is exported and the balance quantity is consumed within the country. India exports about 2,00,000 tonnes of coffee in a year to over 90 countries of the world. It includes re-exported coffee after value addition. The top five export markets for Indian Coffee are Italy, Russian Federation, Germany, Belgium and Spain. A total quantity of 1, 97,171 metric tonnes of coffee was exported from India valued at Rs. 2,242.64 crore during the year 2008-09. This is an all time high in rupee as well as dollar terms (USD 506.14 million). A total quantity of 1,95,716 metric tonnes of coffee was exported from India valued at Rs. 2,067.70 crores during the year 2009-10.

There are 70 processing centres licensed by the Coffee Board and about 300 registered exporters. Consequent to liberalization, the marketing of coffee is carried out entirely by the private sector.

India's domestic coffee consumption is mainly restricted to the four southern states of Tamil Nadu, Karnataka, Andhra Pradesh and Kerala and the per capita consumption is only 75 gm. Another unique feature of Indian coffee industry is that, it produces both Robusta (65%) and Arabica (35%). The Indian coffee industry employs nearly 1 million people directly and indirectly. The highest per capita consumption is in Tamil Nadu (450 gms).

As export is function of production, commensurate to the increase in production from 1990 onwards, the coffee exports also rose steadily both in terms of volume and value during the last 20 years.

**Table 1.27:** Performance of coffee exports

| Year | Quantity (MT) | Unit Price (Rs. / MT) | Value (Rs. Crore) |
|---|---|---|---|
| 2008-09 | 1, 97,171 | 113,741 | 2,242.64 |
| 2009-10 | 1,95,716 | 105,648 | 2,067.70 |

## Rubber

Rubber plantations occupy 0.4 percent of the gross cropped area in India, and accounts for 0.19 percent of the country's GDP. Rubber-based manufactured products sector is one of the few sectors for which India has positive trade balance. Exports of rubber products from the country during 2004-05 were Rupees 31,990 million whereas the imports were only Rs. 9260 million. In export of NR, though India is a new entrant, the country holds 1.1 per cent of the global share compared to the global share of only 0.8 percent in the total merchandise exports of India. One of the noteworthy developments during the QR-free regime has been India emerging as an exporter of NR. Export of natural rubber from India rose from the low level of 6995 tonnes in 2001-02 to 55311 tonnes during 2002-03 and 75905 tonnes during 2003-04. During 2005-06, though the financial incentive for NR exports was withdrawn and the international prices remained unfavourable for about three months, 73,830 tonnes was exported. Table 1.28 shows the annual trend in exports of NR and value thereof during the period from 1991-92 to 2005-06. Due to presence of a strong manufacturing base, the natural rubber is also imported in India. The details of production, consumption, import and export are given in Table. 1.28 and 1.28a. In 2008-09 and 2009-10, export of Natural Rubber stood at US$ 97.95 million and US$ 52.76 million respectively.

**Table 1.28**: Quantity and Value of Natural Rubber Exported

| Export Year | Quantity in tonne | Value in Rs. Crore |
|---|---|---|
| 1991-92 | 5834 | NA |
| 1995-96 | 1130 | 6.67 |
| 2000-01 | 13356 | 37.36 |
| 2005-06 | 73830 | 458.29 |

**Table 1.28a:** Export and Import of Natural Rubber (Tonnes)

| Year | Production | Consumption | Import | Export |
|---|---|---|---|---|
| 2009-10 | 8,31,400 | 9,30,565 | 1,77,130 | 25,090 |
| 2010-11 | 8,61,950 | 9,47,715 | 1,88,337 | 29,851 |

## Tea

Until 1987-88, tea was the most significant item of agricultural export from India accounting for 20.7% of agricultural exports. Disintegration of the then USSR and loss of other hard currency markets had adversely affected the Indian export over the last 20 years. India's world ranking as an exporter has come down from number one to number four, in the face of stiff competition from Sri Lanka, Kenya, and China. Presently, tea export constitutes about 0.5% of total commodity exports and about 5% of total agro-export. (Chadha and Choudhary, 2007).

The quantity, value and export price of tea for last five years is given in Table 1.29. India has been importing 11 to 33 million kg tea at value of Rs. 66-145 Crore and at a price of Rs. 44.61 to 60.53 Rs./kg and most of this has been re-exported .

**Table 1.29:** Quantity, Value and Export Price of Tea

| Year | Quantity (Million kg) | Value (Rs. Crore) | Value (Million US $) | Unit price (Rs./kg) |
|---|---|---|---|---|
| 2005-06 | 196.67 | 1793.58 | 405.15 | 91.20 |
| 2006-07 | 218.15 | 2045.72 | 451.79 | 93.77 |
| 2007-08 | 185.32 | 1888.68 | 469.64 | 101.91 |
| 2008-09 | 190.64 | 2381.79 | 518.04 | 124.94 |
| 2009-10 | 213.43 | 3038.69 | 633.06 | 142..37 |

# *Chapter 2*
# Status of Mechanization of Agriculture

Agriculture needs timely operations and agricultural jobs are very strenuous. Due to higher level of education in the rural area youth is migrating to cities where they get less strenuous jobs and better wages. The Government of India has also introduced the Mahatma Gandhi National Rural Employment Guarantee Scheme as per which the rural unemployed person is entitled for 100 days work at the rate as per minimum wages act. Due to this the availability of labour has reduced in the villages.

The mechanization of Indian agriculture is the need of the day. India is producing largest number of tractors 6 lakhs (Table 2.3) in the world. However, these are only used for seed bed preparation and seed sowing to a limited extent and for farm transport. Very few tractor operated sprayers / dusters, weeders and harvesters are available. The tractor is highly under utilized as this is used for 300-400 hours per year. The tractors of 31-40 hp are mostly popular in India and on an average the expenditure per day is about Rs. 2500 per day and for economical consideration any tractor operated implement has to work equivalent to 20 to 25 labours. The irrigation and threshing operation of crops is mechanized to a great extent using stationary threshers and electric or diesel operated pumpsets. The cultural practices followed for cultivation does not permit entry of tractor in the field after crop germination. The necessary cultivation practices are to be followed for mechanization of weeding, spraying and harvesting and necessary machinery needs development.

## 2.1 History of Agriculture Mechanization in India

Before independence bullock drawn country plough, blade harrow (bakhar) wooden peg tooth harrow and wooden plank were used for pulverization, compaction and smoothening of seed bed. The bullock cart was used for farm transport and Persian wheel was used for irrigation. The handle tools like khurpi, sickle, spade, crowbar, kodali etc., were used for weeding, harvesting, irrigation and other miscellaneous farm operations. The chaff cutter was also being used in north India for cutting fodder for animals. Most of the bullock drawn implements and hand tools were fabricated locally by carpenter and blacksmith. The spade, pickaxe, crow bar and watering buckets were also manufactured by Tata Iron and Steel Company in large scale. Efforts were made to introduce western plough (mould board plough) of 15 cm width by the Allahabad Agricultural Institute.

The mechanization of Indian agriculture started with start of green revolution in Punjab. The cultivation of dwarf varieties of wheat and hybrid varieties of maize and application of fertilizers at subsidized rates, increased the income of the farmers. The start of electrification of villages in early 1960's replaced the Persian wheel and bore wells were drilled inside the open well. Electrical centrifugal pump sets were installed inside the open wells for irrigation of the farms. Later on the centrifugal pump sets were installed in open pits as its suction capacity was only 8 to 10 meters. Wherever, electricity was not available, diesel operated pump sets were used. As the water table went down during 1980 and 90's the submersible pump sets replaced the centrifugal pump sets.

During early 1960's the animal drawn disc harrow were introduced in Punjab. The chaff cutter was modified as drummy thresher by providing projections/ beaters on fly wheel and enclosing it in a casing and screen was provided in the lower half of casing. The blower was also provided to separate the straw (bhusa) from grains. The large scale adoption of threshers for wheat started during early 1970's when the thresher (conch or ghughu type) developed by Prof. Mason of Allahabad Agriculture Institute was fabricated by a large number of fabricators in Moga city of Punjab and nearby areas. The axial flow paddy threshers also became popular after 1975. Later on cutter type threshers/Haramba thresher and multi-crop threshers were adopted.

The combine harvesters were imported from West Germany in large number by Punjab State Agricultural Co-operative Marketing Federation Ltd., during late 1960's. These were immediately adopted by farmers due to unpredictable rains during harvesting season and shortage of labour during wheat harvesting season. Some private entrepreneurs and graduates in agricultural engineering also purchased the combine harvesters for custom hiring. These combine harvesters were also deployed for paddy harvesting and proved a big commercial success. Now wheat and paddy is being harvested by combine harvesters in large areas in almost all the states from North, South and West India. There are 13 manufacturers of combine harvesters in India now and out of these 12 manufacturers are in Punjab. The thresher manufacturers started the manufacturing of self propelled and tractor ride type combine harvesters.

Steam tractors were introduced in India by S. Joginder Singh, Agriculture Minister of Punjab in 1919 to reclaim waste land and eradicate kans. Tractors with internal combustion engines were introduced in India after independence to reclaim the marshy lands in Tarai of UP for settlement of displaced people who migrated from Pakistan at time of independence. Tractors were imported in CKD (completely knocked down) condition and assembled, marketed, operated and serviced by trained Indian technicians. The indigenous manufacture of tractors started in India with foreign collaboration in 1961 by Eicher Tractors Ltd., and Tractors and Farm Equipment Ltd. Today, tractors are manufactured by 13 firms. Swaraj and Sonalika tractors have been designed and manufactured in India based on the design of Central Mechanical Engineering Research Institute (CMFRI), CSIR, Durgapur. The tractor operated implements like mould board plough, disc plough, disc harrow and cultivators became popular for seed bed preparation. The seed-cum-fertilizer drills were introduced during late 1960's and became very popular. The manufacture of the tractor operated implements was reserved for small industries till 31st March 2000 and mostly tillage implements, land levellers and seed-cum-fertilizer drills were manufactured. The tractor operated trailors were also manufactured for farm transport. However, the tractor operated weeders and sprayers were not manufactured as the cultivation practices did not permit the entry of tractor in field with crop. The tractor operated rotavator for seed bed preparation was introduced in 1975 and has become lately popular in North India. The manufacture of tractor operated reaper for harvest of wheat and paddy started in 1975.

The animal drawn disc harrow and seed drill were introduced in late 1960's, however, these did not become very popular as the conventional implements were easily available. The tractor operated seed planters were also introduced with seed drill, however, these did not become popular.

The tractor operated potato planters and potato diggers became popular after 1978 as the fabrication started in Jalandhar (Punjab) based on Central Potato Research Jalandhar Institute design. The tractor operated post hole diggers were introduced in 1975.

Today India is biggest manufacturer of tractors in the world and 6,07,568 tractors were sold during 2011-12. The manufacture of power tiller started in 1961 and there were 12 manufacturers at a time. However, these did not become popular due to slow and inconvenient in operation. The power tillers are used in some areas in West Bengal for paddy cultivation.

Mango, Sapota, Guava, Orange and Lime harvesters were developed at Dr Balasaheb Sawant Konkan Krishi Vidyapeeth, Dapoli, Indian Institute of Horticultural Research, Bangalore, Central Institute of Post Harvest Technology, Ludhiana and Central Institute of Subtropic Horticulture, Lucknow during early 1990's. The coconut climbers were designed and commercialized in 1990's in Kerala. The vegetable seedling transplanters developed at CIAE and IIHR are also becoming popular. The air blast sprayer is also becoming popular for spraying mango orchards in UP and grape orchards in Maharashtra.

## 2.2 Farm Power Sources in Agriculture

The main power sources at farm are Human Power, Draught Animal Power, Tractive Power (Tractors and Power Tillers) and Electro-Mechanical Power (Diesel

Engines and Electric Motors). The number of different power sources and the power supply per hectare of cultivated area in Kilowatt per hectare (kW/ha) is given in Tables 2.1 and 2.1A. The cultivated area, production and power availability per hectare for different states is given in Table 2.2. The power availability in India has increased to 0.32 kW/ha to 1.44 kW/ha from 1965-66 to 2004-05. The cropping intensity has also increased 114% to 135.6% in the corresponding duration.

**Human Power**

Digging, clod breaking, sowing, transplanting, inter culture, spraying, harvesting, threshing, cleaning and grading are performed by human power using traditional tools. Improved tools have been designed and are available commercially to improve the efficiency of the workers. The agricultural workers comprise of small cultivators and agricultural labourers. The number of agricultural workers increased from 125.7 million to 244.5 million from 1971-72 to 2010-11(Table 2.1 and 2.1A). The human power availability increased from 0.045 to 0.087 kW/ha (almost doubled) from 1971-72 to 2005-06 (Table 2.1).

Though the availability of the human power per unit area increased, however, the level of literacy also increased to about 70% in India. The literate labour does not want to do the tedious work with conventional tools, he likes to use technology and work with machines, which require less manual effort. About 31% of the agricultural workers are women, hence the efficient tools and machinery are to be developed suitable for men and women workers.

**Draught Animal Power**

Traditionally, draught animals are used for field operations, transport and agro-processing. There are about 79% small and marginal farmers who have limited land holding and resources and over 57% of farming area is commanded by the draught animals. Field operations in hilly regions are performed by animal and human power only. With introduction of tractors, electrical motors and diesel engines the number of draught animals has reduced from 78.4 million in 1971-72 to 51.3 million in 2010-11. The power availability has also reduced from 0.133 to 0.120 kW/ha during 2005-06 (Tables 2.1 and 2.1A). Animals are still used to do 24.7% of tillage and 20.6% of sowing work in India (Table 2.7).

**Tractors**

Tractors are being manufactured by 26 manufacturers in India as per Central Farm Machinery Training and Testing Institute, Budni, M.P. The sale of tractors have increased from about 80 thousand in 1986-87 (Table 2.3) to about 4.40 lakh per year during 2009-10 (Table -2.5) and 6.08 lakh during 2011-12 (Table 2.3). The production of tractors increased from about 80,000 in 1986-87 (Table 2.3) to about 4.33 lakhs during 2009-10 (Table 2.5). The tractors in the range of 31–40 hp are more popular and manufactured by 9 firms with a annual production of about 1.98 lakhs per year during 2009–10 (Tables 2.4 & 2.5).

India is largest manufacturers of the tractors in world and about 20 to 30 thousand tractors are exported every year. Different sizes of tractors are manufactured in India ranging from less than 25 hp to more than 45 hp but most popular range is 31-40 hp. The states of Punjab, Haryana and Western UP have contributed largely in mechanization. The number of tractors in India has increased from 0.119 million to 3.929 million from 1971-72 to 2010-11(Tables 2.1 & 2.1A). The tractor power availability for cultivated land has increased from 0.02 to 0.70 kW/ha during this period. However, the tractors are used only 300-400 hours per year due to non availability of suitable implements. It is noticed that the area sown per tractor was 2162 hectare per tractor during 1965-66 and it has reduced to 50 hectare per tractor during 2004-05 (Table 2.2). The state-wise sales of tractor is given in Table 2.6. During 2009-10, maximum number of tractors 73,680 were sold Uttar Pradesh followed by Andhra Pradesh 34,690, Maharashtra 34,313 and Madhya Pradesh 33,329. The lowest sale of tractor was in Kerala 587 followed by Uttarakhand 2327 during 2009-10 (Table 2.6). The level of mechanization of different operations in India is given in Table 2.7. The tractor is mostly used for tillage *i.e.* 15.6% of tillage work in India is done by tractors. The animals are still doing 24.7% of tillage work (Table 2.7). The sowing with tractor operated seed drill is 8.3 % of area under sowing and the harvesting is done to the extent of 0.56% by tractor operated reapers (Table 2.7).

The operations of weeding, spraying etc. are not performed by the tractor as the matching implements are not available.

**Power Tillers**

The manufacture of power tillers started during 1961. Once there were 12 manufacturers and now there are only 2 manufacturers. The number of power tillers increased from about 16 thousands in 1971-72 to about 1.81 lakhs during 2005-06 (Table -2.1) and 2.43 lakhs during 2010-11 (Table -2.1A). Power Tillers did not become popular as these were slow in operation, the operator has to walk behind the power tiller and the operation is tedious. Currently about 18,000 power tillers are produced every year. However, the power tiller is the most versatile power source to work in fruit orchards.

**Electrical Motors and Diesel Pump sets**

These are stationery power sources and are used for irrigation, threshing and processing industry. The use of electrical motors increased from about 1.5 million during 1971-72 to about 12 million during 2005-06 (Table -2.1) and 17.5 million during 2010-11( Table 2.1A). The electric motors are preferred over diesel engines due to hassle free operation and low cost of electricity for agricultural purposes. The number of diesel engines increased from about 1.4 million to 7.4 million during the period 1971-72 to 2005-06 and 9.2 million during 2010-11 (Table 2.1A).

**Farm Power Availability and Productivity**

The farm power availability and average productivity of food grains for different states is given in Table 2.10. Punjab had 3.5 kW/ha farm power availability and food grain productivity was 4032 kg/ha followed by Haryana with 2.25 kW/ha power availability and productivity of about 3088 kg/ha. Orissa and Chhattishgarh had 0.60 kW/ha and the productivity was only about 799 kg/ha. On all India basis we have 1.35 kW/ha power availability and about 1723 kg/ha food grain productivity.

## 2.3 Agricultural/Farm Machinery in India

### Animal drawn Implements

The availability of animal and power operated implements in India is given in Table 2.8. The number of steel ploughs increased from 6.5 millions to 24.1 millions from 1976-77 to 2010-11. The number of seed drills and planters was 4.9 millions during 1976-77 and increased to 11.4 million during 2010-11(Table 2.8). The animals still contributed to 24.7% of tillage and 20.6% sowing operations (Table 2.7). The level of mechanization in different agricultural operations is given in Table 2.7. The tractor operated implements are mostly for tillage / seed bed preparation and seed drilling.

### Tractor drawn Implements

The trend of tractor / power drawn implements is given in Table 2.9. The number of mould board and disc ploughs have increased from about 25 thousand to 13.19 lakhs from 1972-73 to 2010-11. The number of harrows and tillers have increased from about 56 thousand and 82 thousands respectively during 1972-73 to 11.56 lakhs and 11.04 lakhs respectively during 2010-11. The seed planters have increased from about 9 thousand to 87 thousand during 1972-73 to 2010-11.

The number of tractor operated combines have increased from about 450 to 3500 during 1971-72 to 1991-92. There are 48 tractor and self propelled combines manufacturers in India today (mostly in Punjab).

The number of tractor operated sprayers and weeders has not increased correspondingly and still most of the operations are done manually by backpack hand operated or petrol engine operated sprayer or mist blower and need attention.

### Threshers

The number of stationery threshers mostly for wheat and paddy has increased from 2.1 lakh to 70.4 lakhs from 1972-73 to 2010-11. The threshing operation of these crops is mostly mechanized in India.

## 2.4 Manufacturers of Farm Machinery in India

The fabrication of agricultural machinery was reserved for small industries till 31st March 2000, hence mostly these are manufactured in small towns. The traditional animal drawn implements are fabricated by village craftmen and organized small scale industries also. The tractors, power tillers and sprayers are fabricated by medium and large scale industries. Now the manufacturing of agricultural machinery has been opened up for all kind of manufacturers. The status of farm machinery industry in India and the regions having concentration of the farm machinery manufacturing units is given in Tables 2.11 and 2.12 respectively. The Northern Region has maximum number of manufacturers followed by Western Region.

## 2.5 Research and Developments (R & D) in Agricultural Machinery in India

A large number of implements for mechanization of different crops have been developed by state agricultural universities and institutes of Indian Council of Agricultural Research under All India Co-ordinate Project of ICAR on Farm Implements and Machinery. Yet the system for end to end (seed bed preparation to harvesting and threshing) mechanization is yet to be followed in Indian agriculture. The cultural practices for mechanized operations of weeding, spraying and harvesting of field crops are to be followed.

**Centres of All India Co-ordinate Project on Farm Implements and Machinery**

| | | |
|---|---|---|
| CIAE, Bhopal | - | Central Institute of Agricultural Engineering, Bhopal - 462 038 |
| TNAU, Coimbatore | - | Tamil Nadu Agricultural University, Coimbatore- 641 003 |
| PAU, Ludhiana | - | Punjab Agricultural University, Ludhiana - 141004 |
| MPKV, Rahuri | - | Mahatma Phule Krishi Vidyapeeth, Rahuri - 413 722 |
| ANGRAU, Hyderabad | - | Acharya NG Ranga Agri. Univ., Rajendranagar, Hyderabad-500 030 |
| NEH, Barapani | - | ICAR Res. Complex for NEH Region, Barapani-793 103 |
| IISR, Lucknow | - | Indian Institute of Sugarcane Research, Lucknow- 226 002 |
| BAU, Ranchi | - | Birsa Agril. University, Ranchi – 834 006 |
| HAU, Hisar | - | CCS Haryana Agril. University, Hisar-125044 |
| JNKV, Jabalpur | - | Jawahar Lal Nehru Krishi Vishwa Vidyalaya, Jabalpur – 482 044 |
| IGFRI, Jhansi | - | Indian Grassland and Fodder Research Institute, Jhansi- 284 003 |
| GBPUAT, Pantnagar | - | G B Pant University of Agri. & Technology, Pantnagar- 263 145 |
| IIT, Kharagpur | - | Indian Institute of Technology, Kharagpur- 721 302 |
| KAU, Tavanur | - | Kerala Agricultural University, Tavanur- 679 573 |
| MPUAT, Udaipur | - | Maharana Pratap University of Agri. & Tech., Udaipur- 313 001 |
| AAU, Jorhat | - | Assam Agril. University, Jorhat-785 013 |
| Dr. PDKV, Akola | - | Dr. Panjabrao Deshmukh Krishi Vidyapeeth, Akola – 444 104 |
| NDUAT, Faizabad | - | Narendra Deva Univ. of Agri. & Tech. Ambedkarnagar – 224 122 |
| UAS, Raichur | - | Univeristy of Agril. Sciences, Raichur- 584 102 |
| RAU, Pusa | - | Rajendra Agricultural University, Pusa, Samastipur-848 125 |

contd...

*contd...*

| | | |
|---|---|---|
| AADU, Allahabad | - | Sam Higginbottom Institute of Agriculture, Technology & Science (Deemed University) Allahabad – 211 007 |
| HPKV, Palampur | - | CSK Himachal P Krishi Vishvavidyalaya, Palampur – 176 062 |
| OUAT, Bhubaneswar | - | Orissa University of Agri. and Tech. Bhubaneshwar – 751 003 |
| CAU Gangtok | - | Central Agri. University, Ranipool, Gangtok – 737 135 (Sikkim) |
| **Other Institutes** | | |
| IIHR, Bangalore | - | Indian Institute of Horticultural Research, Bangalore – 560 089. |
| CISH, Lucknow | - | Central Institute of Sub-tropical Horticulture, Lucknow – 227 107. |
| DBSKKV, Dapoli | - | Dr. Balasaheb Sawant Konkan Krishi VidhyaPeeth, Dapoli – 415 712 |
| CMERI, Durgapur. | - | Central Mechanical Engineering Research Institute, Durgapur- 713 209 |

The implements developed by these institutions for different operations are given below:

## 2.5.1 Tillage and Seedbed Preparation

### Animal Drawn Tillage Implements

1. **Ridger Plough-** It is a double mould board plough for ploughing and making ridges in light soils of hilly terrain. It is an improvement over local plough. The width of cut is 150 mm and depth of cut is 140 mm.
2. **Bose Plough-** It is a mould board plough suitable for upland paddy and dry land cultivation.
3. **Tool Carrier-** It is a multipurpose frame with a pair of pneumatic wheels and a common frame for hitching the attachments like iron plough, cultivator, sweep, bund former and bed former.
4. **Disc Harrow-** It is a single acting double gang type disc harrow suitable for secondary tillage operation. It has 2 gangs of 3 discs each.
5. **Naveen Bakhar Blade-** It is a V-shaped blade which is modified version of conventional straight blade of harrow (bakhar). It can be attached to the local bakhar frame easily. It is suitable for primary and secondary tillage operations in black cotton soil. Width and depth of cut is 460 mm and 60 mm respectively.
6. **Channel cum Bund Former-** It is a bullock drawn implement suitable for making field channels to manage the irrigation water effectively. It can

prepare the bunds of 250-550 mm base width and 80 to 150 mm top width and 80 to 150 mm height. The channel of upto 300 mm top width and depth can be prepared.

7. **Puddler-** Puddler is suitable for puddling operation under wet land condition for paddy cultivation.

**Tractor Drawn Tillage Implements**

1. **Bed & Furrow Former-** This is capable of forming alternate beds and channels. The beds are suitable for planting crops like sorghum, maize and cotton. This bed and furrow system is ideal for efficient irrigation management.

2. **Basin Lister-** It is an attachment fitted behind 9-tine tractor mounted cultivator. It is used for water harvesting, and conservation of soil and moisture by making pits in soil.

3. **Pulverising Roller Attachment-** It is an attachment to commercially available tractor operated tined cultivators. It is suitable for puddling as well dry seed bed preparation. It consists of Helical blades fitted around a shaft.

4. **Spiked Clod Crusher-** It is used as a combination tillage tool with tractor drawn harrow or cultivator. It is suitable for breaking and segregation of clods for seed bed preparation after paddy harvest. It can also be pulled by a pair of bullocks.

5. **Combine Tillage Tool-** It has a blade in front which loosens the soil. It also has an operated rotary shaft fitted with 'L' shaped blades which rotate at 110 rpm making it suitable for seed bed preparation in one pass under irrigated soil conditions.

## 2.5.2 Sowing and Planting Equipments

**Manually Operated Sowing and Planting Equipments**

1. **Manually Operated Drill-** This machine is suitable for sowing rape seed and mustard. The crops like wheat, green gram, ground nut and maize can also be sown by changing fluted roller.

2. **Naveen Dibbler-** It is suitable for sowing pea, soya bean, sorghum and maize in small plots or hilly terrain. It can also be used for gap filling.

3. **Rotary Dibbler-** The dibbler is operated by pushing it manually into the soil in the direction of travel. The jaws enter into soil and automatic dropping of seeds take place. It is suitable for dibbling maize, soybean, sorghum, pigeon pea and Bengal gram.

4. **Adjustable Row Marker-** It is a marker suitable for marking rows at different spacing according to crop for sowing seeds.

**Animal Drawn Sowing and Planting Equipments**

1. **Seed-cum-fertilizer drills-** These animal drawn implements are suitable for sowing wheat, rapeseed, mustard, pearl-millet, cotton, guar, bajra, gram, methi and groundnut.

2. **Seed Planter-** These implements maintain seed to seed distance and are suitable for planting groundnut, pea, sunflower, safflower, soybean, pigeon pea, bengalgram , cotton, maize and sorghum.
3. **Potato Planter –** This animal drawn implement is a rotating sector type planter suitable for making ridges and planting potato in the ridges.
4. **Seeding Attachment-** It is an animal drawn seeding device attachable to wooden or iron country ploughs or multipurpose tool carrier. It consists of a seed box, a spike type ground wheel, a variable aperture and rubber agitator type seed metering mechanism, a furrow openers and a mounting-cum-depth adjusting clamp.
5. **Birsa Seed Drill-** It is a bullock drawn implement suitable for sowing paddy, ragi, wheat, linseed, gram, safflower and other small sized seeds.

**Tractor Drawn Sowing and Planting Equipments**

1. **Seed-cum-fertilizer Drill**- These tractor drawn implements are suitable for sowing wheat, rapeseed, mustard, pearl-millet, cotton, guar, bajra, gram, groundnut.
2. **Zero-till Drill-** Tractor mounted zero-till seed-cum-fertilizer drill was developed to sow wheat directly in rice-harvested fields without preparing the seedbed. It is a nine-row/11-row unit consisting of fluted rollers for seed metering and agitators over adjustable openings for fertilizer metering.
3. **Seed, Fertilizer Drill-cum-Planter-** It is a planting attachment to the commercially available tractor operated seed-cum-fertilizer drills suitable for sowing maize, soybean, groundnut, cotton etc. It maintains seed to seed distance.
4. **Inclined Plate Planter-** The 6-row tractor mounted inclined plate planter has been adopted for sowing intercrop on broad beds and a bed shaping/forming attachment has been added as an integral part in the refined design of the planter.
5. **Pneumatic Planter-** It is for precision planting works based on the air suction principle. Six row (2 and 4 optional) unit, can be operated with a 26kW (35 hp) tractor. The air suction principle and release of the seed near the ground provides lower variation in seed planting.
6. **Ridge Planter for Winter Maize-** It is a tractor operated implement suitable for planting crops like maize on one side of ridges. In this, a planting unit is mounted on the commercially available ridger.
7. **Broad Bed Former cum Seed Planter-** It is a tractor mounted implement which forms raised bed of 1.5 m width with trenches at both ends and sows seeds. It is suitable for sowing maize, cotton, groundnut etc.
8. **Ridger Seeder-** It forms ridges and furrows and carries out sowing either on the ridge, one side of the ridge or in the furrows as desired. It is suitable for planting raya, gram and bajra seeds.

## 2.5.3 Fertilizer Applicators

1. **Fertilizer Broadcaster-** It is a manually operated device suitable for broadcasting granular fertilizers like Urea and DAP.
2. **Granule Applicator –**It is manually operated float type implement suitable for placement of urea super granules in root zone of low land paddy crop.

## 2.5.4 Equipment for Weeding and Intercultivation

### Manually Operated Weeding and Intercultivation Implements

1. **Grubber-** It is a manual, pull type implement suitable for weeding and inter-culture of upland row crops in black soil regions.
2. **Peg Type Dry Land Weeder-** It is manually operated push-pull type of weeder, suitable for weeding and mulching in row crops with 225-450 mm row-to-row spacing like groundnut, sunflower, safflower, soybean and sorghum.
3. **Wheel Hoe-** It is manually operated push-pull type of Single Wheel Hoe. It is suitable for weeding crops having row spacing of 300 mm or more such as sorghum, soybean, maize and pigeon pea.
4. **Star Weeder-** It is manual push-pull type weeder suitable for weeding operation in all types of row crops. It consists of two rows, of star wheels and a blade behind these.

### Animal Drawn Weeding and Intercultivation Implements

1. **Multipurpose Hoe-** It is suitable for weeding and interculture operation in groundnut, sunflower, safflower, sorghum, soyabean etc.
2. **Animal drawn Sweeps-** It is an animal drawn implement with adjustment for row-to-row spacing and depth. It removes the shallow rooted weeds and provides a soil mulch.
3. **Weeding Attachment-** It is an attachment of CIAE multipurpose tool frame. Sweep type tines are mounted on a steerable tool bar behind the tool frame. It is suitable for interculture and weeding operations in row crops.

### Self-Propelled Weeding and Intercultivation Implement

1. **Self-Propelled Power Weeder-** The self-propelled power weeder has been found to be an efficient and cost effective machine.

## 2.5.5 Plant Protection Equipments

1. **Low Volume Disc Sprayer-** It is battery operated knapsack sprayer suitable for spraying in crops like paddy, cotton, groundnut, pulses and vegetable.
2. **CDA Crop Sprayer-** It is a tractor mounted sprayer suitable for application of herbicides, fungicides and insecticides in row crops. It has spinning discs for atomization of chemical.
3. **Orchard Sprayer-** It is a tractor mounted air assisted rotary atomizer type sprayer suitable for spraying of fungicide on small berries and mango trees of upto 6m height.

4. **High Clearance Sprayer-** It is a self propelled unit suitable for spraying on tall crops like cotton. This four wheel riding type machine consists of two rear steered wheels, two front lugged wheels powered with a 20 hp diesel engine through a gear box. Tank, hydraulic pump and boom is fitted with 15 nozzles.

5. **Aero Blast Sprayer-** Tractor operated aero blast sprayer is useful for spraying on horticultural trees and tall crop such as cotton, sunflower, pigeon pea etc. The machine consists of a tank of 400 litres capacity, pump, fan, control valve, filling unit, spout adjustable handle and spraying nozzles to release the pesticide solution in to a stream of air blast produced by the centrifugal blower. The air blast distributes the chemical in the form of very fine particles throughout its swath width, which is on one side of tractor.

## 2.5.6 Harvesting Equipments

### Manually Operated Harvesting Tools

1. **Improvised Sickles-** It is serrated blade sickle suitable for harvesting wheat, rice and grasses. The wooden handle has a bend at the rear for better grip and to avoid hand injury during operation.

### Engine Operated (Self Propelled)/ Power Tiller Operated Harvesters

1. **Vertical Conveyor Reaper-** It is an engine operated (self propelled), walking type harvester suitable for harvesting and winowing cereals and oilseed crops. The engine power is transmitted to cutter bar and conveyor belts through belt-pulleys.

2. **Vertical Conveyor Reaper Windrower-** It is a power tiller front mounted, walking type reaper windrower. The reaping attachment consists of cutter bar crop conveyor belt, crop row dividers and star wheels. The cutter bar and conveyer belt are driven by engine through belt-pulley and safety clutch. It is suitable for harvesting and windrowing of erect rice crop.

### Tractor Operated Harvesters

1. **Vertical Conveyor Reaper Windrower-** It is a tractor front mounted reaper suitable for harvesting and windrowing of rice and wheat crops. The reaper consists of cutter bar, crop conveyor belts, crop divider and star wheels. Lifting and lowering is done by tractor hydraulic.

## 2.5.7 Threshing Equipment

### Manually Operated Equipments

1. **Pedal Operated Thresher-** It consists of wire-loop type threshing cylinder operated by foot pedal. It is suitable for threshing rice.

2. **Sunflower Thresher-** It is a pedal operated hold on type sunflower thresher suitable for separating seeds from sunflower heads. It consists of a threshing wheel having spokes and a blower fan operated by foot pedal through chain and sprocket. Four holes are provided on the cover for holding sunflower heads to the threshing wheel by four persons.

**Electric Motor or Diesel Engine Operated Threshers**

1. **Multicrop Thrersher-** It consists of spike tooth or rasp bar cylinder (speeds 7-21 m/s), aspirator type blower and sieve shaker. It is suitable for threshing wheat, maize, sorghum, rice, gram, pigeon pea, soybean, mustard, sunflower, safflower and linseed crops.
2. **Semi-Axial Flow Multicrop Thresher-** It consists of spike-tooth cylinder, aspirator blower and sieve shaker. It works on axial flow principle but crop flow is restricted by inserting a semi-circular plate between cylinder and thrower. It is suitable for threshing wheat, soybean, sorghum, maize, pigeon pea, rice, sunflower and safflower crops.
3. **Axial-flow Multicrop Thresher-** It is a modified version of IRRI Axial Flow Paddy thresher. It consists of peg type cylinder, aspirator and sieve shaker. The top cover is provided with louvers for axial movement of crop while threshing. The louvers are covered with a ribbed casing.
4. **Pigeon Pea Thresher-** It is a high capacity thresher suitable for operation by 7.5 kW/10 hp electric motor or 35 hp tractor. It consists of an automatic feeding mechanism. Crop bundles are fed one by one from feed end and threshed bundles are discharged through other end after threshing.

### 2.5.8 Agricultural Processing Equipments

1. **Cleaner/Grader (CIAE, Bhopal)-** A dust free cleaner (900 kg/h) was developed at CIAE Bhopal. The dust emission could be reduced to the acceptable limits (10 mg/$m^3$) by using this cleaner.
2. **Cleaner for Lighter Seeds (CIAE, Bhopal) -** A cleaner based on the aerodynamic properties of seeds such as berseem, cumin and coriander was designed and developed with four outlets. Seed purity of 98%, 97% and 95% could be achieved for cumin, coriander & berseem, respectively.

### 2.5.9 Miscellaneous Equipments

1. **Seed Treating Drum (APAU, Hyderabad) -** It consists of a drum mounted on a frame at 40° with horizontal and operated manually by a crank handle. It is used for thorough mixing of chemicals like Agroson and Sirason with seeds before sowing as a plant protection measures.
2. **Light Weight Power Tiller (CIAE, Bhopal) -** The light weight power tiller is powered by a 3.75 kW petrol start kerosene run engine. The power tiller is equipped with handle, speed control lever, main clutch lever and two levers fitted on handles for turning the power tiller on either side, one rotary engaging lever and tail wheel for depth adjustment during operation.

## 2.6 Horticultural Machinery

1. **Vegetable Nursery Machinery** (IIHR Bangalore)– The nursery machinery for media sieving, mixing, protray filling, dibbling and sowing has been developed.

2. **Plastic Mulch Laying Machine** (CIAE, Bhopal)- Tractor Mounted Tractor mounted plastic mulch laying machine is developed for mechanization of plastic mulch laying operation to reduce time and labour requirement. It consists of roll holder assembly (with adjustments for height and width), bund or raised bed former assembly, two pneumatic press wheels of 350 mm diameter, two pneumatic wheels of 600 mm diameter for depth control and transport, two scrapers for covering plastic film roll edges with soil.

3. **Seedling Transplanter–** (CIAE, Bhopal and PAU, Ludhiana) It is tractor operated two-row transplanter for seedlings of chilli, brinjal, tomato etc. It is a finger type transplanter. The seedlings are placed manually in fingers/clips, which are released in soil.

4. **Raised bed former cum Transplanter–** (IIHR, Bangalore) It is tractor operated bed former cum transplanter for seedlings of chilli, brinjal, tomato etc. It is a punch type transplanter. The seedlings are placed manually in pits fomed by the punch wheels.

5. **Potato Digger - Animal Drawn –** (PAU, Ludhiana) It is single row animal drawn digger consisting of a V-shaped blade or ridger shaped bottom with welded extension rods on the wings. These rods help in separation of soil and dirt from the potato tubers. This implement was suitable for digging potato tubers after removal of vines from the field.

6. **Potato Digger –** Tractor Mounted – (PAU, Ludhiana) It is a tractor mounted equipment suitable for digging and exposing tubers from two rows. Two V-shaped blades with round bars at the rear are mounted on cultivator frame with provision for adjustment of row to row distance. The lifter rods help in separation of soil and potato tubers.

7. **Potato Digger Elevator -** Tractor Mounted-(PAU, Ludhiana) It is a tractor rear mounted single row unit consisting of digging blade, endless rod chain conveyor and gauge wheels. The elevator is driven by tractor pto. It is suitable for digging and exposing tubers.

8. **Strawberry harvester** - It is hand operated strawberry harvesting cum Holding Tool.

9. **Harvesters for Mango, Sapota, Guava, Lime** (IIHR, Bangalore) The manual harvesting aids have been developed.

10. **Harvester for Kinnow** (CIPHET, Ludhiana) The clipper for harvest of kinnow has been developed.

## 2.7 Horticultural Processing Equipments

1. **Fruit Grader** (CIAE, Bhopal)- A fruit grader of 4 t/h capacity which grades the fruits on the basis of size was developed.

2. **Steam Blancher for Vegetables** (CIAE, Bhopal)- Batch type steam blancher has been developed for cabbage, cauliflower, pea and okra. Two minutes of blanching of cabbage and four minutes of blanching okra could effect enzyme inactivation (peroxidase).

3. **Cassava Chipper** (TNAU, Coimbatore) - The tubers fed through the guides of the feed hopper reach the chipping disc and get sliced by the knives. The chips are discharged through the inclined chips outlet.

4. **Turmeric Boiler** (TNAU, Coimbatore) - A farm level turmeric-boiling unit has been developed and popularized which reduces water and fuel requirement. This is a wood fired boiler. Rhizomes are boiled by the steam liberated from the boiling water.

5. **Banana Hand Cutter** (CIPHET, Ludhiana)- It is useful to de-stem the hand from the banana bunch. It replaces the use of sickle, which was much labour and time consuming. It is safe for human and banana hands and can be used for faster work.

6. **Fruit Grader** (CIPHET, Ludhiana)- It grades all spherical commodities (sphericity above 0.8) in 5 to 8 grades, has a capacity of 1.0 to 1.2 t/h, requires 1.0 hp electric motor, and 3-4 persons to operate. It is commercially available.

7. **Tomato Grader** (CIPHET, Ludhiana) - Tomato is graded on the basis of size into four grades i.e. 25-40 mm, 40-55 mm, 55-70 mm and > 70 mm. It works on the principle of gravity and has mobility from one field to another.

8. **Arecanut Dehusker** (TNAU, Coimbatore)- It is a manually operated unit (capacity 160 kg per 8h, cost Rs 3000/ where four persons can dehusk freshly harvested mature green areca nuts simultaneously. The technology has been commercialized

9. **Fruit and Vegetable Washing Machine** (PAU, Ludhiana) - The stainless steel portable mechanical washing machine is suitable for a wide range of fruits and vegetables (carrot, potato, radish, turnip, ginger, okra, tomato, spinach, kinnow and pears). This machine could also be successfully used for breaking garlic bulbs and peeling potatoes.

10. **Rasper for Tuber Starch Extraction** ( CTCRI, Trivandrum) - An industrial rasper has been developed by Trivandrum centre for extracting starch from tubers such as cassava or tapioca, sweet potato, dioscorea, curcuma and arrowroot.

11. **Garlic Processing Machines** (MPKVV, Udaipur)- An improved garlic bulb breaking machine, a garlic clove flaking machine and a garlic peeling machine for batch type and continuous type have been developed by to remove the drudging and mechanize these operations.

12. **Tender Coconut Punch and Cutter** (CPCRI, Kasargod)- A hand-operated tender coconut punch machine has been developed to save time and drudgery involved in cutting and preparing the tender nut prior to drinking its water, by using a straw through the punched hole. A coconut cutter has also been developed to split open the tender nut into two halves.

13. **Snow Ball Tender Coconut making Machine** (CPCRI, Kasargod)- Technology and equipment for making snowball tender nut (SBTN) has been developed to

facilitate consumption of tender coconut water along with the tender coconut Kernel (capacity 300 snow balls/8h).

14. **Technology for Apricot Kernel Oil** (YSPUAF Solan)- The process for extraction of apricot oil was standardized by Solan centre, comprising of mechanical decortication of apricot fruit stone, separation of kernels, followed by recovery of oil by expeller and filter press. Oil yield is about 122 litres from 1000 kg apricot stones.

15. **Solar-cum-biomass-cum-electrical Dryer for Coconut** (CPCRI Kasargod)- A coconut dryer has been developed to dry 1500 nuts/batch in 3 days . The dryer can use solar energy, electrical heaters and combustion of coconut husk and coconut shell, individually or in combination, as the source(s) for heating the air.

16. **Pomegranate Aril Extractor** (CIPHET Ludhiana)- This is a mechanical and continuous system for processing of whole pomegranate of any size, shape and variety; and for recovery of clean, whole and undamaged arils.The machine is capable to process the whole pomegranate with extraction capacity of 90-94% and with little damage of arils that is about 1-2%.

17. **Cassava Chipper** (TNAU, Coimbatore)- The tubers fed through the guides of the feed hopper reach the chipping disc and get sliced by the knives. The chips are discharged through the inclined chips outlet. Suitable for other tuber crops like carrot, potato, etc.

18. **Centrifuge for Sterilised Pepper** (TNAU, Coimbatore)- Continuous type centrifuge consists of a conical basket rotated at high speed. Due to high centrifugal force, the water in the sterilized pepper is discharged through the perforations in the screen and drained through the outer cylinder.

19. **Cashewnut Decorticator** (CMERI Durgapur) - The Decorticator splits the nuts between two horizontally mounted blades, specially shaped to suit the contour of the raw nut. The outer shell is conveniently separated by engaging a pair of horizontally mounted blades.

20. **Palm Oil Twin Screw Press** (CMERI Durgapur)- Twin Screw Press for extraction of Palm Oil has been developed.

**Table 2.1:** Population of Power Sources and their Power Availability in India

| Year | Agriculture workers | | Draught animal | | Tractor | | Power tiller | | Diesel Engine | | Electric motor | | Total power |
|---|---|---|---|---|---|---|---|---|---|---|---|---|---|
| | Numbers (Lakhs) | Power kW/ha | Numbers (Lakhs) | Power kW/ ha | Numbers (Lakhs) | Power kW/ ha | Numbers (Lakhs) | Power kW/ ha | Numbers (Lakhs) | Power kW/ ha | Numbers (Lakhs) | Power kW/ ha | Power kW/ ha |
| 1971-72 | 1257 | 0.045 | 784 | 0.133 | 1.19 | 0.02 | 0.16 | 0.759 | 14.43 | 0.053 | 15.36 | 0.041 | 0.293 |
| 1975-76 | 1338 | 0.048 | 775 | 0.135 | 2.07 | 0.04 | 0.23 | 1.110 | 20.75 | 0.078 | 20.64 | 0.056 | 0.358 |
| 1981-82 | 1468 | 0.051 | 762 | 0.128 | 5.13 | 0.09 | 0.32 | 1.562 | 30.61 | 0.112 | 32.03 | 0.084 | 0.467 |
| 1985-86 | 1611 | 0.057 | 754 | 0.129 | 7.47 | 0.14 | 0.40 | 1.971 | 37.42 | 0.139 | 41.92 | 0.111 | 0.578 |
| 1991-92 | 1852 | 0.065 | 741 | 0.126 | 12.44 | 0.23 | 0.60 | 3.020 | 48.00 | 0.177 | 60.19 | 0.159 | 0.760 |
| 1995-96 | 2010 | 0.071 | 733 | 0.124 | 17.34 | 0.32 | 0.82 | 4.098 | 55.28 | 0.203 | 74.64 | 0.196 | 0.918 |
| 2000-01 | 2226 | 0.079 | 723 | 0.122 | 26.00 | 0.48 | 1.22 | 6.112 | 64.66 | 0.238 | 95.25 | 0.250 | 1.175 |
| 2005-06 * | 2464 | 0.087 | 713 | 0.120 | 38.19 | 0.70 | 1.81 | 9.035 | 74.32 | 0.273 | 118.66 | 0.311 | 1.500 |

*Note* : **Projection.*

*Sources:* 1. *Power Availability in India Agriculture, 2000, Central Institute of Agricultural Engineering, Nabi Bagh, Berasia Road, Bhopal.*

2. *TERI Energy Data Directory & Yearbook, 2005-06, Tata Energy Research, New Delhi.*

**Table 2.2**: Cultivated Area, Production and Power Availability in India.

| Year | Cropping Intensity (%) | Net shown area (Million hectare) | Gross shown area (M ha) | Irrigated area (M ha) | Productivity* (tonne/hectare) | Net shown area per tractor (ha) | Power available (kW/ha) | Power per unit production (kW/t) |
|---|---|---|---|---|---|---|---|---|
| 1965-66 | 114 | 136.2 | 155.3 | 27.8 | 0.636 | 2162 | 0.32 | 0.50 |
| 1970-71 | 118 | 140.8 | 165.8 | 31.1 | 0.872 | 838 | 0.41 | 0.47 |
| 1975-76 | 120 | 142.2 | 171.0 | 34.6 | 0.944 | 487 | 0.48 | 0.551 |
| 1980-81 | 124 | 140.3 | 173.3 | 38.7 | 1.023 | 264 | 0.55 | 0.54 |
| 1985-86 | 127 | 141.0 | 178.8 | 42.1 | 1.184 | 17 | 0.73 | 0.62 |
| 1990-91 | 130 | 142.2 | 185.5 | 48.0 | 1.468 | 119 | 0.87 | 0.59 |
| 1995-96 | 131 | 142.8 | 186.8 | 53.0 | 1.499 | 84 | 1.05 | 0.70 |
| 2000-01 | 131.6 | 141.2 | 185.7 | 54.83 | 1.630 | 57 | 1.29 | 0.79 |
| 2004-05 | 135.6 | 141.0 | 191.2 | 56.21 | 1.650 | 50 | 1.44 | 0.87 |

*Sources : Data Book -2007 - Indian Agricultural Statistics Research Institute, (ICAR), New Delhi*

**Table 2.1A**: Projected Population of Power Sources in India **.

| Year | Agriculture workers (Lakhs) | Draught animal (Lakhs) | Tractor (Lakhs) | Power Tiller (Lakhs) | Diesel Engine (Lakhs) | Electric Motor (Lakhs) |
|---|---|---|---|---|---|---|
| 2010-11 | 2447 | 513 | 39.29 | 2.43 | 81.34 | 174.88 |
| 2013-14 | 2531 | 486 | 43.62 | 2.96 | 83.68 | 486.00 |
| 2020-21 | 2711 | 423 | 52.18 | 4.42 | 89.14 | 423.00 |
| 2024-25 | 2830 | 387 | 55.36 | 5.39 | 92.26 | 387.00 |

*** Sources : Tyagi KK, Jagbir Singh, KK Kher and VK Jain and Surendra Singh (2010), Status and Projection Estimates of Agricultural Implements and Machinery in India, Agricultural Engineering Today, 34(4): 5-14*

**Table 2.3.** Year-wise Production and Sale of Tractors and Power Tillers

(In Numbers)

| Year | Production | | Sale | |
|---|---|---|---|---|
| | Tractors | Powertillers | Tractors | Powertillers |
| 1986-87 | 80369 | 3325 | 80164 | 3209 |
| 1989-90 | 121624 | 5334 | 122098 | 5442 |
| 1990-91 | 139233 | 6228 | 139831 | 6316 |
| 1991-92 | 151759 | 7580 | 150582 | 7528 |
| 1992-93 | 147016 | 3648 | 144330 | 8642 |
| 1993-94 | 137352 | 9034 | 138796 | 9446 |
| 1994-95 | 164029 | 8334 | 164841 | 8376 |
| 1995-96 | 191329 | 10147 | 191329 | 10147 |
| 1996-97 | 222769 | 11000 | 222769 | 11000 |
| 1997-98 | 260815 | 12200 | 254279 | 12200 |
| 1998-99 | 261609 | 18840 | 262351 | 14880 |
| 1999-00 | 278556 | 16891 | 273181 | 16891 |
| 2010-11* | 545128** | 55000** | 545128 | 55000 |
| 2011-12* | 607658** | 39900** | 607658 | 39900 |

*Sources:*

1. *Tractors Manufacturers Association, Lodhi Road, New Delhi.*
2. *Indian agriculture in Brief, 27 edition, Directorate of Economics and Statistics, Ministry of Agriculture, Govt. of India.*
3. *Agricultural Research Data Book 2003, IASRI, New Delhi.*

   **Federation of Indian Chambers of Commerce and Industries, Tansen Marg, New Delhi.*

** *Estimates (equal to sale)*

**Table 2.4:** Horse Power-wise Tractor Production by Corporate Sector.

| Year | ≤ 20 hp | 21-30 hp | 31-40 hp | 41-50 hp | >50 hp | Total |
|---|---|---|---|---|---|---|
| 2000-01 | 251 | 54293 | 129654 | 33169 | 17235 | 235602 |
| 2001-02 | 380 | 46164 | 109416 | 35084 | 16208 | 207252 |
| 2002-03 | 216 | 34626 | 88090 | 26902 | 6779 | 156613 |
| 2003-04 | 518 | 42959 | 94450 | 40421 | 13285 | 191633 |
| 2004-05 | 975 | 48760 | 125993 | 51893 | 21456 | 249077 |
| 2005-06 | 1190 | 51392 | 152660 | 67506 | 23332 | 296080 |
| 2006-07 | 1537 | 60014 | 170417 | 91223 | 29177 | 352368 |
| 2007-08 | 1755 | 49792 | 155340 | 98351 | 40524 | 345762 |
| 2008-09 | 2451 | 49086 | 152558 | 88596 | 50829 | 343520 |
| 2009-10 | 3752 | 64896 | 197682 | 105842 | 61035 | 433207 |

*(in numbers)*

*Note : hp: Horse Power*

*Sources :*

1. *Agricultural Engineering Data Book, Jan. 2009, Central Institute of Agricultural Engineering, Nabi Bagh, Berasia Road, Bhopal.*
2. *Information received from Tractor Manufactures Association, Chennai.*

**Table 2.5:** Number of Tractor Manufacturing Companies, Production and Sale of Different Category of Tractors

| Tractor Category | Manufacturers | 2006-07 | | 2007-08 | | 2008-09 | | 2009-10 | |
|---|---|---|---|---|---|---|---|---|---|
| | | Production | Sale | Production | Sale | Production | Sale | Production | Sale |
| ≤ 20 hp | 2 | 1537 | 1495 | 1755 | 1714 | 2451 | 2329 | 3752 | 3761 |
| 21-30 hp | 7 | 60014 | 58610 | 49792 | 51675 | 49086 | 49988 | 64896 | 66027 |
| 31-40 hp | 9 | 170417 | 172236 | 155340 | 158401 | 152558 | 158099 | 197682 | 202134 |
| 41-50 hp | 9 | 91223 | 91650 | 98351 | 94972 | 88596 | 85483 | 105842 | 105821 |
| >51 hp | 6 | 29177 | 28790 | 40524 | 39539 | 50829 | 51111 | 61035 | 62487 |
| Total | | 352368 | 352781 | 345762 | 346501 | 343520 | 347010 | 433207 | 440230 |

*Sources:*

1. *Information received from Tractor Manufactures Association, Chennai.*
2. *Agricultural Engineering Data Book, Jan. 2008, Technical Bulletin No.CIAE/2008/132.*

**Table 2.6:** State-wise Annual Sale of Tractors

| State | 1994-95 | 1995-96 | 1998-99 | 2000-01 | 2002-03 | 2005-06 | 2007-08* | 2008-09* | 2009-10* |
|---|---|---|---|---|---|---|---|---|---|
| Andhra Pradesh | 7333 | 10888 | 10052 | 17520 | 10987 | 22444 | 43758 | 38417 | 34690 |
| Assam | 355 | 428 | 450 | 540 | 502 | 770 | 1296 | 1588 | 3578 |
| Bihar | 3708 | 5312 | 13120 | 17344 | 13114 | 12295 | 13104 | 17572 | 29047 |
| Chhattisgarh | - | - | - | - | - | - | 6570 | 8790 | 10174 |
| Gujarat | 13966 | 17388 | 23238 | 11546 | 8186 | 23060 | 25255 | 20179 | 24346 |
| Haryana | 17585 | 18950 | 22361 | 16560 | 11313 | 14572 | 23627 | 22573 | 28715 |
| Himachal Pradesh | 106 | 188 | 499 | 395 | 704 | 758 | 891 | 907 | 1203 |
| Jammu & Kashmir | 401 | 367 | 704 | 850 | 1402 | 1290 | 1328 | 1115 | 1614 |
| Jharkhand | - | - | - | - | - | - | 2897 | 3643 | 5667 |
| Karnataka | 6196 | 8671 | 6822 | 11261 | 6244 | 25291 | 17230 | 13835 | 23927 |
| Kerala | 646 | 1024 | 539 | 669 | 135 | 697 | 493 | 374 | 587 |
| Madhya Pradesh | 21175 | 23069 | 34686 | 21901 | 21027 | 28136 | 18242 | 24306 | 33329 |
| Maharashtra | 9847 | 12143 | 16522 | 14420 | 4412 | 16916 | 29815 | 25610 | 34313 |
| Orissa | 1106 | 1623 | 2012 | 17344 | 2953 | 6672 | 4993 | 5099 | 7940 |
| Punjab | 25123 | 25027 | 30550 | 20879 | 14976 | 12051 | 18062 | 20027 | 28742 |
| Rajasthan | 12390 | 16854 | 24589 | 14692 | 10722 | 27949 | 29456 | 25763 | 31799 |
| Tamil Nadu | 8736 | 11380 | 7442 | 9931 | 4822 | 18278 | 16894 | 14609 | 15819 |
| Uttarakhand | - | - | - | - | - | - | 2479 | 1869 | 2327 |
| Uttar Pradesh | 31717 | 32996 | 60342 | 65253 | 39586 | 45652 | 39682 | 51513 | 73680 |
| West Bengal | 984 | 1442 | 3188 | 3385 | 2157 | 4335 | 5354 | 6069 | 9153 |
| Export & others | 3396 | 3579 | 5053 | 7449 | 11940 | 30514 | 45075 | 43125 | 39681 |
| **Total** | **164770** | **191329** | **262169** | **251939** | **168182** | **291680** | **346501** | **347010** | **440331** |

*Note: * Information received from Tractor Manufactures Association, Chennai.*

*Sources :*

1. *Agricultural Engineering Directory-1998 (Orange Book-ISAE, New Delhi) & Hand Book of Agriculture (ICAR), 2006.*
2. *Souvenir of 42nd ISAE Annual Convention & Symposium Feb. 1-3, 2008.*

**Table 2.7:** Level of Mechanization of Different Agricultural Operations

| Operation | Percentage |
|---|---|
| Tillage | 40.2 |
| • Tractor | 15.6 |
| • Animal | 24.7 |
| Sowing with seed drill/ Seed-ferti-drill | 28.9 |
| • Tractor | 8.3 |
| • Animal | 20.6 |
| Irrigation | 37 |
| Thresher | 52.2 |
| • Wheat | 47.8 |
| • Thresher Paddy and others | 4.4 |
| Harvesting | 0.93 |
| • Reaper | 0.56 |
| • Combine | 0.37 |
| Plant protection | 34.2 |

*Source: Proceedings of Tractor and Farm Machinery Manufacturer's Meet-2007.*

**Table 2.8:** Population of Manual Operated Animal Drawn Implements (In Lakhs) in India

| Implements | 1976-77 | 1980-81 | 1990-91 | 2000-01 | 2010-11* | 2013-14* | 2020-21* | 2024-25* |
|---|---|---|---|---|---|---|---|---|
| **Manual operated** | | | | | | | | |
| Sprayers & dusters | 15.52 | 15.52 | 17.72 | 19.99 | 21.50 | 22.14 | 25.19 | 28.66 |
| **Animal operated** | | | | | | | | |
| Wooden plough | 410.31 | 423.39 | 433.18 | 440.48 | 447.78 | 449.97 | 455.08 | 458.00 |
| Steel plough | 65.16 | 66.52 | 113.87 | 177.17 | 240.67 | 259.72 | 304.17 | 329.57 |
| Puddlers | 20.55 | 22.17 | 45.33 | 76.23 | 107.13 | 116.40 | 138.03 | 150.39 |
| Seed drills & planters | 48.63 | 54.67 | 72.94 | 93.65 | 114.52 | 120.66 | 134.63 | 142.36 |
| Sugarcane crushers | 7.77 | 7.05 | 7.46 | 8.32 | 9.31 | 9.61 | 10.31 | 10.71 |
| Bullock carts | 126.70 | 134.70 | 149.30 | 162.26 | 173.49 | 176.63 | 183.60 | 187.38 |

* *Projected nos.*

*Source : Tyagi KK, Jagbir Singh, KK Kher and VK Jain and Surendra Singh (2010), Status and Projection Estimates of Agricultural Implements and Machinery in India, Agricultural Engineering Today, 34(4): 5-14 .*

**Table 2.9**: Actual and Projected Population of Power Operated Implements, Sprayers and Threshers (In Lakhs ) in India

| Implements | 1972-73 | 1980-81 | 1990-91 | 2000-01 | 2010-11* | 2013-14* | 2020-21* | 2024-25* |
|---|---|---|---|---|---|---|---|---|
| **Plough** | **0.25** | **1.17** | 3.94 | 8.14 | 13.19 | 14.74 | 18.31 | 20.28 |
| Levellers | 0.49 | 2.97 | 9.15 | 16.80 | 22.23 | 23.43 | 24.58 | 23.92 |
| Harrows | 0.56 | 1.65 | 4.62 | 8.41 | 11.56 | 12.60 | 15.02 | 16.41 |
| Tillers | 0.82 | 2.60 | 6.62 | 9.17 | 11.04 | 11.70 | 13.25 | 14.13 |
| Planter | 0.09 | 0.28 | 0.50 | 0.69 | 0.87 | 0.92 | 1.03 | 1.09 |
| Sprayers & dusters | 0.35 | 1.08 | 2.56 | 5.02 | 7.26 | 7.96 | 9.61 | 10.55 |
| Threshers | 2.06 | 8.09 | 21.04 | 46.68 | 70.36 | 77.75 | 95.02 | 104.88 |

** Estimated nos.*

*Source : Tyagi KK, Jagbir Singh, KK Kher and VK Jain and Surendra Singh (2010), Status and Projection Estimates of Agricultural Implements and Machinery in India, Agricultural Engineering Today, 34(4): 5-14 .*

**Table 2.10:** Farm Power Availability and Average Productivity of Food Grains in India in 2001.

| Name of the State | Farm Power availability (kW/ha) | Foodgrain productivity (kg/ha) |
|---|---|---|
| Punjab | 3.50 | 4,032 |
| Haryana | 2.25 | 3,088 |
| Uttar Pradesh | 1.75 | 2,105 |
| Andhra Pradesh | 1.60 | 1,995 |
| Uttaranchal | 1.60 | 1,712 |
| West Bengal | 1.25 | 2,217 |
| Tamil Nadu | 0.90 | 2,262 |
| Karnataka | 0.90 | 1,406 |
| Kerala | 0.80 | 2,162 |

*Contd...*

*Table 2.10: contd...*

| | | |
|---|---|---|
| Assam | 0.80 | 1,443 |
| Bihar | 0.80 | 1,622 |
| Gujarat | 0.80 | 1,169 |
| Madhya Pradesh | 0.80 | 907 |
| Himachal Pradesh | 0.70 | 1,500 |
| Maharashtra | 0.70 | 757 |
| Rajasthan | 0.65 | 884 |
| Jharkhand | 0.60 | 1,095 |
| Jammu & Kashmir | 0.60 | 1,050 |
| Orissa | 0.60 | 799 |
| Chhattisgarh | 0.60 | 799 |
| **All India** | **1.35** | **1,723** |

*Source: Proceedings of Tractor and Farm Machinery Manufacturer's Meet-2007.*

**Table 2.11**: Status of Farm Machinery Industries in India

| S. No. | Equipment manufacturers | No. of units |
|---|---|---|
| 1. | Agricultural tractors | 13 |
| 2. | Power tillers | 2 |
| 3. | Earth movers | 3 |
| 4. | Pumps | 600 |
| 5. | Sprinkler set | 35 |
| 6. | Drip irrigation system | 35 |
| 7. | Plant protection equipment | 300 |
| 8. | Combines | 48 |
| 9. | Reapers | 60 |
| 10. | Threshers | 6000 |
| 11. | Seed drills | 2500 |
| 12. | Ploughs, cultivator and harrows | 5000 |
| 13. | Tractors parts and accessories | 546 |
| 14. | Earth moving machinery and parts | 188 |

*Contd...*

*Table 2.10: contd...*

| | | |
|---|---|---|
| 15. | Diesel oil engines | 200 |
| 16. | Rice processing machinery | 300 |
| 17. | Sugarcane crusher | 50 |
| 18. | Chaff cutter | 50 |
| 19. | Dairy and food industries | 500 |
| 20. | Village craftsmen | 1 million |

*Source: Data Book on mechanization and Agro-Processing since Independence, 1997, CIAE, Bhopal.*

**Table 2.12:** Regions having Concentration of Agricultural Machinery Manufacturers

| |
|---|
| **Northern region** |
| Ludhiana, Moga, Jalandhar, Goraya, Batala, Hoshiyarpur, Karnal, Panipat, Faridabad, Delhi, Agra, Ghaziabad, Meerut, Rudrapur, Muzaffarnagar, Lucknow, Kanpur, Fatehpur and Allahabad. |
| **Western region** |
| Bombay, Pune, Nagpur, Ahmed Nagar, Sangli, Kolhapur, Sholapur, Ahmedabad, Baroda, Anand, Junagarh, Bhopal, Indore, Dewas, Bina, Khurai, Raipur, Vidisha and Gwalior. |
| **Southern region** |
| Hyderabad, Guntur, Anantpur, Kakinada, Coimbatore, Madurai, Chennai, Salem, Palghat, Ernakulam, Kochin and Bangalore. |
| **Eastern region** |
| Calcutta, Vardhaman, Durgapur, Bhubaneshwar, Sambhalpur, Patna, Ranchi, Dhanbad and Muzaffarpur. |

*Source : Status of Farm Mechanization in India 2006 - Indian Agricultural Statistics Research Institute, (ICAR), New Delhi.*

# *Chapter 3*

# Seed Bed Preparation

The seed bed is prepared to create favourable soil conditions for seed placement, germination and growth. The basic operations for opening the cultivable land are known as primary tillage and are done by ploughing. The harrowing and cultivator operations are known as secondary tillage operations. Levelling, clod breaking/planking, bund and bed forming are known as tertiary operations. The main functions accomplished by these operations are:

1. Preparation of the seed bed by pulverizing the soil
2. Destroying weeds by burying them and to prevent their growth
3. Adding humus and fertility to the soil by burying and mixing the vegetation and manures
4. Destroying breeding places, eggs and larvae of insects
5. Increasing air circulation in the soil for plant growth
6. Retaining more moisture from rain
7. Preventing soil erosion by water and wind
8. Leveling land for irrigation and other operations.

The tillage operations are generally performed by animal or power tiller or tractor operated implements based on the farm holding and level of mechanization in the area. Even though slow and labour intensive the animal drawn implements are still quite popular in whole of India except in the states of Punjab, Haryana, Western U.P., some parts of Maharashtra and coastal Andhra Pradesh. Power tiller implements are becoming popular in some parts of West Bengal and Assam. The tillage is the main operation done by tractors in India. Almost all kind of tractor tillage implements are available throughout the country. In Punjab and Haryana most of the area is cultivated by tractor operated implements. The various animal, power tiller and tractor drawn tillage implements used in India are detailed in this chapter.

## 3.1 Animal Drawn Implements

### Country Plough

The country or indigenous plough (*Desihal*) is the most popular implement for seed bed preparation. The type, size and shape of the plough varies throughout the country (Figs. 3.1 to 3.4) based on the soil type and type of animals used. The plough generally consists of beam and body and is hitched to animals by yoke. The shoe, land side and handle are attached to body. The share is supported on shoe and is made of mild steel. All the parts except share are made of wood. The ridger type country plough in which share and wings are made of mild steel was mostly used in Punjab for sandy soils. It cuts a trapezoidal slice of soil of 10 to 15 cm depth and pulverizes it. At least three passes / operations are required to complete seed bed preparation and plough man has to walk about 60 to 70 kilometer and takes 25 to 35 hours to plough one hectare of land.

Fig. 3.1 Country Plough (Desi Plough)
http://www.flickr.com/photos/iqbalkhatri

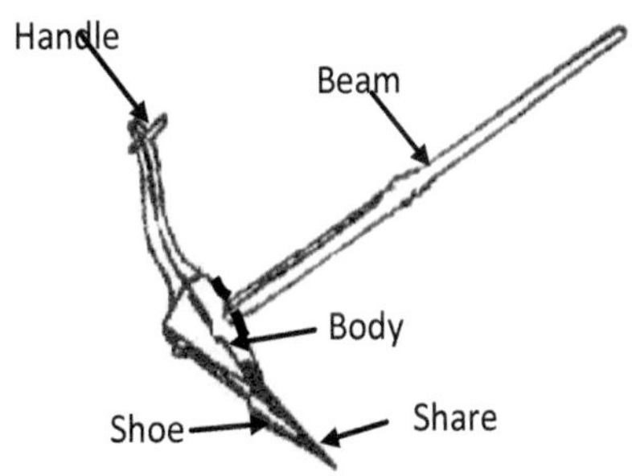


Fig. 3.2 Country Plough (Desi Plough)
Micheal A.M. & Ojah T.P., Agricultural Engineering Vol:I

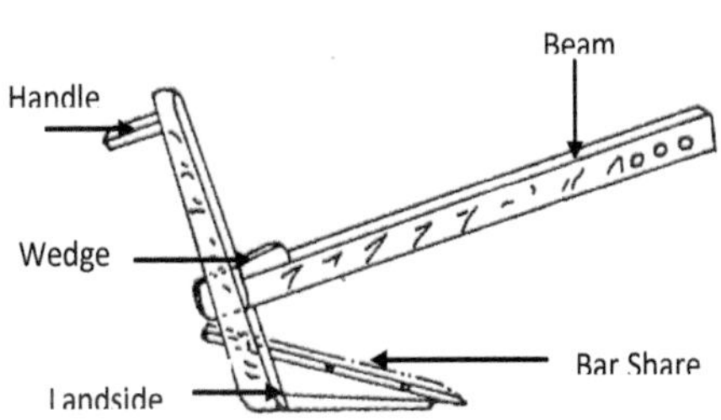


Pandey, M.M., et.al., Farm Machinery Research Digest, Central Institute of Agricultural Engineering, Bhopal.

Kisan Plough
Mamta Iron Works,Billimora
http://mamtaironworks.tradeindia.com

Fig. 3.3 Country Plough

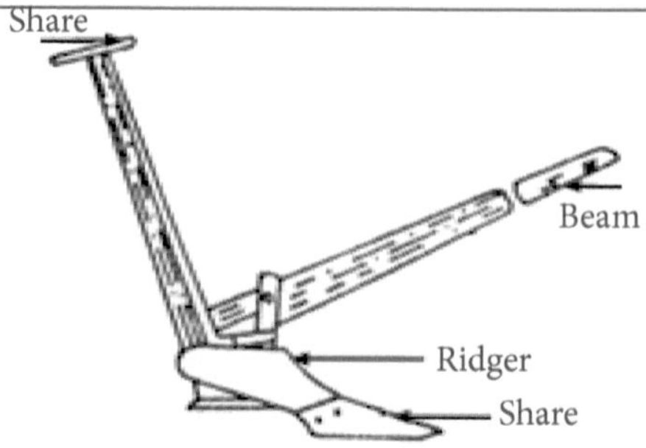


Fig. 3.4 a Ridger Type Country Plough
Pandey, M.M., et.al., Farm Machinery Research Digest, Central Institute of Agricultural Engineering, Bhopal

Fig. 3.4 (b) Ridger Plough
K.S.Worldwide Exports Pvt. Ltd. Nichi Mangli, Ludhiana, http://www.ksexport.co.in

**Mould Board Plough**

The mould board plough (Figs. 3.5 and 3.6) cuts a rectangular slice of soil and inverts and pulvarizes the same. The weeds and green manure are turned and buried and so humus is added to the soil. The plough consists of share for cutting the soil slice, mould board for inverting and pulverizing the soil, land side for balancing the plough (by taking side thrust) and handle for controlling the same. All these parts are attached to the frog (for assembling) and then to the wooden beam for hitching to the pair of animals through yoke. The plough may have single or double handle. The size of plough is determined by the width of share and may vary from 10 to 20 cm based on the type of soil and type and size of animals used. The depth of ploughing is 10 to 15 cm. The field capacity of mould board plough varies from 0.01 to 0.03 hectare per hour (ha/h).

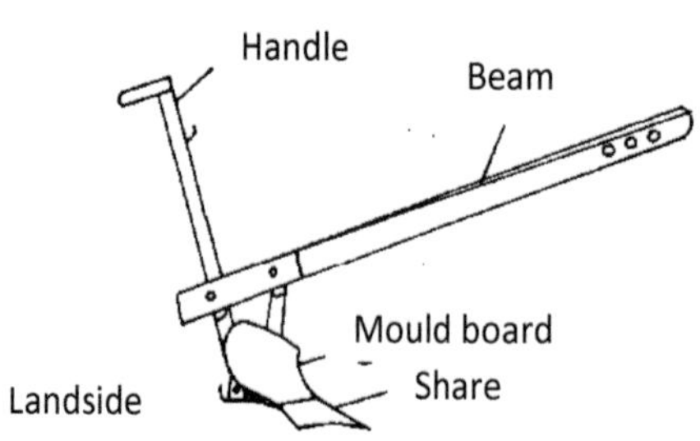


Fig. 3.5 Mould Board Plough
Pandey, M.M., *et.al.*, Farm Machinery Research Digest,Central Institute of Agricultural Engineering, Bhopal.

Fig. 3.6 Mould Board Plough
Mamta Iron Works, Billimora http://mamtaironworks.tradeindia.com

**Blade Harrow**

Blade harrow (*bakhar*) (Fig. 3.7) is a very popular implement for primary tillage in black cotton soils which are very crumbling type. It is also used after country plough to chisel out uncut portions left after ploughing. The blade width varies from 40 to 105 cm. based on the use and area and the field capacity varies from 0.05 to 0.12 ha/h.

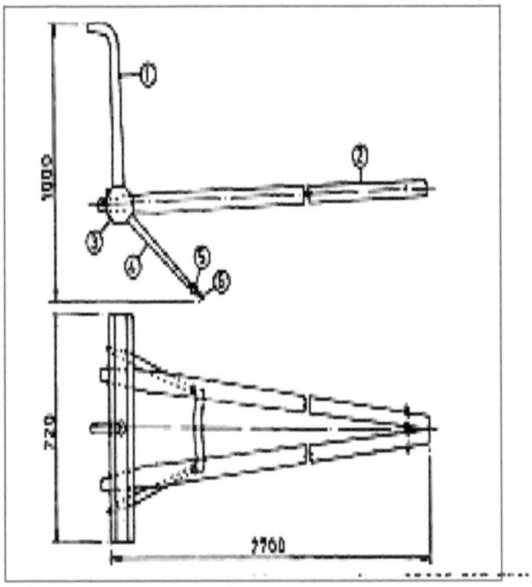

Fig. 3.7(a) Traditional Blade Harrow
Pandey, M.M., et.al., Farm Machinery Research Digest, Central Institute of Agricultural Engineering, Bhopal

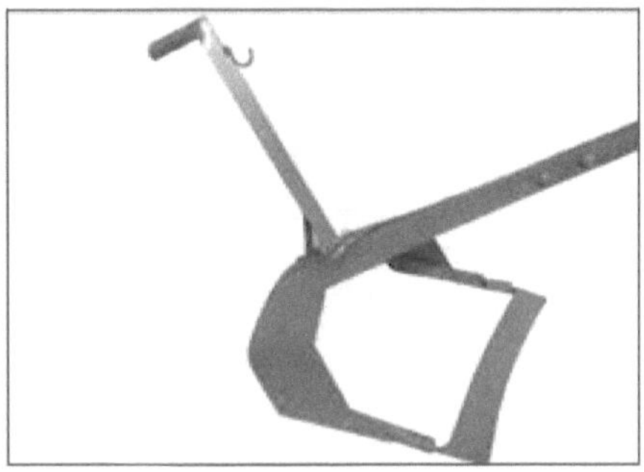

Fig. 3.7(b) Traditional Blade Harrow
Mamta Iron Works, Billimora
http://mamtaironworks.tradeindia.com

**Cultivator**

Tined cultivators (Fig. 3.8) are extensively used in U.P., Haryana and Rajasthan under loamy soil conditions. The cultivator gives about 2 times more output than conventional plough. Three to five rigid tines are fitted on a frame with 5 cm size (width) reversible shovels. Tines are generally fixed at 20 to 25 cm spacing. Depth of operation is 10 to 15 cm. The output varies from 0.06 to 0.10 ha/h. In Karnataka the cultivator has 7-8 wooden pegs of about 5-6 cm dia, 20-25 cm length in one row for tillage.

Fig. 3.8(a) Three Tined Cultivator
Rajasthan Mechanical Works Ltd.Jaipur
http://www.brahmpuri.com

Fig. 3.8 (b) Three Tined Cultivator
Mamta Iron Works,Billimora
http://mamtaironworks.tradeindia.com

### Disc Harrow

Disc harrows (Figs. 3.9 and 3.10) are widely used for secondary tillage in Punjab, Haryana and Uttar Pradesh. The disc harrow may have 4 or 6 discs and these are fixed in two gangs, opposite to each other and a tine is provided at the centre, to cover the entire width. A seat is provided on the frame for operator and the weight of the operator helps in penetration of the implement. The depth of operation is 10 to 12 cm and it gives an output of about 0.12 ha/h of fine seed bed.

Fig. 3.9 Animal Drawn Disc Harrow
K.S.Worldwide Exports Pvt. Ltd. Nichi Mangli, Ludhiana, http://www.ksexport.co.in

Fig. 3.10 Animal Drawn Disc Harrow
Pandey, M.M., et.al., Farm Machinery Research Digest, Central Institute of Agricultural Engineering, Bhopal

## 3.2 Power Tiller Drawn Tillage Implements

Power tillers of 2 to 12 horse power are available. The power tillers of more than 5 hp can be used for tillage operations for preparing the seed bed for vegetable and ornamental crops.

### Rotavator

The power tiller is generally supplied along with a rotavator (Figs. 3.11 and 3.12). This can do primary and secondary tillage by opening the cultivable land and pulverizing the same to prepare the seed bed. The rotavator can also be used to pulverize the soil after ploughing by mould board plough. The depth of operation for 8 hp power tiller is upto 10 cm in unploughed land and upto 15 cm in ploughed field. The field capacity for rotavating in unploughed field is about 0.04 ha/h and the speed of operation is 1.0 to 1.2 kilo meter per hour (km/h). The speed of operation in ploughed field or for second operation of rotavating is 1.4 to 1.7 km/h and the field capacity is 0.06 to 0.08 ha/h. The fuel consumption for operation in unploughed field is about 18 litres/ha and for subsequent operation in ploughed / rotavated fields is 10 to 14 litres /ha.

Fig. 3.11 Power Tiller with Rotavator
http://www.vsttillers.com
VST Tillers Tractors Ltd

Fig. 3.12 Power Tiller Rotavator working in the field
http://www.afftis.or.jp
Agriculture, Forestry and Fisheries Technical Information Society, Japan

**Mould Board Plough**

The power tiller is generally supplied with a single bottom MB plough (Fig. 3.13) or single bottom reversible MB plough (Fig. 3.15). Power tillers are also supplied with 2 bottom reversible MB plough (Figs. 3.14 and 3.15). For ploughing with MB plough the field has to be first divided into portions and the head land is used for idle traveling or turning at the end of the field. The plough can be right handed or left handed. In reversible MB plough (Fig. 3.14) one bottom is right handed and another bottom is left handed. The plough bottoms are reversed at the end of pass at the time of turning. Power tiller wheel can be accommodated in the furrow created by the previous operation. The MB plough leaves the ridges and furrows the field equal to the number of portions in which it was divided or ploughed. The reversible plough leaves the field level (without furrow and ridges). The reversible plough (Figs. 3.15 and 3.16) has regular share while the mould board is made of slats. It can be converted into right or left plough as per requirement at the end of the pass at time of turning for the subsequent pass. Two bottom reversible ploughs (Fig. 3.15) are also available. The depth of ploughing for a 8 hp power tiller reversible plough is upto 15 cm and it is operated at a speed of 2.2 to 2.4 km/h. The field capacity varies from 0.04 to 0.06 ha/h and the fuel consumption 14 to 19 litres/ha.

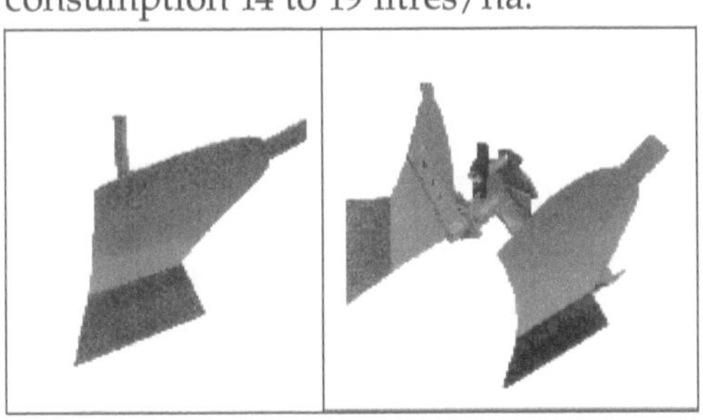

Fig. 3.13 Single Bottom M.B Plough and Reversible Single Bottom M.B Plough

Fig. 3.14 Power Tiller with Two Bottom Reversible M.B. Plough

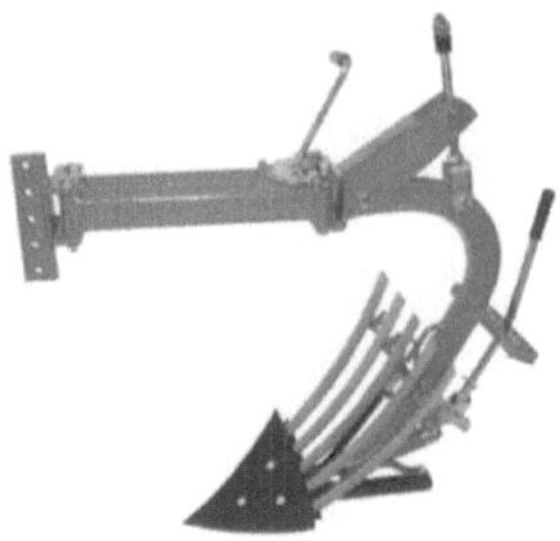

Fig. 3.15(a) Single Bottom Reversible Plough
http://www.tracmaster.co.uk/BCS-Two-Wheel-Tractors.html
P. Tuckwell Ltd, UK

Fig. 3.15(b) Two Bottom Reversible Plough

Fig. 3.16(a) Two Bottom Reversible Disc Plough
http://www.dftractor.com
Changzhou Dongfeng Agricultural Machinery Group Co., Ltd

Fig. 3.16(b) Single Bottom Reversible Plough working in field
Indian Institute of Horticultural Research, Bangalore

## Cultivator

Five tined cultivators (Fig. 3.17) are generally available in India. Six tined cultivator with a rotor pulverizer is also available. These are used for pulverizing the soil after operation of the mould board plough. The speed of operation varies from 1.4 to 2.0 km/h for the first and second operation after ploughing and depth of operation is about 10 cm.

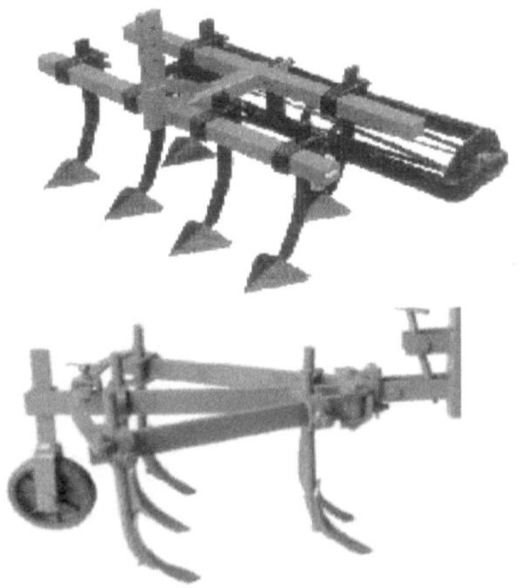

Fig. 3.17 Six Tined and Five Tined Cultivators
Conservation Technology & Machinery Ltd, Richmond, North Yorkshire, UK, http://www.ctm-ltd.co.uk

Fig. 3.18 Five Tined in operation
Indian Institute of Horticultural Research, Bangalore

## 3.3 Tractor Drawn Tillage Implements

### Mould Board Plough

The tractor mounted mould board ploughs are extensively used in central India. The two bottom plough (Fig. 3.19 and 3.20) is generally suitable for tractors of 20 to 35 hp and three bottom plough (Fig. 3.21) for 35 to 45 hp tractor. Mould board plough can plough 15 to 30 cm deep and operates at a speed of 4 to 6 km/hr. The field capacity for a 2 bottom plough is 0.25 to 0.30 ha/hr and for a 3 bottom plough is 0.35 to 0.40 ha/hr. The fuel consumption is around 15 litres/ha.

A circular coulter can also be fitted in some of the ploughs to cut the deep roots of weeds for proper turning, inversion and giving a deep cut in front for proper defining of cutting width. The jointer is also fitted in some of the ploughs to give a narrow and shallow cut in front of the plough to help in ploughing and to cut the weeds for proper inversion.

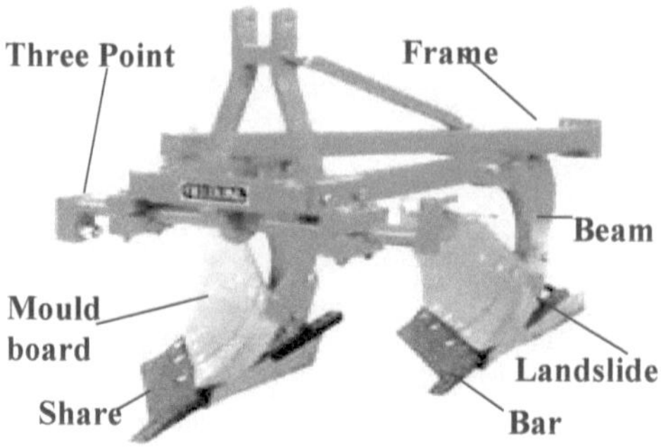


Fig. 3.19 Mould Board Plough Two Bottom
Arson Agro Industries, Moga, Punjab, India

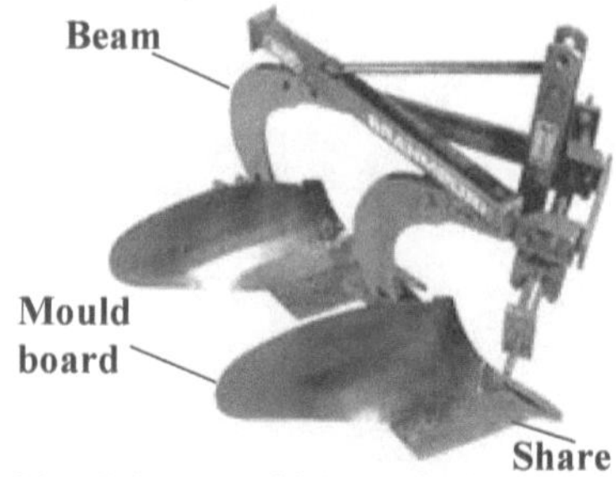


Fig. 3.20 Mould Board Plough Two Bottom
Rajasthan Mechanical Works Ltd.,Jaipur.
http://www.brahmpuri.com

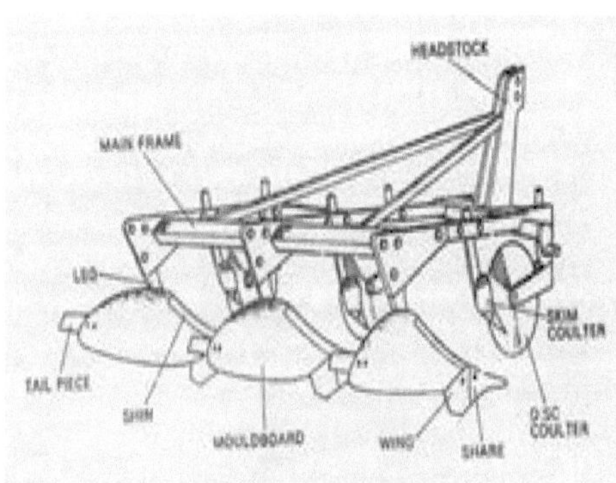


Fig. 3.21 Tractor operated mounted type mould board plough
Training manual on Mechanization of Vegetable crops, IARI, New Delhi

## Reversible Mould Board Plough

In this plough one set each of right hand and left hand bottom are fixed to the frame. During ploughing at the end of the field after turning set of bottoms is turnover so that the tractor wheel can be accommodated in the furrow created previously. In this way continuous ploughing can be done from one end of the field to the other end without dividing the field into blocks for ploughing. The set of the mould boards are fixed at either 180° (Fig. 3.22) or 90° (Figs. 3.23 and 3.24). The turn over / change of bottoms at the end of the field can either be done manually, mechanically or hydraulically by the handle provided on the plough or tractor.

Fig. 3.22 Reversible Mould Board Plough Two Bottom (180°)
Eros Metal Works (P) Limited, Nagpur, India
www.erosgroup.co.in and National Research Center For Soyabean, Indore

Fig. 3.23 Reversible Mould Board Plough
Single Bottom (90°)
Popular Steel Works, Kolhapur, India.
http://www.popularpsw.com/framep.htm

Fig. 3.24 Reversible Plough (90°) working in field
Popular Steel Works, Kolhapur, India.
http://www.popularpsw.com/framep.htm

**Disc Plough**

This plough is used where conditions are very hard, rough and not suitable for mould board plough. It can be 2 to 4 bottom plough with disc diameter of 560, 610, 660 or 710 mm. Discs are mounted on the frame through standard. These rotate and cut the soil slice, pulverize and invert it. The disc angle and tilt angle is adjustable depending upon the soil condition. The scrappers are provided to scrap the sticky soil on disc. Furrow wheel is provided to take side thrust. A two bottom plough requires 25 to 30 hp tractor, while a three bottom 35 to 50 hp tractor and four bottom 50 hp tractor based on the size of discs. The field capacity is generally more than the mould board plough for same width and depth of ploughing.

In reversible disc plough (Fig. 3.28) the set of discs is hinged to the main frame. The set can be rotated to make these to turn the soil to right or left side. The reversing mechanism can be hand or mechanically or hydrautically operated. The set of the discs is rotated / reversed at the end of field / pass. This gives better level ploughed field and the field capacity also increased.

Fig. 3.25 Disc Plough – Two Bottom
Rajasthan Mechanical Works Ltd., Jaipur.
http://www.brahmpuri.com

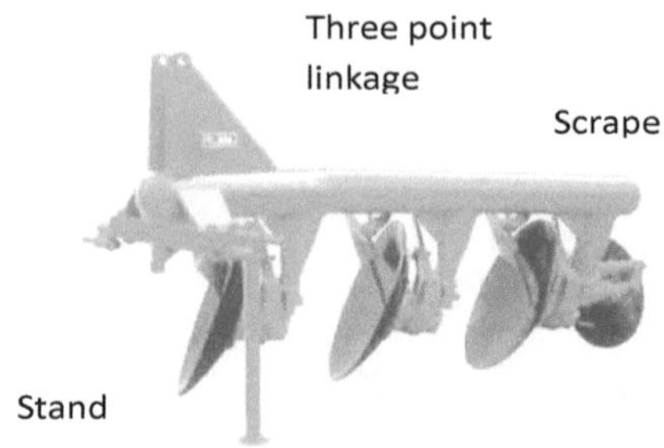


Fig. 3.26 Disc Plough – Three Bottom
Govt. of India, Department of Animal Husbandry &
Dairying http://dahd.nic.in

Fig. 3.27 Disc Plough – Four Bottom
Rajasthan Mechanical Works Ltd.,Jaipur.
http://www.brahmpuri.com

Fig. 3.28 Reversible Disc Plough
Rajasthan Mechanical Works Ltd.,Jaipur.
http://www.brahmpuri.com

Fig.3.29 Single Action Tractor Mounted Disc Harrow

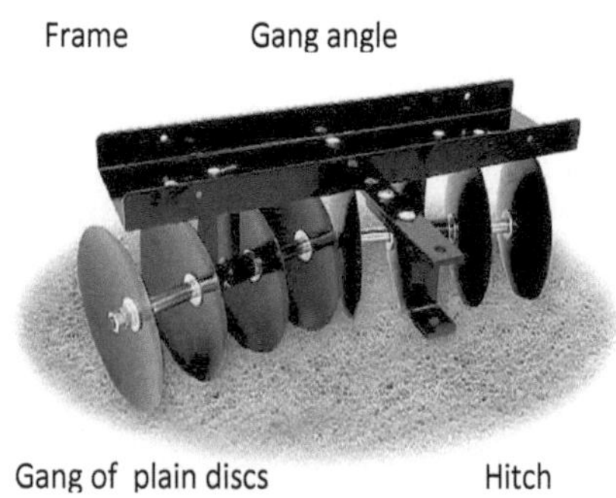


Fig. 3.30 Single Action Trailed Type Disc Harrow

**Disc Harrow**

Disc harrow is used for preparing the seed bed after ploughing by mould board or disc plough. A single action disc harrow has two opposed gangs of disc blades, both throwing soil outwards from the centre of the tilled strip (Figs. 3.29 and 3.30). The gang angle is generally 20° to 26° for cutting the soil strip, inverting and pulverizing. The single action disc harrows are not very popular in India.

The double action trailed type disc harrows are offset type and are very common in North India and are generally used for primary tillage. These harrows are attached to the tractor hitch. The mounted type of disc harrow is supported by three point linkage or hydraulic system. These are generally used for secondary tillage operation after ploughing and are lighter than trailed type disc harrows. The mounted disc harrows have 10 to 16 discs while trailed type have 18-24 discs.

Fig. 3.31 Double Action Tractor Drawn Disc Harrow Trailed Type

Fig. 3.31 Trailed Type Disc Harrow with Transport Wheels

Fig. 3.33 Offset Disc Harrow Mounted Type

Fig. 3.34 Offset Disc Harrow Mounted Type (With transport wheels)

The disc harrows in India generally have two gang, is an assembly of discs all rotating together with a common shaft or bolt through their centres. The discs can be plain (Fig. 3.30) or notched/scalloped (Fig. 3.29). In double action disc harrow, front gang of discs can be notched (Fig. 3.33). The size / diameter of discs is 51 or 56 cm and these are spaced at a spacing of 18-24 cm by using spacer spools. The gang is attached to the frame through cast iron journal bearings. For adjusting the depth of penetration, the gang angle adjustment is provided.

Double action disc harrow has one right hand gang (a gang that moves the soil to the right) and one left hand gang, operating in tandem. Each gang has an angle of 20° with respect to the rear axle of the tractor. For transporting the trailed type disc harrows the gang angle is reduced to zero and then these are transported on the road. In this way the discs do not penetrate in road / soil. For bigger sizes of disc harrow trailed or mounted the transport wheels are provided (Figs. 3.32 & 3.34) for faster transport on road and lesser load on tractor hydraulic system.

The disc harrow cultivates the soil to about 10 cm depth in ploughed land and the speed of operation is 5 to 6 km/h. The field capacity for a 35 hp tractor is about 0.6 ha/h and 50 hp tractor 0.9 ha/h respectively in ploughed field. The fuel/diesel consumption is around 10 litres/ha.

**Cultivator**

Commonly called tiller, the cultivator is a secondary tillage tool generally used after ploughing. In light sandy soils of Punjab, the cultivator (Figs. 3.35 to 3.37) is used for primary tillage also. Cultivators break the soil and lift and pulverize it. These uproot the trash and grasses, which are brought to the surface and get dried and killed in the sun. These are used in heavy soils after ploughing to avoid compaction and preparing seed beds where deep tilth is needed. The cultivator is also suited for inter cultivation / weeding and mulch forming operations.

The cultivator may have 7, 9, 11, 13 or more tines, which are mounted in two rows one behind the other with centre distance of 23 cm. The cultivators of bigger sizes are folding type for easy transport on road. The centre to centre distance of tines is also variable as the necessary holes are provided on the frame at an interval of 2.5 cm.

The cultivator may have fixed tines (Fig. 3.35) or spring loaded tines (Figs. 3.36 & 3.37). For general tillage operations the cultivator is fitted with 6.3 cm wide reversible shovels. The shovels for different kind of operations are shown in Fig. 3.38. For black cotton soils, the V type shovel is generally used. Nine tine cultivators are generally used with 35 hp while thirteen tine cultivators with 50 hp tractors. The field capacity for 35 hp tractor is 0.8 ha/h and 50 hp tractor is 1.1 ha/h while the speed of operation is 5-7 km/h. The fuel/diesel consumption is 4.3 to 5.0 litres/ha.

**Rotavator**

Rotavator is a very effective tillage implement for preparing a fine seed bed, weed control and mixing of soil with crop residues (Figs. 3.39 to 3.42) and is very popular in North India. The soil slices are cut and pulverized by the rotating blades and subsequently pulverized by impact on rear flap. It is mounted to tractor by 3 point linkage hydraulic system and the depth of operation is controlled by the hydraulic system. Some rotavators have wheels fitted to them for depth control. About 30 to 60 "L" shaped blades (bent left or right) are fitted on the main rotor shaft which rotates at about 180-215 rpm. The rotor shaft with blades is rotated by the pto shaft through pto drive shaft, bevel gear box, counter shaft and chain or gear drive system. The rotavators are available from 1.3 to 2.5 metre working width. These rotavate the soil upto a depth of 22 cm. The tractor power requirement of rotavator is 40 hp per metre of operating width. The rotavator with 1.3 meter may be fitted with 30 blades and needs about 30 to 45 horse power. The rotavators of 1.55 and 1.80 metre operating width may be fitted with 36 and 42 blades and need about 45 to 65 hp and 65 to 80 hp respectively. There is hardly any traction load requirement for rotavator, while the other tillage implements work on traction developed by the rear wheels of the tractor.

Fig. 3.35 Cultivator – Fixed Tyne
Modern Industries, Industrial Estate, Hoshiarpur, India – 146001
http://www/moderninds.com

Fig. 3.36 Cultivator – Spring loaded Tyne
Rajasthan Mechanical Works Ltd.,Jaipur.
http://www.brahmpuri.com

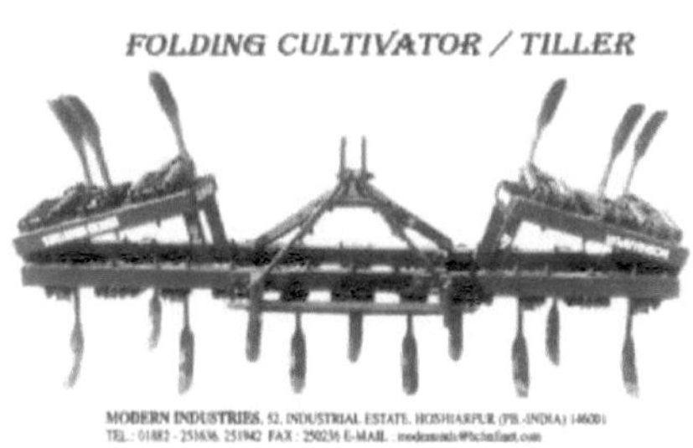


Fig. 3.37 Cultivator – Folding Type
Modern Industries, Industrial Estate, Hoshiarpur, India – 146001
http://www/moderninds.com

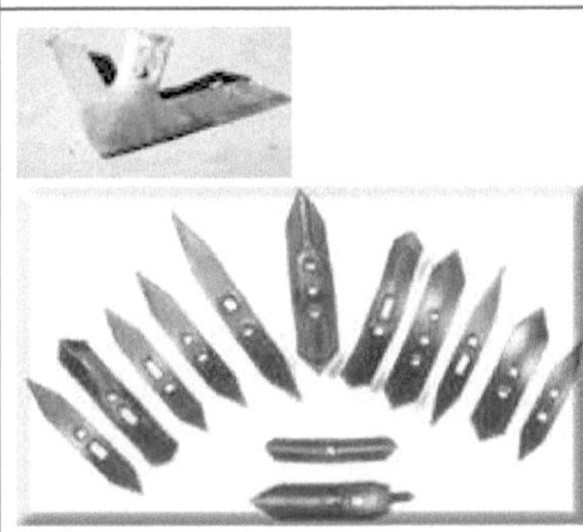

Fig. 3.38 Types of Shovels
Modern Industries,Industrial Estate, Hoshiarpur, India – 146001
http://www/moderninds.com

Rotavator is not suitable in soils with stones as there is a chance of breaking of blade or universal joint cross of pto drive shaft, when the blade hits the stone. The override clutch is generally provided in pto drive shaft to take care of such situation. The farmer has to store some blades and cross of universal joints in stock for replacement due to breakage.

A rotavator of 1.25 meter working width has the field capacity of 0.18 to 0.26 ha/h when working with 35 hp tractor. The working depth is 10 to 12 cm and the fuel consumption is 3.1 to 4.3 litres/h when operating in silt clay loam and black cotton soils.

Fig. 3.39 Rotavator
Modern Industries,Industrial Estate, Hoshiarpur, India – 146001
http://www/moderninds.com

Fig. 3.40 Rotavator
Govt. of India, Department of Animal Husbandry & Dairying
http://dahd.nic.invvvt

Fig. 3.41 Rotavator

Fig. 3.42 Rotavator in operation

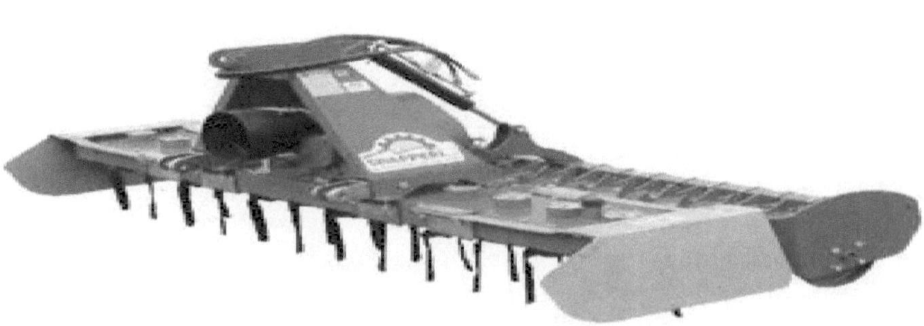

Fig. 3.42 (a) Rotary harrow
http://www.spapperi.it
SPAPPERI S.r.l., Zona Artigianale, San Secondo (PG), Italy,

**Power / rotary harrow**

The power harrows have 8 to 12 sets of rotary blades. The blades are fitted on vertical shafts. These blades are rotated through pto shaft. These are tractor mounted implements (Fig. 2.42a) and have working width of 1.3m and more. One roller is also provided for breakage of clods.

**Roller / Clod Crusher**

Spiked Clod Crusher (Fig. 3.43) is suitable for breaking and segregation of clods for seed bed preparation after paddy harvest. This can also be used as a combination tillage tool with disc harrow or cultivator. The length of roller is 1.5 metre, diameter 35 cm and it weight about 150 kg. It can be drawn by a pair of bullocks or a tractor. The field capacity is 0.4 to 0.6 ha/h, when operating with a tractor at the speed of 4.0 to 4.3 km/h. The clod size is reduced from an initial of 16 to 18 cm to final size of about 1.4 cm when working in silt clay loam soils. Grooved roller (Fig. 3.44) or plain stone roller can also be used for clod crushing. These can also be used for final shaping of beds formed by the bed former.

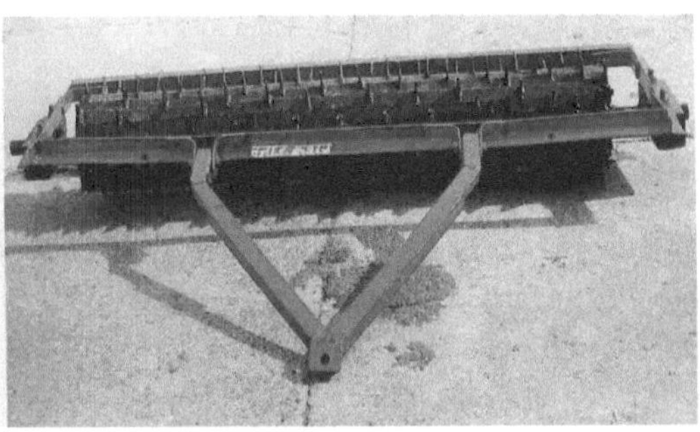

Fig. 3.43 Tractor Drawn Spiked Clod Crusher
Pandey, M.M., *et.al.*, Farm Machinery Research Digest, Central Institute of Agricultural Engineering, Bhopal.

Fig. 3.44 Animal drawn Roller
Pandey, M.M., Farm Machinery Research Digest, CI A E

## 3.4 Tractor Drawn Pulverizing Roller Attachment

This is an attachment to commercially available tractor operated cultivator for breaking the clods. The roller is attached with springs to the frame of the cultivator. The roller is made by ms flats by fitting in a helical cylinder form. The length of the roller is 1.85 m and the diameter 35 cm. The field capacity is 0.62 to 0.75 ha/h when working at a speed of 3 to 5 km/h. The depth of operation is 10.0 to 12.5 cm while fuel consumption is 3.5 to 4.0 litres/hr when working as an attachment to the cultivator.

***Patella* Harrow (Plank)**

*Patella* harrow (*Suhaga*) is used to break clods and compact the top layer of soil. In addition to pulverization, this helps in reducing soil moisture loss by reducing aeration in soil. The patella harrow can be tractor or animal drawn (Fig 3.45 and

3.46. For animal drawn *patella*, the operator has to stand on *patella* for providing necessary weight for clod breaking. The field capacity can be 0.4 to 0.6 ha/h for tractor operation. Hooks for removing and collection of weeds from the soil can also be provided with the harrow. The handle is also provided to lift the hooks and leave the weeds on soil surface for drying.

Fig, 3.45 Animal Drawn *Patella* Harrow
Pandey, M.M., *et.al.*, Farm Machinery Research Digest, CI A E

Fig, 3.46 Tractor Drawn Patella Harrow
http://www.indiamart.com/gurunanak-agriculture-implements
Guru Nanak Agricultural Impl. Mrs, Jalandhar

**Sub Soiler / Chisel plough**

The chisel plough is a pointed curved bar type implement for breaking the hard pan which is formed below normal ploughing due to plough sole and compaction and due to movement of tractor. Hard pan restricts the natural drainage and nutrient availability through soil structure. The animal drawn chisel plough (Fig. 3.47) can break the pan up to a depth of 30 cm. The field capacity is 0.2 ha/h when operating with a pair of bullocks at speed of 2.1 km/h with 90 cm space between successive passes. The tractor drawn chisel plough can have one or more tines (Figs. 3.48 to 3.50) based on the horse power of the tractor. It can break the hard pan up to a depth of 0.50 meter. One tine plough needs 30 to 40 hp tractor and two tine plough needs 60 to 80 hp tractor.

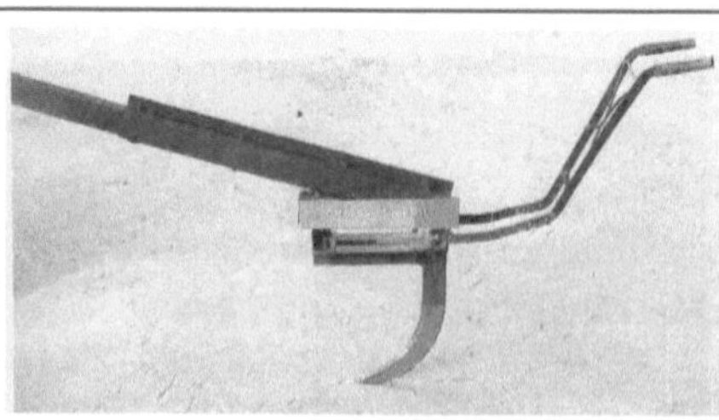

Fig. 3.47 Animal Drawn Chisel Plough – Single Tine
Pandey, M.M., *et.al.*, Farm Machinery Research Digest,
Central Institute of Agricultural Engineering, Bhopal.

Fig. 3.48 Tractor Drawn Chisel Plough – Single Tine Modern Industries, Industrial Estate, Hoshiarpur, India – 146001
http://www/moderninds.com

Fig. 3.49 Tractor Drawn Chisel Plough – Single Tyne
http://arsonagro.com
Arson Agro Industries, Moga

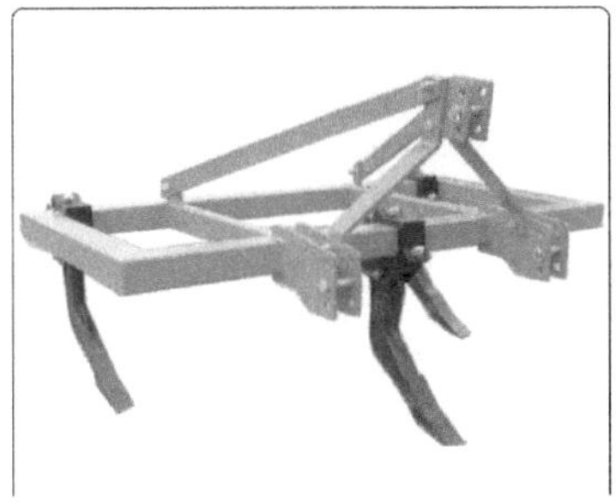

Fig. 3.50 Three Tine Chisel Plough
http://www.aeco.co.in
Agra Engineering Company, Mumbai

## 3.5 Earth Moving Machinery

The soil may need leveling in most cases and terracing in some cases. For heavier works; the dozers (Fig. 3.51) may be required which are track type tractors with horse power varying from 50 to 250 hp. The tine for loosening of the soil may also be fitted with the dozer. Tractor operated bucket loader (Fig 3.52) is able to perform most of the farm leveling operations and terracing. Tractor operated bucket loader cum back hoe loader (Fig. 3.53) commonly known as JCB can be used for most of the earth moving works, including pit digging, channel forming, terracing, etc. The size of back hoe loader/bucket may vary from 0.2 $m^3$ to 1.0 $m^3$ based on the size of tractor. The bucket can have 180°C swing and dig upto 5 meter deep. For heavy duty this equipment may also be of track type.

The terracer blade (Figs 3.54 and 3.55) is the most common implement for field leveling, grading and terracing. It is hydraulically mounted implement. The length of the blade is 1.8 meters or 2.4 meters for 35 hp or 50 hp tractors respectively. The height of terracer blade is around 0.4 meter and the blade angle is adjustable. The wheel (Fig. 3.55) is fitted to adjust the depth of the cut for fine leveling of the field. The terracer blade can be of offset type (Fig 3.55). It can be reversed for backfilling.

The low cost land levelers (Figs. 3.56 & 3.57) are also popular for land leveling and terracing. The land leveler (Fig. 3.57) can also be reversed by using the alternate hitch for land filling.

The triplane (Fig. 3.58) is used for final grading of the field after ploughing and harrowing. The frame length of the triplane varies from 12 to 16 metres and blade width 5 to 8 metres. This is a trailed type implement in which height and angle of the leveling blade is adjusted by using the hydraulic system of the tractor.

The laser guided leveling system (Figs. 3.59 and 3.60) utilizes laser rays for depth control along with hydraulic and electronic control circuitry for accurate leveling and grading. The laser system consists of laser transmitter, laser receiver and controls the depth of the cut i.e., raises or lowers leveller blade. The operator need not worry for raising or lowering blade as it is done automatically. The implement has to be used in properly ploughed and harrowed land. It's field capacity can be 0.17 to 0.38 ha/h.

Fig. 3.51 Dozer
http://india.cat.com/cda
Caterpillar Corporate Headquarters, Peoria, Illinois USA

Fig. 3.52 Tractor operated Bucket Loader
Kishan Agro Industries, Rajkot, Gujarat
www.kishanagro.com

Fig. 3.53 Tractor operated Excavator (Backhoe) cum Bucket Loader
Bull Machines Pvt Ltd,Coimbatore http://www.bullindia.com/tractor_attachments

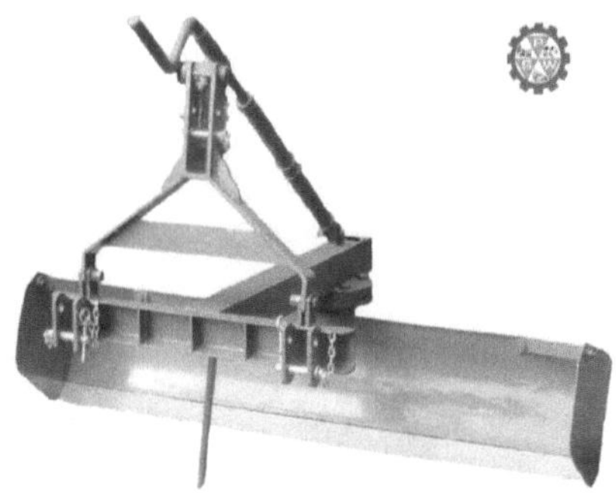

Fig. 3.54 Terracer Blade
Popular Steel Works Kolhapur
http://popularpsw.com/framep.htm

Fig. 3.55 Terracer Blade with Rear Wheel
Mahindra & Mahindra Ltd.
http://tractor.mahindraworld.com/leveller.htm

Fig. 3.56 Land Leveler
Rajasthan Mechanical Works Ltd., Jaipur.
http://www.brahmpuri.com

Fig. 3.57 Land Leveler with Hitch on both Sides
Popular Steel Works Kolhapur
http://popularpsw.com/framep.htm

Fig. 3.58 Triplane (16' x 40')
(SWECO)D.B. West Auctioneers
http://www.westauction.com

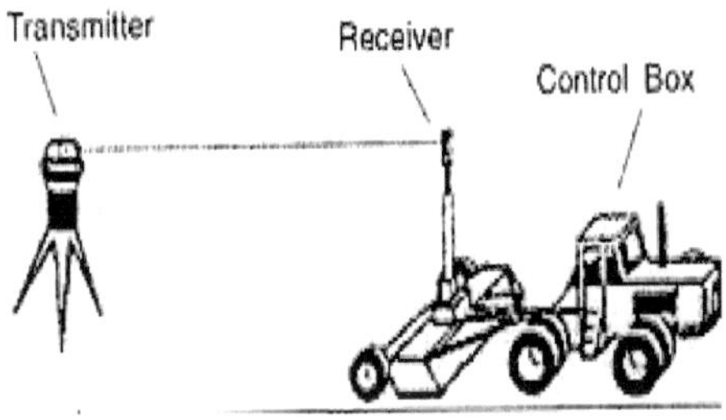



Fig. 3.59 Laser Guided Grader
CIAE Bhopal

Fig. 3.60 Laser Guided Grader
Kevin McNamara & Son P. Ltd. http://www.tenniscourts.com.au/buyers_guide.html

Fig. 3.60(a) Laser Guided Grader
http://laserlevelsz.net

Fig. 3.60(b) Laser Guided Grader
SPL Technologies (P) Ltd. New Delhi
http://spltech.in

**Pit Hole Digger**

For planting of fruit and other perennial trees pits of 0.3 m to 1.0 meter diameter are required to be dug. Similarly, pits of 0.10 to 0.30 meter diameter are required to be dug for fencing and for erecting bower for grapes and other crops. The power tiller operated pit hole digger (Fig. 3.61) has been developed at the Central Institute of Agricultural Engineering, Bhopal and is available commercially. It can dig pits of maximum diameter of 30 cm and up to a depth of 45 cm. Around 25 pits can be dug in an hour and the fuel consumption is about 0.9 litre/hr.

The tractor operated pit hole diggers (Figs. 3.62 to 3.64) can dig pits of up to 0.5 metre diameter and 1 metre depth. These are mounted on 3 point linkage and the auger is rotated by the pto shaft through pto drive shaft and gear box. The pits are always straight and perpendicular to the ground as the parallellogram type of linkages are used to lower the auger. Augers of 10 to 50 cm. diameter are generally available and are operated at 50 to 200 rpm. Conical shape pits can also be dug by using the digger shown in Fig 3.64. The soil has to be properly irrigated and preferably ploughed for efficient operation of digger. Fifty centimeter diameter digger needs about 50 hp tractor and 100 cm diameter digger needs about 100 hp tractor. It takes about one minute to dig a pit.

Fig. 3.61 Power Tiller Operated Pit Hole Digger
Varshney, A.C., Suresh Narang, Central Institute of Agricultural Engineering, Bhopal

Fig. 3.62 Tractor Operated Pit Hole Digger
Modern Industries,Industrial Estate, Hoshiarpur, India – 146001 http://www/moderninds.com

Fig. 3.63 Pit Hole Diggers
Northern Tool and Equipment, Minnesota, USA http://www.northernstores.com

Fig. 3.64 Pit Hole Digger – Conical

The tractor operated back hoe/JCB/excavator can also be used for digging the pits. These can also be used for filling and compacting the pits, during planting of the seedlings.

**Trencher**

Trenches are required for planting of grape and pineapple as also to conserve the rain water. The deep trencher (Fig. 3.65) can be used for digging trench with a top width of 60 cm with bottom width of 30 cm with up to 90 cm depth. The trencher (Fig. 3.66) is suitable for preparing wide trenches for pineapple planting. A depth control wheel can also be provided with the trencher.

The trencher needs a tractor of more than 45 hp and the working speed is 3 to 4 km/h. The trencher can work in properly culivated/ploughed and harrowed field with optimum moisture content.

Fig. 3.65 Trencher for Deep Trench – Tractor Operated Shiva Industries Mandipet Tumkur Karnataka,India,572101

Fig. 3.66 Trencher for Wide Trench –Tractor Operated AGRIPAK INTERNATIONAL *Karachi* http://www.brain.net.pk/~agripak/ag10.html

The vertical trenchers (Fig. 3.67) are required to dig the trenches for laying the main and sub main pipe lines of the drip irrigation system. These dig trench of upto 200 mm width and pipes of upto 15 cm diameter can be installed in these trenches. The depth of trench is 54 cm and above. The tractor power required is 65 hp and above and tractor speed 0.35 to 0.65 km/h. The hydraulic system of the tractor rotates the trencher wheel for trenching.

The ditcher (Fig. 3.68) is required for digging the trench with top width of 75 cm and bottom width of 23 cm for planting the grafts of grape. The depth of the ditch/trench is 50 cm and above. These are rotary trenchers operated with tractor hydraulic system with tractor requirement of 55 hp and above. The tractor speed is 0.8 – 1.2 km/h. These are fitted to Cat-II type 3 point linkage system of the tractor. The hydraulic system of the tractor rotates the trencher wheels of the ditcher.

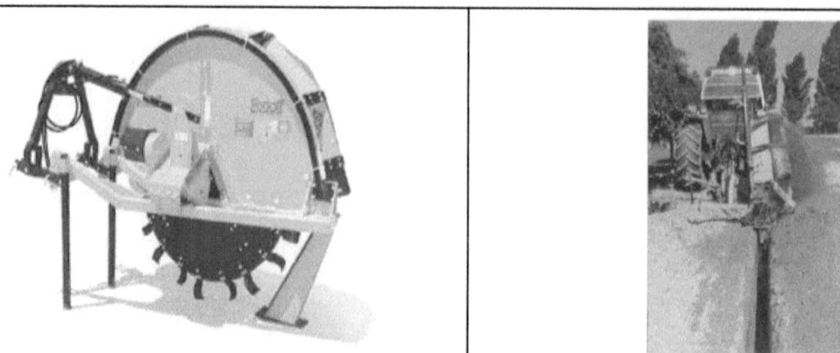

Fig. 3.67 Tractor operated vertical trencher

Dondi Spa, Italy

Fig. 3.68 Tractor operated ditcher

Dondi Spa, Italy

**Ridger**

Most of the vegetable seeds and seedlings are grown on ridges. A two bottom power tiller drawn ridger is shown in Fig. 3.69. The wing adjustment can be provided in each bottom for adjusting the ridge height and top and bottom width. The frame of ridger is generally provided with holes at a distance of 5 cm to adjust the spacing in ridges.

The two bottom ridger has a field capacity of 0.09 to 0.15 ha/h when operating at a speed of 1.8 to 2.0 km/hr with an 8 hp power tiller. It gives ridge spacing 60 cm with the bottom width of the ridge being 30 cm, top width 7 cm and height 15 cm.

The single bottom reversible plough is modified as a single bottom ridger by modifying the share. The single bottom ridger had a field capacity of 0.05 ha/h when working at a speed of 1.75 km/h and the fuel consumption was 0.47 litres/hr and 9.9 litres/ha. The ridge spacing is 53 cm, with bottom width of 32 cm, top width of 7 cm and ridge height of 15 cm.

The tractor operated ridgers generally have 2 or 3 bottoms The shape of wing/mould board varies as per the soil condition and the adjustment for wing width is also provide in most of the cases. The ridges are made by the tractor and then the seeds or seedlings are planted manually in most cases on the ridges. The ridger is also an integral part for the ridge planter/seed drill/transplanter. The two bottom and three bottom ridgers may need 35 and 45 hp tractors respectively, when working at a speed of 5 to 7 km/h. The tied ridger creates small dams in furrows formed by ridger. A wide wheel with 10 to 12 cm high projection creates the depressions and check, dams in furrow. As a result the free flow of the rain water is checked and is conserved in the soil.

**Bed Former**

When vegetable crops are grown on raised beds in double row system, it facilitates to installation of the drip irrigation system at a lower cost. There is also saving of water, compared to irrigation by flooding. Even fruit, vegetable and ornamental crops are also generally raised on beds.

Fig. 3.69 Bed Former– Tractor Drawn- Mould Board Type with Chain
Central Institute of Soil Science, Bhopal

Fig. 3.70 Bed Former – Tractor Drawn- Ridger Type
Indian Institute of Horticultural Research, Bangalore

Fig. 3.71 Bed Former (Furrow Former ) – Tractor Drawn
Pandey, M.M., et al., Farm Machinery Research Digest,
Central Institute of Agricultural Engineering, Bhopal.

Fig. 3.72 Bed Former– Tractor Drawn-in operation
Indian Institute of Horticultural Research, Bangalore

Fig. 3.73

Fig. 3.74 Roller (Grooved) for Bed Finishing
Singh, D.V. Natonal Research Centre for Soyabean, Indore

It is easy to mechanize various operations in crops raised on the bed, as the channel acts as a passage for the movement of tractor wheels. The width of the raised bed is kept equal to the tread width/spacing between rear wheels of the tractor. The bed former (Fig. 3.69) has two mould boards, one left hand and one right hand to raise the bed level and prepare the channel. The beds formed are leveled by the heavy link chain provided in between the two mould boards. The bed former is a tractor mounted implement. The disc type bund former (Fig. 3.85) can also be used to make beds. The bed former (Fig. 3.70) has two ridger bottoms fixed at required distance to form the bed and channel. The beds so formed can be finished by the roller as shown in Fig. 3.73. The roller can have the exact shape of the bed in some cases for finer finish and compaction.

The bed former (Figs.3.71 & 3.72) can make 2 beds of 45 cm width each and 15 cm height. The total working width is 2.25 metres. The field capacity is 0.75 to 1.00 ha/hr, when working with 35 hp tractor at a speed of 3.2 km/hr.

The tied bund former shown in Fig. 3.74 is used to create the small check dams by the projections fitted on the wheels. The wheels run in the channels formed by the bed former. The rain water collected in between small check dams is

## Bund and channel former

The tractor drawn bund former has two sheets 40 cm high and 1.25 meter in length. These sheets collect the soil and form the bund behind the tractor in between rear wheels. The width and height of the bund can be adjusted by adjusting front and rear spacing between the sheets.

The bund former is tractor drawn and makes bunds of 25 to 100 cm base width and 12 to 35 cm height. If the bunds are formed side by side the channels are of 10 to 45 cm width and 12 to 35 cm depth. The operating speed is 2.4 km/h when working with 20 hp tractor.

The channel cum bund former is animal drawn. It has four sheets of 30 cm height and 64 to 78 cm length. The width and height of the bund and channel is adjustable. The bunds can have 25 to 55 cm base width and 8 to 15 cm height. The channel can have a width of 10 to 30 cm. The operational speed is 1.5 km/h.

The tractor operated bund former is fitted with 2 discs to collect the soil at the centre and make the bund. It is suitable for tractors of 25 hp and above. The disc type bund former can also be used to make beds. Tractor operated channel former is a modified the bottom ridger the centre bottom width is increased to from a channel .

Fig. 3.75 Channel cum Bund Former
Pandey, M.M., *et al.*, Farm Machinery Research Digest
Central Institute of Agricultural Engineering, Bhopal.

Fig. 3.76 Bund Former
http://mamtaironworks.tradeindia.com
Mamta Iron Works, Billimora Gujarat

## Tractor Drawn Bund Former and Channel Former

Fig. 3.77 Bund Former
Pandey, M.M., *et al.*, Farm Machinery Research Digest
Central Institute of Agricultural Engineering, Bhopal.

Fig. 3.78 Bund Former-Disc Type
Agra Engineering Company, Mumbai
http://www.aeco.co.in

Fig. 3.79 Channel Former
Guru Nanak Agricultural Impl. Mrs, Jalandhar
http://www.indiamart.com/gurunanak-agriculture-implements

# *Chapter 4*
# Crops Planting

Crop planting is placing seeds, tubers or cuttings in the soil at a required depth and spacing. This can be achieved by one of the following methods.

1. **Broad casting :** It is random scattering of the seeds over the seed bed. This system is mostly used for cultivation of crops like onion and carrot primarily under rain fed condition.
2. **Drilling :** It is random dropping and covering of seeds in furrows in definite rows. While the seed to seed distance cannot be maintained, the number of seeds per unit length of row or the seed rate per unit area is approximately constant. The seeds of green leafy vegetables and narrow spaced root crops like raddish, carrot, sugar beet and turnip are usually drilled.
3. **Planting :** It is accurate placing of single seeds at equal interval in rows. Most of the vegetable crops like okra (lady's finger), beans, etc. are dibbled or planted manually in the field.

   The seed bed for planting can have the following contours

   i. **Flat planting:** When the crop is planted on flat fields. Green leaves and rain fed onions are generally grown on flat/leveled fields.

   ii. **Ridge planting:** Root or tuber crops like raddish, carrot, potato, beans and okra are generally planted in a single row on ridges.

   iii. **Bed planting:** Planting on beds in paired rows (2 rows per bed) is very common for planting beans and root crops to save irrigation water. This practice is also well suited for drip irrigation of the crop. This practice is also suitable for mechanized cultivation of crops with tractor when bed width is equal to tractor tyre base width.

iv. **Furrow planting:** In this case seeds are put in furrows created in between ridges. This practice is followed under rain fed cultivation.

The application of fertilizer is also one of the operation to be performed along with crop planting. The details of the machines used for various operations are as follows.

## 4.1 Fertilizer Broadcaster

### a) Manual broadcaster

The seeds are generally broadcasted manually for narrowly spaced crops. For attaining better uniformity of seed spacing, the seeds are broadcasted in 2 to 3 passes. Fertilizer is also broadcasted manually. The manually operated broadcaster (Fig. 4.1) is held by the shoulder straps. The broadcaster can hold upto 15 kg of seeds or fertilizer. It consists of hopper generally made of plastic or stainless steel. There is an adjustable opening at the bottom of the hopper to control seed or fertilizer application rate. A horizontal spinning disc is provided below the opening. Spinning disc is rotated by handle through gears for spreading/broadcasting the seed or fertilizer. The on and off control handles are also provided to start or stop the broad casting of the seed/fertilizer by closing the opening at the bottom of the.

Fig. 4.1 Manually operated broadcaster for seeds and fertilizer
Neptune Sprayers www.neptunesprayers.com

Fig. 4.1 (a) Manually operated broadcaster for seeds and fertilizer
Hymatic Agro http://hymaticagro.com

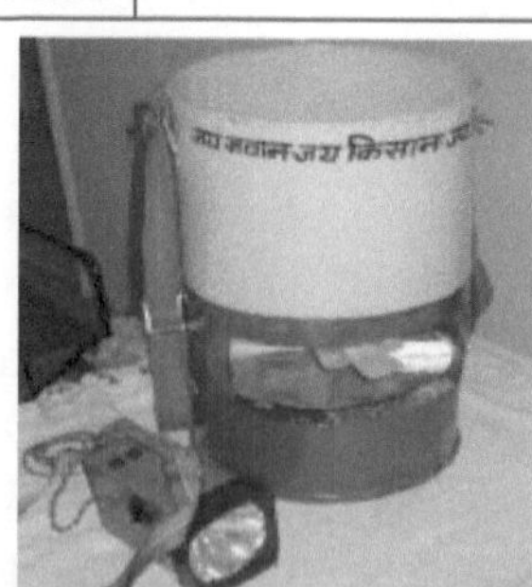

Fig. 4.2 Battery operated broadcaster for seeds/ fertilizer
www.nifindia.org

hopper. In battery operated broadcaster (Fig. 4.2), the spin disc is operated by a micro motor. This innovation by Dharmendra Patidar from Maharashtra a science student has been awarded by National Innovation Foundation of India.

b) **Tractor operated broadcaster**

Spin type broad caster—The spin type broadcaster (Figs. 4.3 and 4.4) have a fertilizer hopper of 400 to 900 litre capacity. The hopper is generally made from plastic or stainless steel sheet. In low cost broadcasters the hopper may be made from mild steel sheet. At the bottom of the hopper, an adjustable opening is provided to control the flow of fertilizer. A horizontal spin disc is provided below the fertilizer opening. The spin disc is rotated by the pto drive shaft through gears. The broadcasters with small capacity hoppers are mounted type on three-point linkage. The broadcasters with higher capacity fertilizer hoppers are usually trailed type. The spreading width is adjustable from 6 meter to 24 m. The broadcasting rate is adjustable by adjusting the opening at bottom of the hopper and the opening can be opened or closed to start or stop the broadcasting. The broadcaster can also be used to broadcast seeds. A tractor of upto 45 hp is sufficient to operate the broadcaster and the broadcasting capacity is upto 3 ha/h. The operating speed of the tractor varies from 3 to16 km/hr.

Fertilizer can also be applied to tree crop rows by using double spin type broadcaster (Figs. 4.4 & 4.5). For this 2 spin discs may be used with openings targeted to the row of trees.

Swing type broadcaster- In these broadcasters (Figs. 4.6 to 4.8) oscillating spout is provided in place of spin disc for spreading the fertilizer or seed. The oscillating spout is operated by the pto drive shaft of the tractor. The spreading width is adjustable and can be varied from 6 m to 24 m. The broadcasters generally have more rate of application at centre and less at the far end. The uniformity of fertilizer spreading in the field can be achieved by overlapping the spreads (Fig.4.9).

Fig. 4.3 Tractor operated fertilizer spreader – spin disc type, www.kenttractors.com

Fig. 4.4 Tractor operated fertilizer spreader – spin disc type – http://www.emdek.com.my

Fig. 4.5 Double disc spin fertilizer broadcaster for applying to row crops. http://www.emdek.com.my

Fig. 4.6 Tractor operated fertilizer broadcaster – swing type, http://www.marketfarm.com

Fig. 4.7 Swing type (Oscillating spout) tractor operated Fertilizer Spreader in operation Kverneland pendulum

Fig. 4.8 Swing type (Oscillating spout) tractor operated Fertilizer Spreader in operation Kverneland pendulum

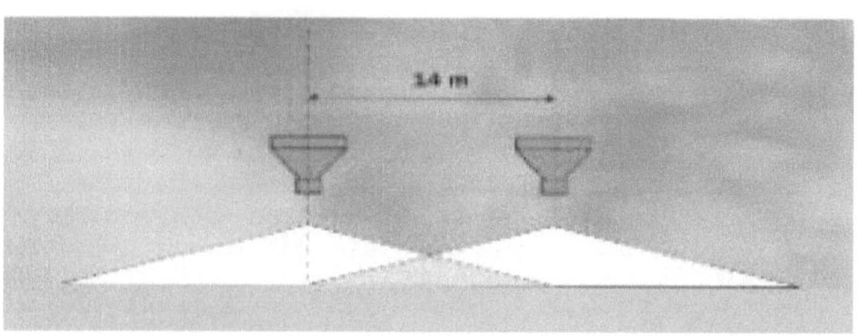


Fig.4.9 Overlapping the spread surface to achieve uniform application of fertilizer
Kverneland pendulum

## 4.2 Manure spreader

To spread the farm yard manure evenly trailor type spreader (Figs 4.10 to 4.17) is mostly used. The trailor (Fig. 4.11 ) is generally tipping type by the hydraulic system of the tractor. One beater drum (Figs. 4.12 and Fig. 4.13) with big projections is provided at the rear of the trailor. The beater drum pulverizes the manure, moves it to the rear and spreads it on the field through the outlet. The size of outlet (Fig. 4.14) can be controlled to adjust the manure application rate. The manure conveyor (Fig. 4.12) is also provided at the bottom of the trailor for pushing the manure towards backside or towards the beater. The beater drum can be rotated by the trailor wheel (Fig. 4.14 and Fig. 4.15) through sprocket and chain or through pto drive shaft (Fig. 4.16.). The application of manure is shown in field Fig. 4.17.

Fig. 4.10 Manure spreader-tipping type trailer with beater drum
http://www.wikco.com

Fig. 4.11 Manure spreader-tipping type trailer with beater drum
D.V.Singh, NRC Soyabean, Indore

Fig. 4.12 Beater drum and manure conveyor of manure spreader
Hamilton Equipment, Ephrata, PA, USA
http://www.honeyspreaders.com

Fig. 4.13 Beater drum of manure spreader
http://www.wikco.com

Fig. 4.14 Sprocket for driving beater drum and outlet adjustment handle
D.V.Singh, NRC Soyabean, Indore

Fig. 4.15 Manure spreader drive to beater drum from ground wheel
http://www.wikco.com

Fig. 4.16 Drive to beater drum from pto shaft
http://www.wikco.com

Fig. 4.17 Spreading FYM in field
http://www.wikco.com Wikco.com Inc.1467 N Grant Ave., Casa Grande AZ 85222 USA

## 4.3 Seed Cum Fertilizer Drill

Seed drills are used to sow the seeds in rows and maintain the number of seeds per unit length. The seed to seed distance is low and the seed spacing is not constant. The simplest form of seed drill is the traditional funnel (pora) fitted behind the country plough (Fig. 4.22). The seeds are dropped in the furrows almost at constant rate by the ploughman and are covered by the soil automatically behind the plough. The uniformity of sowing depends upon the skill of the operator. Two row animal drawn seed drill (Fig. 4.23) has been developed at CIAE, Bhopal and three row seed-cum-fertilizer drill at ANGRAU, Tirupathi and ICRISAT Hyderabad (Fig. 4.24). In these, seed and fertilizer is metered manually and the seed spacing depends upon the skill of the ploughman. The mechanical seed drill has the following parts.

i. Seed hopper – To store the seeds for drilling and control.

ii. Seed metering mechanism – To meter and control the seeds for distribution in rows. Approximately equal numbers of seeds are sown in one meter length. The seed rate can be varied by the seed rate control mechanism.

iii. Seed pipes – To convey the seeds from seed metering mechanism to the furrow.

iv. Furrow opener – To open the furrow of required width and depth for placement of seeds and fertilizers in the soil.

v. Furrow closer – To cover the seeds and fertilizers with soil after sowing

vi. Ground drive wheel – To rotate the seed and fertilizer metering systems, through belt/chain and pulley/sprocket.

vii. Transport wheel – To support the weight of the machine in field and for control of depth of seed sowing.

Generally seed-cum-fertilizer drills are in use in India and these have following additional parts for application of fertilizer in the field.

i. Fertilizer box – To store the fertilizer for drilling in the field.

ii. Fertilizer metering system – To meter and control the fertilizer application.

iii. Fertilizer piper- To convey fertilizer from fertilizer metering system to furrow opener.

**Seed hopper**

Seed hopper stores / holds the seeds for drilling and is generally made of mild steel sheet and properly painted to control the rusting and reaction with the chemicals used for treatment of seeds. The shape of the box is such that the level or quantity of seeds in the hopper does not have any effect on seed rate. When the seed hopper becomes empty after completion of drilling, no seed should be left in corners of the hopper. The light weight seed hoppers can be made from moulded fibre reinforced plastic (FRP).

**Fertilizer hopper**

Fertilizer hopper stores / holds the seeds for drilling and is generally fabricated from mild steel sheet, however, it is preferred to be made out of FRP for resistance to chemical fertilizers.

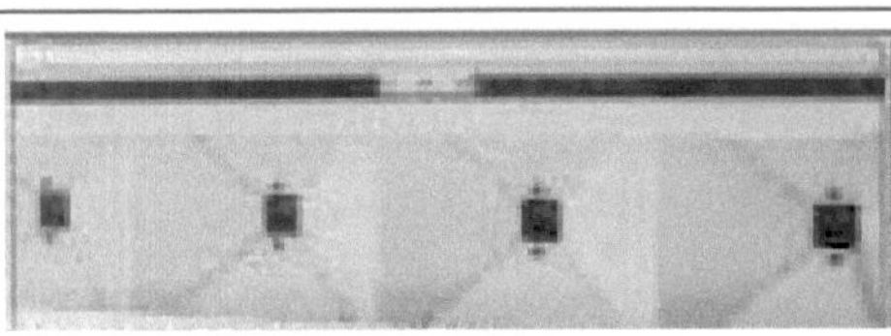

a) Variable opening
http:/www.truaxcomp.com

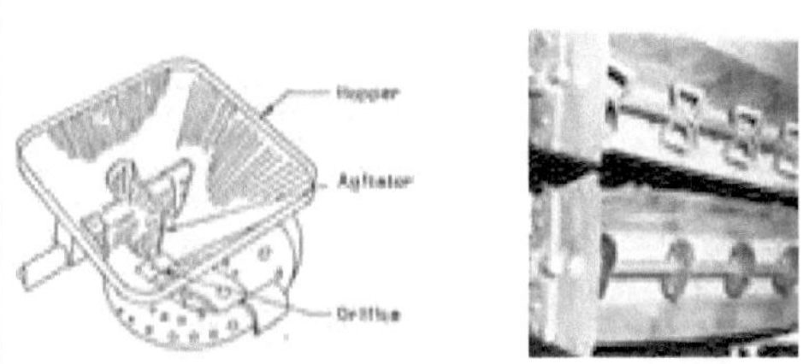

b) Variable opening with agitator
Pandey, M.M *et al.*, Farm Machinery Research Digest, CIAE

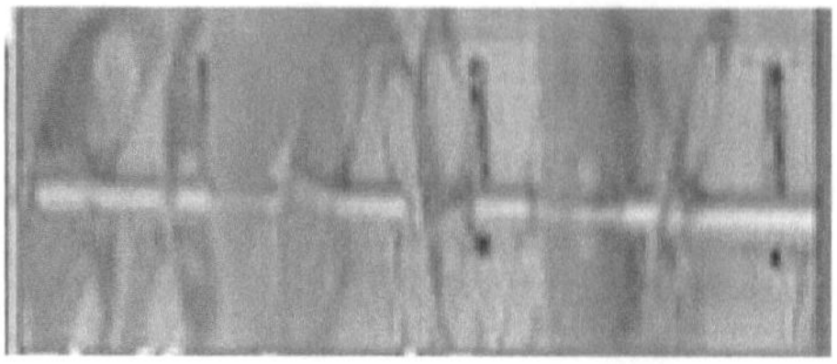

c) Agitator for fluff seeds
http:/www.truaxcomp.com

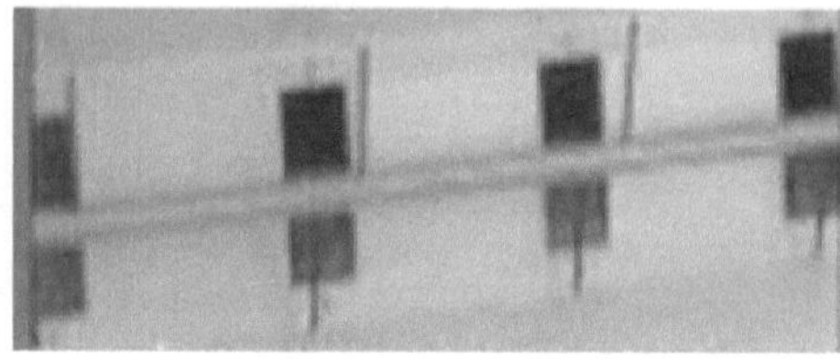

d) Agitator for grain seeds
http:/www.truaxcomp.com

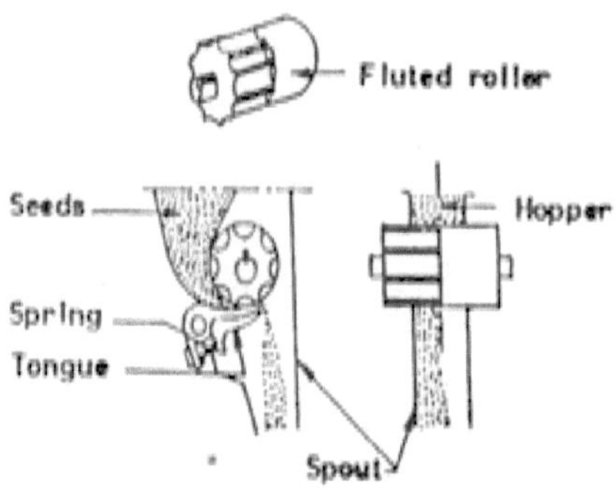

e) Flutted roller
Pandey, M.M *et al.* Farm Machinery Research Digest, CIAE, Bhopal and Mandhar S.C. PAU, Ludhiana

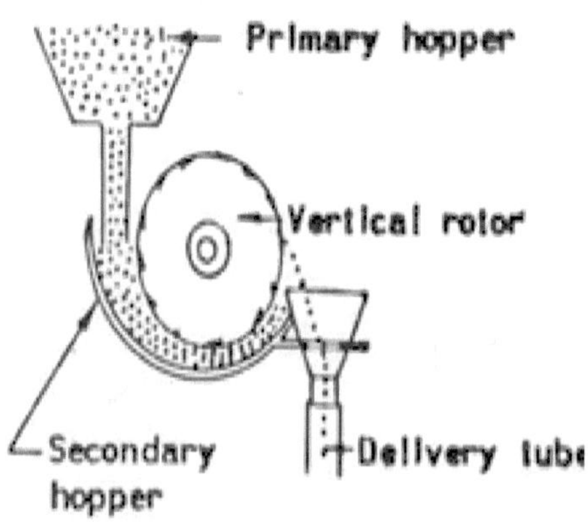

f) Vertical rotor with cells
Pandey, M.M *et al.* Farm Machinery Res. Digest, CIAE & Mandhar SC PAU, Ludhiana

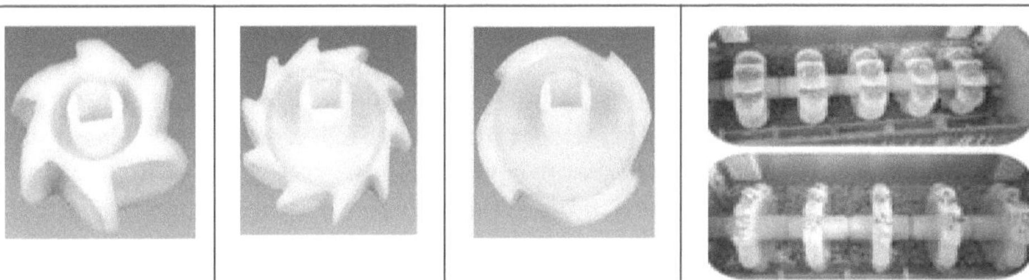

g) Vertical rotors with cells for different size of seeds and seed spacing & seed picking and dropping system

Khedut Agro Engineering http://khedutagro.com

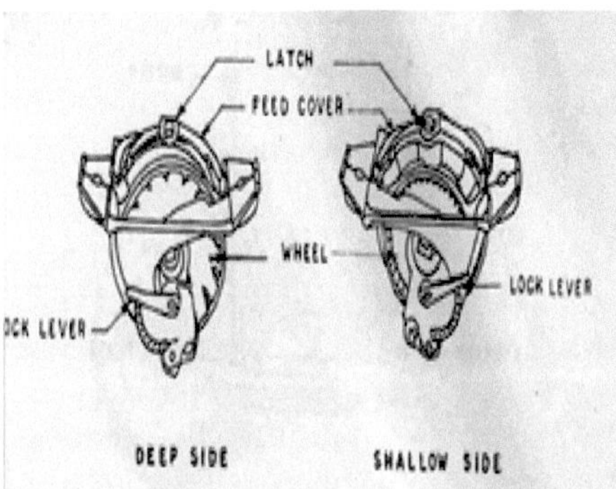


h) Internal double run plate

CIAE, Bhopal

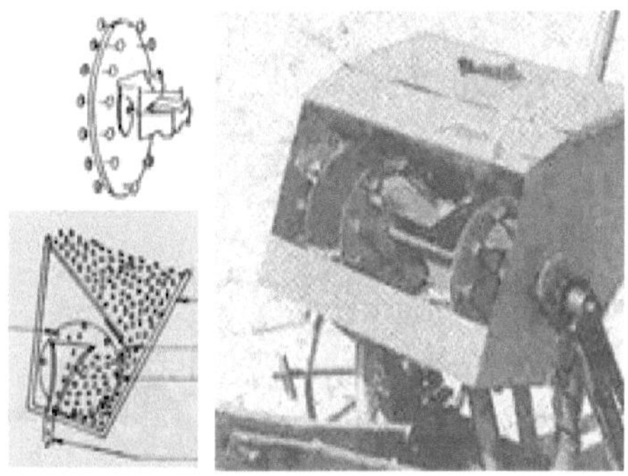

i) Cup feed

Pandey, M.M *et al* Farm Machinery Research Digest, CIAE

j) Rotary drum with holes of seed metering

Pandey, M.M *et al.* Farm Machinery Research Digest, CIAE

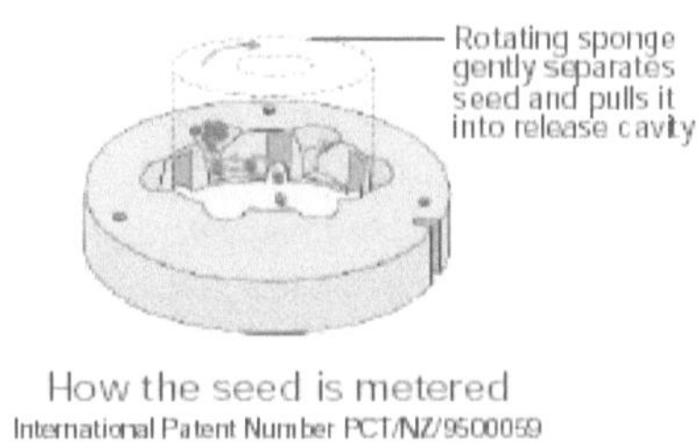


j) Electronic seed metering

http://www.seedspider.com

Fig.4.18 Seed metering devices for seed drill

**Seed Metering Systems**

The size and shape of the seed, the seed rate and the uniformity of seed distribution in row decides the type of seed metering system to be used. The following are the main types of metering systems used in seed drills.

a. **Variable opening:** The seeds are made to flow under gravity, through an opening (Fig. 4.18 a). The size of opening can be adjusted according to seed size and seed rate. The system is fitted at the bottom of the seed hopper and the sides of the seed box are given slope such that no seed is left in the box when it gets emptied. This type of system is most suitable for small seeds like radish and turnip and grasses for turf.

b. **Variable opening with agitator:** The adjustable opening or the disc with different sizes of openings/orifices is provided at the bottom of seed hopper and an agitator is provided above that for smooth flow of seeds through the opening (Figs. 4.18b to 18d). Only one orifice is below the agitator and the orifice is changed for different sizes of seeds and for adjusting seed rate. The size of the orifice or adjustable opening is bigger than the maximum dimension of the seed. Different type of agitators are required for different sizes and shapes of seeds (Figs 4.18 b to 18d). The system is suitable for drilling all kinds of seeds varying from raddish, carrot to french bean. However, there is a variation in seed rate, within the row and among the rows.

c. **Flutted roller**: It consists of a flutted roller rotating about horizontal axis and fitted below the seed hopper (Fig. 4.18e). The seeds are picked by the flutes (grooves) and discharged through the delivery gate to delivery tube. This is a positive feed system and seed rate is adjusted by adjusting the exposed or used length of the flutted rolls. Changing the opening of delivery gate can do the adjustment for different size of seeds. The flutted rollers are suited for medium sizes seeds like okra, pea, etc.

d. **Vertical rotor with seed cells:** In this system of metering the seed hopper is divided into 2 parts/hoppers (Figs 4.18f, 4.18 g and 4.36). The primary hopper is at higher level and the seeds move through pipe to the secondary hopper. In secondary hopper, one horizontal shaft is fitted with vertical rotors which have cells on the periphery to pick up seeds from secondary hopper and drop these in delivery tube (Figs. 4.18f and 4.18). The seed rate can be increased or decreased by raising or lowering the level of seeds in secondary hopper. This is achieved by raising or lowering the level of primary hopper. This system is suitable for drilling bigger seeds like pea, French bean, etc.

e. **Cup feed system:** In this system the discs with cups are provided on the horizontal shaft (Fig. 4.18i). The shaft assembly is fitted inside the seed box. The seeds are picked from the seed box and dropped in the seed tube through the outlet hopper. Cups of different sizes are required for different seed sizes and seed rates.

f. **Double run or internal force disc**: This metering system consists of a wheel or disc rotating on a horizontal shaft fixed below the seed hopper (Fig.4.18h). The two sides of the wheel are provided with seed cells of different sizes. At one time only one side of wheel is used as per requirement of seed size and seed rate. This kind of system has not become popular in India.

g. **Rotary Drum with holes:** In this system the seeds are filled in a drum the drum is provided with holes along the periphery (Fig.4.18j and 4.26). The rotation of seed drum meters the seeds and drops these in the rows. The seed rate can be varied by changing number and diameter of holes along with circumference.

h. **Electronic system:** In this patented system a 12 volt DC geared motor drives a sponge rotor within a vertical cylindrical metering plate. This metering plate has multiple channels in its internal wall, along which seeds are gently separated by the rotating sponge and carried to individual outlets (Figs. 4.18k and 4.39 and 4.40). The seed rate is varied by changing the speed of drive motor by micro processor control. The speed of seed plate or drive motor gets adjusted as per forward speed of tractor through remote control. The sponge rotor can be changed to accommodate different seed sizes and shapes.

i. **Air assisted metering system:** In this case the seeds are metered from the seed box and further distributed in the number of rows by air blown by an air blower (Figs 4.46 & 4.48).

j. **Cone type metering system:** The seeds are fed above a cone and these get distributed equally in all directions (Fig. 4.50). The rate can be adjusted by adjusting the size of annular opening for feeding the seed or size of outlet discharging the seeds to seed pipes.

## Fertilizer metering system

The fertilizer metering system is generally gravity fed. This has a adjustable/ variable opening (at the bottom of fertilizer box) with a agitator above in the form of vertical rotor or horizontal star wheel of revolving plate. The agitator system maintains the continuous flow of the fertilizer through the variable opening. This kind of system is very efficient for granular or crystalline fertilizers. The fertilizer rate is varied by varying the size of the variable opening. Vertical rotors with cells are also used in some of the seed cum fertilizer drills.

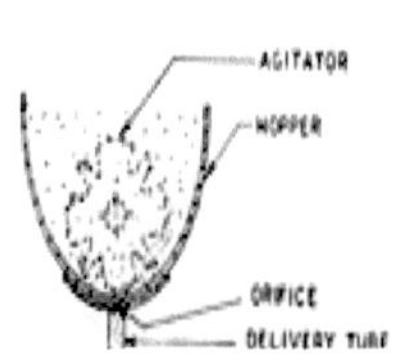


a) Agitator over adjustable orifice

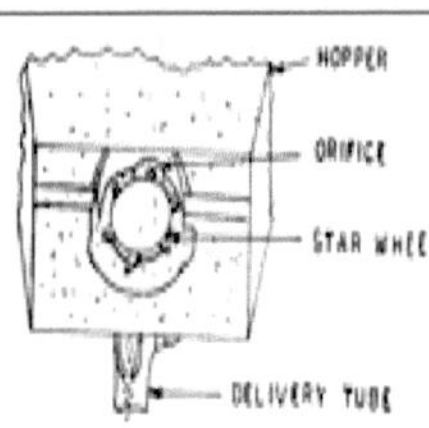


b) Star wheel feed

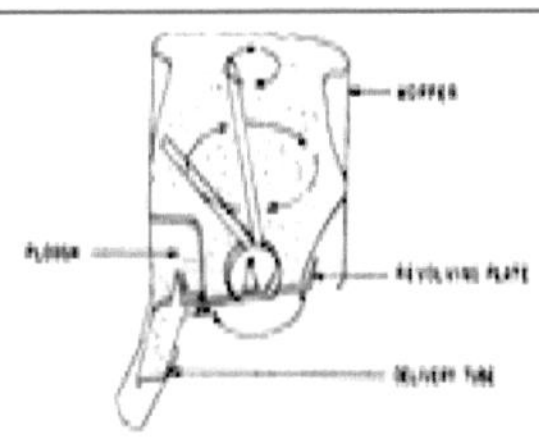

c) Revolving plate feed
Pandey,M.M, Farm Mach Res. Digest

Fig.4.19 Fertilizer metering devices for seed drill

**Ground drive wheel**

The seed and fertilizer metering systems are rotated for continuous flow of seeds and fertilizer. This is done by providing ground wheel with projections in the form of lugs or pegs (Fig. 4.19b and 19c) or pneumatic tyres. This wheel rotates with a slip of 10 to 15% when engaged with soil during the movement of drill in field. The rotation of the ground wheel is transferred to the seed and fertilizer metering systems through V pulleys and belts or chain and sprockets or gears and shafts.

The plain wheels (Fig. 4.19a) provided to the seed drill are transport wheels which control depth of drilling of seed and fertilizer and for balance and support the seed drill in the field.

| | | |
|---|---|---|
| 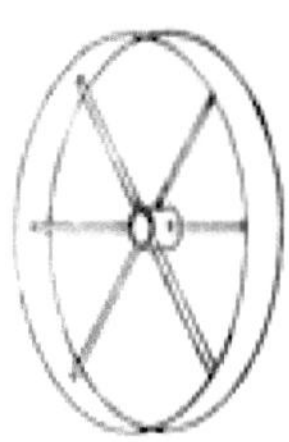 | 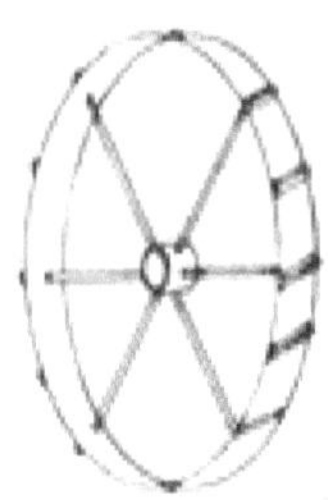 | 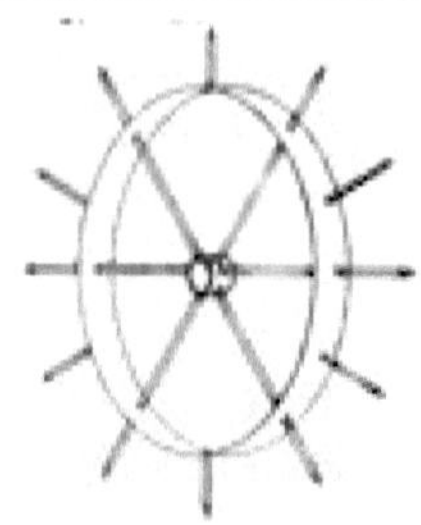 |
| a) Plain Transport wheel | b) Ground drive wheel (with lugs) | c) Ground drive wheel (with pegs)<br>Pandey, M.M, Farm Mach Res. Digest |

Fig.4. 19 Different kinds of wheels of seed drills

**Seed and fertilizer pipes**

These pipes convey the seed and fertilizer from the respective metering system to the furrow openers. These should have sufficient diameter and minimum number of bends for free flow of seed or fertilizer. These are generally made from the transparent plastic pipes to show the blocking of the fertilizer or seeds in the pipe. The blocks are to be immediately cleared while drilling in the field.

| | | | |
|---|---|---|---|
| 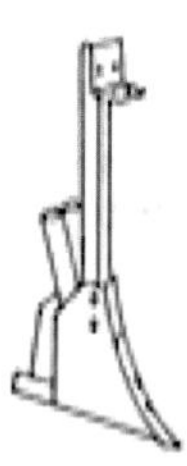 | 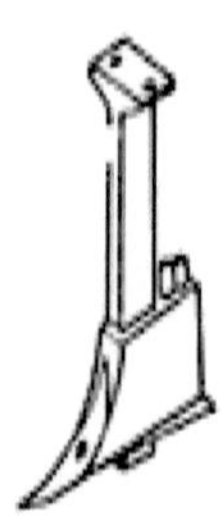 | 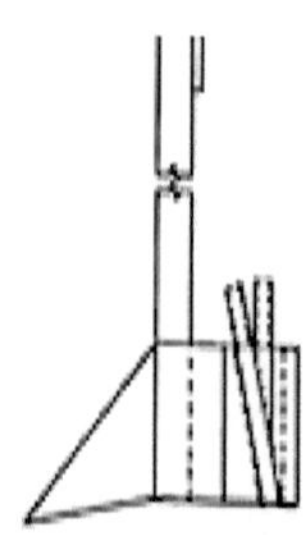 |  |
| a) Shovel Type | b) Chisel Type | c) Shoe Type | d) Curved Runner Type |

| | | |
|---|---|---|
|  e) Single Disc Type | 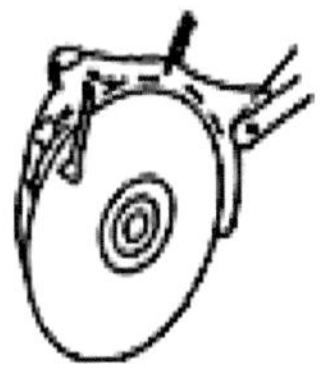 f) Double Disc Type Pandey, M.M Farm Mach Res. Digest |  g) V- shape Roller Type http:/www.truaxcomp.com |

Fig. 4. 20 Different kinds of furrow openers of seed drill

**Furrow openers**

These are used to create furrows and place the seed and fertilizer at the required depth in the field. The fertilizer has to be placed 2 to 3 cm away from the side and 2 to 3 cm deeper than seed. The shovel type furrow opener (Fig. 4.20) is most common. Chisel point type and shoe type furrow openers are used for narrow furrows. The curved runner type, single disc type and double disc openers are used in fields when there is previous crop residue. Rollers with V shaped rings are also used to make furrow for seed sowing. The disc type furrow openers are costly and are not used in India at present.

**Furrow closer**

To cover the seeds with soil and press the soil around seed, furrow closer is provided. This is provided behind furrow opener. This can be a plank, roller, link chains or a 'V' shaped soil closer (Fig. 4.49), etc. These can also be press wheels or V shaped rollers.

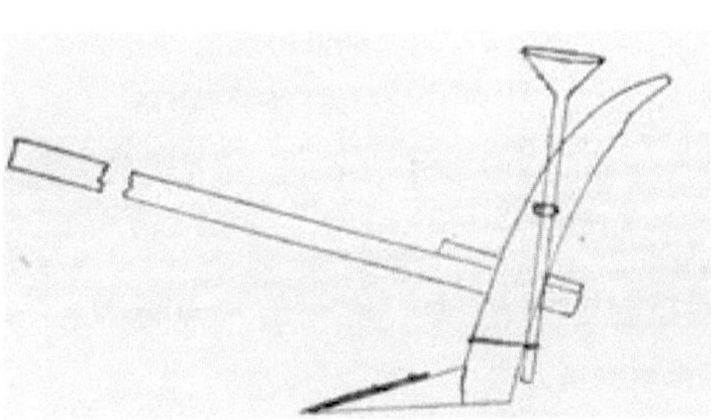

Fig.4.21 Desiplough fitted with funnel for directing seed (Pora)
CIAE, Bhopal

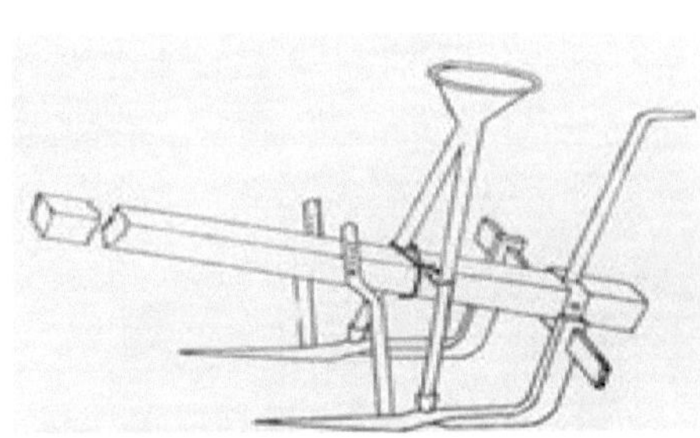

Fig.4.22 Animal drawn cultivator 2-row with seed drilling attachment
CIAE, Bhopal

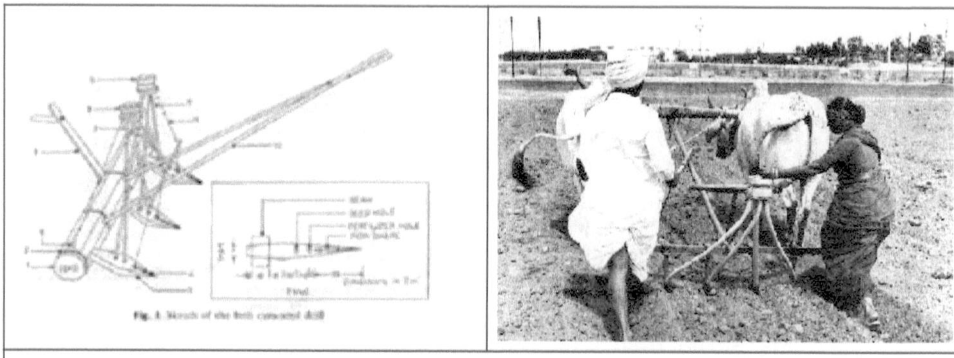

Fig. 4.23 Animal drawn multi row seed cum fertilizer drill
Shaik Kaleemullah, APAU, Tirupathi & ICRISAT Hyderabad

## Technical Details of Some of Seed Drills

A two row drilling funnel for animal drawn plough/cultivator (Fig. 4.22) has been developed CIAE, Bhopal and is modification of traditional seed single row seed drill at (Fig 4.21). While a three row animal drawn seed-cum-fertilizer drill is developed at ANGRAU, Tirupathi and ICRISAT Hyderabad. Two operators have to walk behind the drill to pour seed and fertilizer in the funnels (Fig. 4.23).

The manually operated seed drill (Fig. 4.24) developed at Punjab Agricultural University, Ludhiana is provided with flutted roller for seed metering, and can be used for drilling pea and cowpea seeds by changing the flutted rollers. It places the seeds to a depth of 3 to 5 cm, operates at speed of 3 to 4 km/hr, needs 2 labours to operate and has a field capacity is 0.05 to 0.10 ha/hr. The manually operated pull type, single row seed drill is provided with vertical rotor with seed cells and is suitable for most of the vegetable crops by changing the rotor (Fig. 4.25).

Fig. 4.24 Manually operated seed drill with flutted roller
PAU, Ludhiana

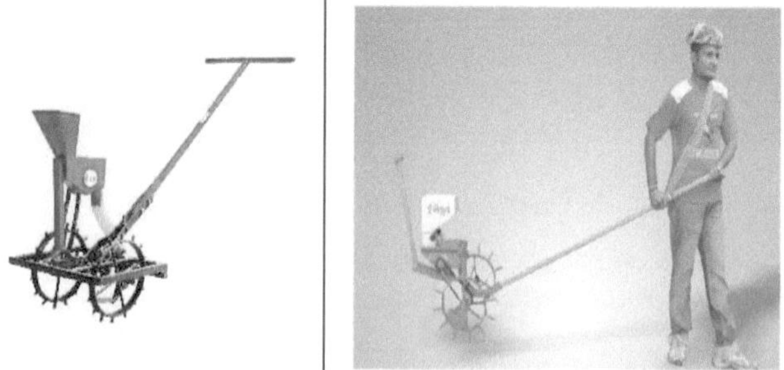

Fig. 4.25 Manually operated seed drill with vertical rotor with cells
Khedut Agro Engineering http://khedutagro.com

The manually drawn drum seeder developed at Indian Institute of Horticultural Research, Bangalore. (Fig. 4.26) for onions has holes along the circumference for row spacing of 10 cm. For varying the seed, rate the diameter of holes or number of holes are to be changed along the circumference. The furrow opener and closer is also provided.

| | |
|---|---|
|  |  |
| Fig. 4.26 Manually operated Onion drum seeder<br>IIHR, Bangalore | Fig. 4.27 Animal drawn seed drilling attachment for country plough<br>IGFRI, Jhansi |
|  |  |
| Fig. 4.27a Animal drawn seed drill<br>Khedut Agro Engineering.<br>Veraval Rajkot. http://khedutagro.com | Fig. 4.28 Animal drawn seed drill<br>Dharti Agro Engineering. Rajkot , Gujarat<br>http://dhartiagro.tradeindia.com |

The seeding attachment to animal drawn country plough has been developed at Indian Grass Land and Fodder Research Institute, Jhansi (Fig. 4.27). The seeding attachment consists of a seed box, variable aperture with rubber agitator metering system and ground wheel with spikes. This system can be used to sow most of the vegetable seeds of small to big size. The operating speed is 1.5 km/h and field capacity is 0.05 to 0.10 ha/h depending on row spacing. The seed drills shown in Fig. 4.27a and Fig. 4.28 are 5 row seed drills with vertical roller with seed cells and rear drive wheel. These are suitable for sowing most of the vegetable seeds by replacing vertical rollers, as for each crop the size of seed cell depend on seed size and seed rate per hectare. The operating speed is 1.5 km/h and field capacity is 0.10 to 0.15 ha/h depending on row spacing.

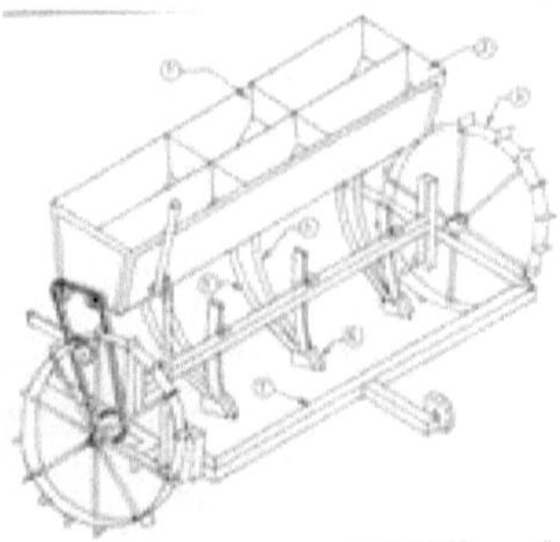

Fig. 4.28a Parts of animal drawn seed cum fertilizer drill
APAU, Hyderabad

Fig. 4.29 Animal drawn seed cum fertilier drill
HAU, Hissar

Fig. 4.30 Animal drawn seed cum fertilizer
APAU, Hyderabad

Fig. 4.31 Animal drawn seed cum fertilizer drill
CIAE, Bhopal

Fig.4.31a Animal drawn seed cum fertilizer drill
Indira Gandhi Agricultural University, Raipur, CG, India http://igau.edu.in

Fig.4.31b Animal drawn seed cum fertilizer drill
Rajasthan Mechanical Works Ltd.,Jaipur
http://www.brahmpuri.com

| | |
|---|---|
| 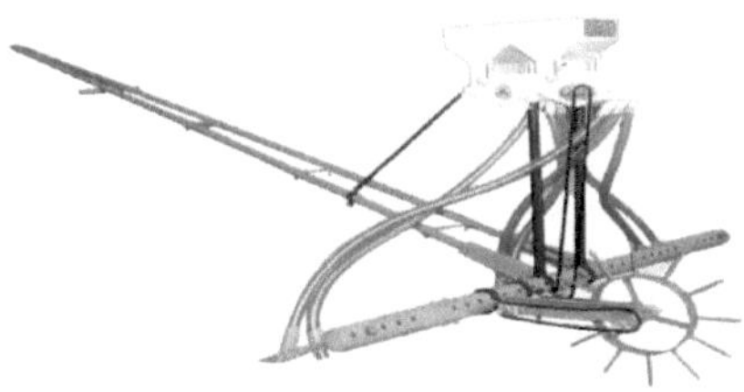 | 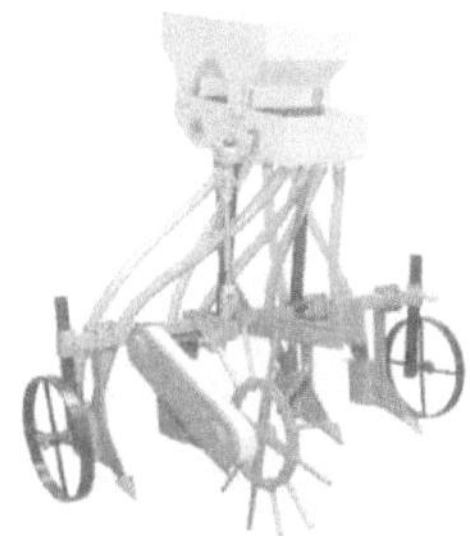 |
| Fig. 4.31(c) Animal drawn seed cum fertilizer drill<br>*Dharti* Agro Engineering. Rajkot - 360024, *Gujarat*<br>http://dhartiagro.tradeindia.com | Fig. 4.31(d) Animal drawn seed cum fertilizer drill<br>**Mamta Iron Works**, Mahadev Nagar, Gujarat http://mamtaironworks.tradeindia.com/ |

A large number of animal drawn seed cum fertilizer drills have been developed at Agricultural Universities and ICAR Institutes and are available commercially (Figs. 4.29 to 4.31a). Most of these are fitted with flutted rollers for seed metering and some models have variable opening or aperture with rubber agitator or cup feed system for seed. The fertilizer metering system is generally a variable opening with agitator. The row to row spacing is adjustable depending upon the requirement of the crop. The front box of the seed drill is fertilizer hopper and the rear box is seed hopper. The suitability for different crops depends upon the type of seed metering used in the seed drill. These are generally 3 row seed drills with working width varying mg from 0.6 meter to 1.5 meter and the operating speed 2 to 3 km/h. The field capacity varies from 0.10 ha/h to 0.25 ha/h. The seed drills shown in Figs. 4.31c and 4.31d are fitted with vertical rollers with cells for seed and fertilizer metering.

The power tiller operated seed cum fertilizer drill (Fig. 4.32) developed at Indian Institute of Horticultural Research, Bangalore has variable opening with rubber agitator for seed metering and mild steel rotor for fertilizer metering. The drive wheel is provided with mild steel flat lugs to rotate the metering systems through V-belt and pulley. The lever system is provided to vary the fertilizer and seed metering apertures/openings. The machine has been tested for sowing radish, cowpea, okra, peas, dolichos and French beans. The performance was satisfactory with average seed spacing for radish being 11.2 cm, cowpea 15.3 cm, okra 15.6 cm, peas 9.8 cm, dolichos 31.7 cm and French bean 28.9 cm. About 60 to 70% seeds are sown at required spacing within 0.5 to 1.5 times targeted spacing, 20 to 30% with spacing less than 0.5 times of targeted spacing and 5 to 10% with spacing more than 1.5 times the targeted spacing. Its speed of operation is 2.4 km/h and the field capacity 0.2 ha/h. The seeds are placed at a depth of 5.2 cm while fertilizer at a depth of 6.0 cm.

Fig. 4. 32 Power Tiller drawn seed-cum-fertilizer drill
IIHR, Bangalore

Fig. 4 33 Power Tiller drawn seed-cum-fertilizer drill
CIAE, Bhopal

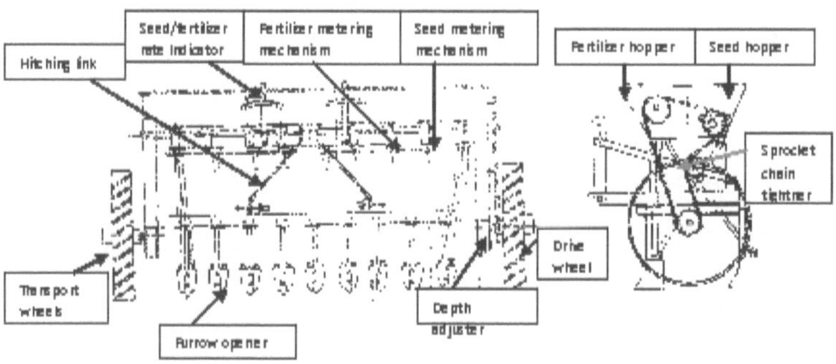


Fig.4.34 Parts of tractor drawn seed-cum-fertilizer drill
Mandhar S.C. PAU, Ludhiana

The power tiller operated seed cum fertilizer drill (Fig. 4.33) developed at Central Institute of Agricultural Engineering (CIAE), Bhopal, is for five rows and is fitted with flutted rollers for seed metering. The drive wheel is spike type and is mounted in front part of the machine, below the frame. The machine can sow okra, dolichos, cowpea and peas. The field capacity is about 0.2 ha/h.

The tractor drawn seed cum fertilizer drill available in North India. 4.34. The seed drill is provided with flutted rollers for seed metering and variable opening with agitators for fertilizer metering. One ground wheel is provided on one side and power is transferred to seed and fertilizer metering systems through sprokets and chain. One transport wheel is also provided. Front portion of the box is fertilizer hopper and rear portion is seed hopper. The seed and fertilizer metering systems with index plates are provided. Boot type furrow openers are provided to drill the seed and fertilizer side by side. The seed and fertilizer application rate adjustment system with adjustment handles and index plate for seed and fertilizer rates are also provided.

Most of the commercially available tractor drawn seed cum fertilizer drills are having 9 or 11 rows. These are suitable for tractors of 35 hp and above. The field capacity is 0.7 to 1.0 ha/h when working at speed of 5 to 8 km/hr.

The seed cum fertilizer drill (Fig. 4.35 ) with 9 row has 2 side wheels, while the lugged wheel is the drive wheel for seed and fertilizer metering systems, the plain wheel is a transport wheel. It has flutted rollers for seed metering and vertical discs with cells for fertilizer metering. The fertilizer application rate is varied by varying the height of fertilizer hopper. It has chisel type furrow opener and these are no / zero till dill and no furrow closers are provided.

National Agro Industries, Ludhiana
www.nationalagroinds.com

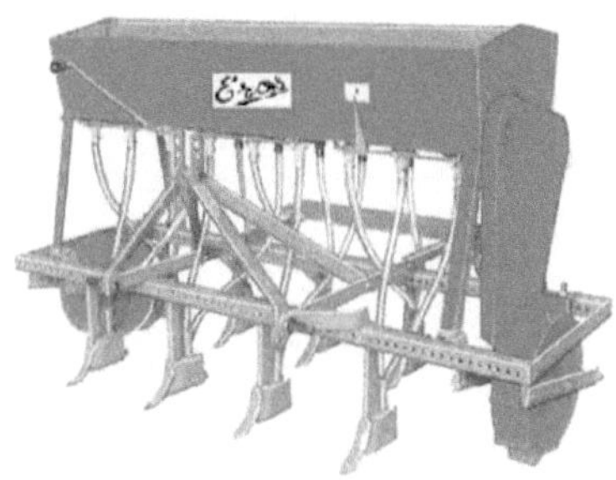

Eros Metal Works (p) Ltd., Nagpur, Maharashtra,
http://www.erosmetalworks.com

Fig. 4.35 Tractor drawn seed-cum-fertilizer drill – with side drive wheel

Fig. 4.36 Tractor drawn seed-cum-fertilizer drill
Dasmesh Mechanical Works – Amargarh, Punjab, http://www.dasmesh.net

The seed cum fertilizer drill (Fig. 4.36) is for sowing 11 rows It has 2 side wheels for transport and one drive wheel (lugged) in front. It has flutted rollers for seed metering and vertical discs with cells for fertilizer metering. The fertilizer application rate is varied by varying the height of fertilizer hopper. It has shovel type furrow opener while vertical angle iron projections are provided as furrow closer.

The seed cum fertilizer drill (Fig. 4.37) is provided with a handle to stop dropping of the fertilizer through variable opening during transport and at turning points in the field. The raised bed seed-cum-fertilizer drills are shown in (Figs.4.41 to 4.44). The seed drill-cum-rotavator (roto till drill) (Fig. 4.45) does the secondary tillage and sows the crop.

Fig. 4. 37 Seed-cum-fertilizer (no till ) drill with drive wheel in front
National Agro Industries, Ludhiana
http://www.nationalagroinds.com

Fig. 4. 38 Tractor operated seed-cum-fertilizer drill with ridger
Dept. of Animal Husbandry & Dairying, GOI
http://dahd.nic.in

Fig. 4.39 Tractor operated seed drill with electronic metering system
http://www.seedspider.com

Fig. 4.40 Tractor operated seed with electronic
metering system http://www.seedspider.com

Fig.4.41 Seed-cum-fertilizer drill with tridger attachment
Modern Industries, Hoshiarpur, Punjab

Fig.4. 42 Raised bed seed-cum-fertilizer drill
National Agro Industries, Ludhiana

Fig. 4.43 Raised bed seed-cum-fertilizer drill
CIAE, Bhopal

Fig. 4.44 Raised bed seed-cum-fertilizer drill
National Agro Industries, Ludhiana

Fig. 4.45 Roto till drill
Bir Singh & Sons, Samrala, Punjab

Fig. 4.46 Air assisted power tiller and tractor operated seed drill
TNAU, Coimbatore

The tractor drawn air assisted seed drills are shown in Figs. 4.46 to 4.48. The machine are fitted with seed box, bulk seed metering system, air blower, seed distributors, blower air pipes and furrow openers. The seeds are stored in seed hopper and the bulk metering for the required number of rows is done by one metering system only. The air blower supplies the air for seed distribution to different rows. These get distributed through the distributor and get conveyed to furrow through air-cum-seed pipes.

The seed drill (Fig. 4.20g) is provided with 3 seed boxes. The metering system for seeds is variable opening. The seed box for small seeds (legumes) is not provided with agitator (Fig. 4.18a). The seed box for fluffy seeds is provided with auger type agitator (Fig. 4.18c) and the seed box for grains is having rotating rods as agitators (Fig. 4.18d). The roller type having V-shaped rings, the furrow closer is a notched roller, it compacts the soil around seed and create little pockets on the soil surface to trap soil moisture.

The electronic seed drill is shown in Figs. 4.39 & 4.40, and experiment plot seed drill with cone metering system is shown in Fig.4.50.

Fig. 4.47 Tractor operated air assisted seed drill

http://www. flexicoil.com

Fig. 4.48 Tractor operated air assisted seed drill

http://www. flexicoil.com

Fig. 4.49 Raised bed seed drill with furrow closer

NRC for Soyabean, Indore

Fig. 4.50 Plot seed drill with cone type seed metering system

CIAE, Bhopal

## 4.4 Seed Planter

Presently the seed of the crops like okra and bean are dibbled manually in India. By using seed planter these seeds can be sown in rows and seed to seed distance can also be maintained. Necessary adjustments for different types of seeds, row and seed spacing can also be made. In seed drill, the bulk metering of seeds is done, while in seed planter, individual seed metering is done. The remaining systems are almost similar in seed drill and seed planter. Due to non availability of seed planters in India some time the seeds which require planting are drilled and the seed to seed distance is not maintained.

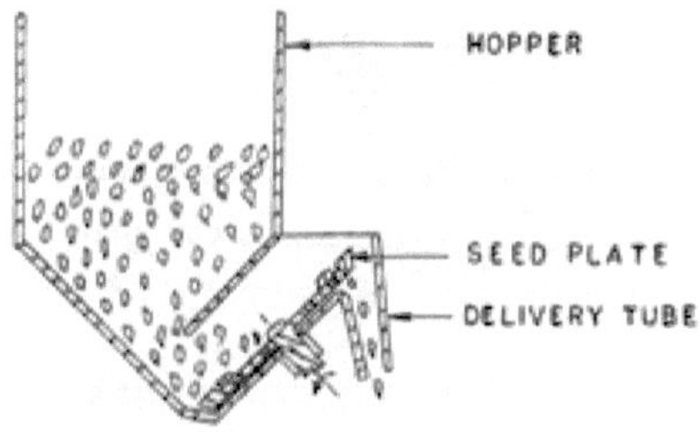


Fig. 4.51 Inclined plate seed metering system
CIAE, Bhopal

Fig. 4.52 Different types of seed plate for different seeds and seed spacing
IIHR, Bangalore

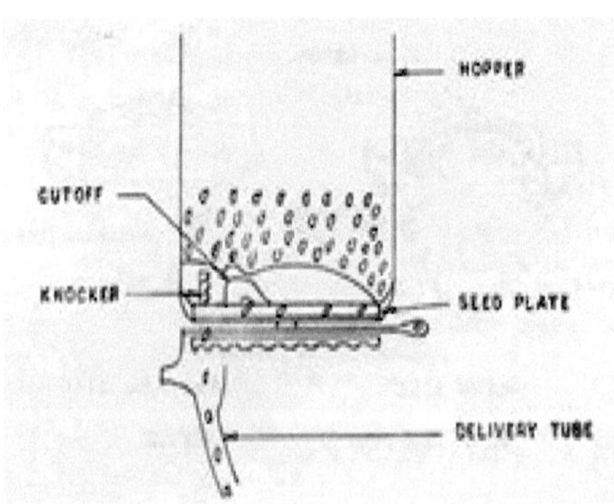


Fig. 4.53 Horizontal seed plate metering mechanism
CIAE, Bhopal

Fig. 4.54 Seed metering (1) horizontal plate, fertilizer metering (2) variable opening with agitator
IIHR, Bangalore

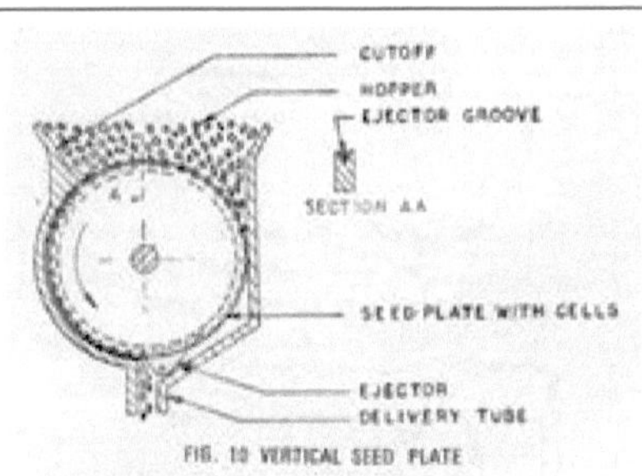


Fig. 4.55 Vertical seed plate metering mechanism
CIAE, Bhopal

Fig. 4.56 Manual planter with vertical seed plate metering
Hare Krishna Rural Life http://hkrl.com/gardening.html

Fig. 4.57 Vacuum seed metering mechanism
http://www.vesconite.com/industry/agric/monosem_planters.htm

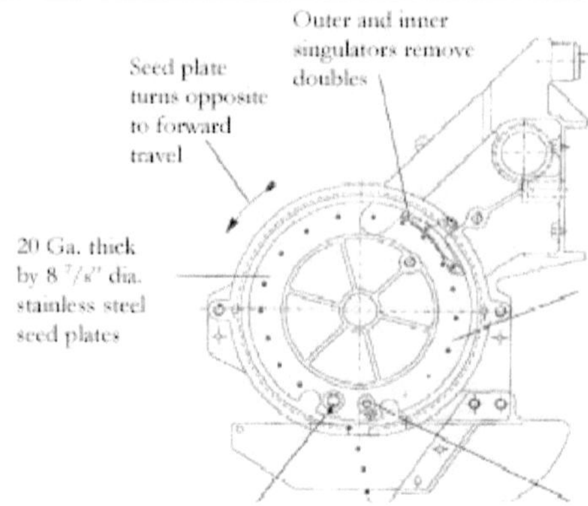


Fig. 4.58 Systematic operation of vacuum seed metering system
http://www.gearmore.com/vacplant.pdf

## Seed Metering Systems

The following seed metering systems have been designed.

a) **Inclined seed plate:** Individual seeds are picked up by a seed plate with cells fitted at the bottom of the seed hopper (Figs. 4.51 & 4.66). The size of cells is just enough to accommodate one seed. Different seed plates with different size seed cells are required for different crops. The plates with more number of cells are required for close spacing of seeds (Fig. 4.52). The inclined seed plate is fixed at the bottom of the seed hopper in a separate chamber. It rotates through the drive system with the rotation of drive wheel. The individual seeds are picked up and dropped through the outlet, maintaining the required seed spacing in the row.

b) **Vertical seed plate with cells:** The seed plate in some of the planters is fixed vertically (Figs. 4.56 & 4.62). In this case seeds fall through the side opening. In some cases, vertical rollers (Figs. 4.55 & 4.68) with cells are provided at the bottom of seed hopper and the cut off for dropping of excess seeds is provided. In some of the seed planters, the vertical rotor with cells in the secondary hopper (Figs. 4.18f and 4.18g) or cup feed (Fig. 4.18i) systems are used.

c) **Horizontal seed plate:** The seed planter is fitted with a horizontal plate with seed cells (Figs. 4.53 & 4.54) at bottom. Only one seed is accommodated in the cell. The seed plate is rotated through the drive system by drive wheel. There is additional seed cut off mechanism before seed enters the seed drop mechanism. The seed cut off mechanism is a cover over seed plate; it does not allow the additional seeds to enter the seed drop mechanism. The seed cell emptying is also ensured by a spring loaded ejector mechanism on seed cell. The seed falls through the outlet into the seed pipe.

d) **Vacuum seed metering:** The system consists of a rotating seed plate with seed picking holes (Figs. 4.57 & 4.58). The seeds are picked by vacuum and dropped at required spacing by the cut off of the vacuum. The vacuum is created by a high-speed blower or a vacuum pump, which is connected to vacuum chamber. The seed plate rotates over a chamber, which has vacuum chamber and atmospheric pressure chamber. The seeds are picked from the seed hopper, when the seed opening is on vacuum chamber. As the seed opening rotates and comes over the vacuum cut off or atmospheric pressure chamber, the seeds fall from the seed plate into the seed pipe and are carried to the furrow for planting.

**Technical details of some of the seed planters**

The Naveen dibbler (Fig. 4.59) has been developed at Central Institute of Agricultural Engineering, Bhopal. This is fitted with a seed hopper and vertical cell type roller for seed metering. The planting operation is accomplished by pushing the jaws into the soil at desired depth and forward movement of the dibbler to open the jaw to meter and release the seed into soil. The metering system (roller) can be changed for different crops like peas, dolichos and beans. The field capacity is 0.02 to 0.03 ha/h. The dibbler reduces the human drudgery of bending and sowing the seeds.

Fig. 4.59 Seed dibbler,
CIAE, Bhopal

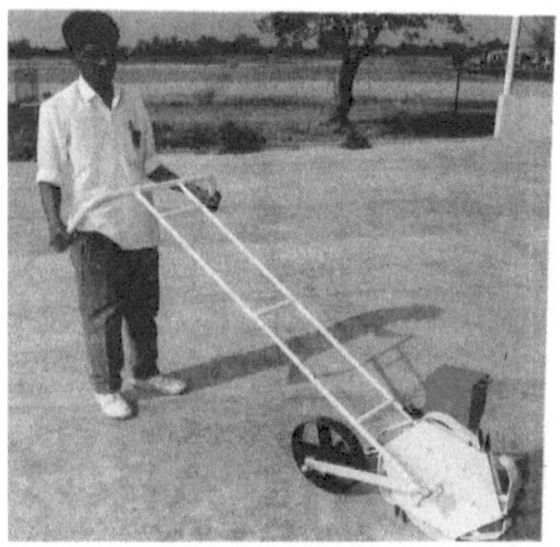

Fig. 4.60 Rotary dibbler
CIAE, Bhopal

Fig. 4.61 Manual vegetable seed planter
Lambert Products. Inc
http://www.thstore.com

Fig. 4.62 Manual drawn vegetable seed planter with vertical metering plate
http://www.bizrate.com/buy/noncat_prod_details__oid--31788120.html

The rotary dibbler (Fig. 4.60) developed at CIAE, Bhopal has vertical cell type wooden roller as metering mechanism. The roller rotates due to forward moment of the dibbler and meters the seed. The jaws enter into the soil and drop the seed at the required depth. The field capacity of the dibbler is 0.03 to 0.04 ha/h.

The manually operated seed planters are shown in (Figs. 4.56, 4.61 & 4.62). These are fitted with line marker for nextline. The planter (Fig. 4.56) has vertical cell plate fitted in seed hopper. The size of cells and number of cells in a plate vary as per requirements of seed size and spacing. The planter is suitable for most of the vegetable crops.

Fig. 4.63 Animal drawn planter
CIAE, Bhopal

Fig. 4.64 Animal drawn seed drill
MPKV, Pune

The bullock operated seed planter (Fig. 4.63) developed at CIAE, Bhopal has the fertilizer drilling attachment also. The three row planter has inclined seed plates for metering seeds and flutted rollers for fertilizer metering or seed drilling. The row spacing is 45 cm and coverage is 1.35 meter. The operating speed is 2.3 km/h and field capacity 0.12 to 0.15 ha/h.

Animal drawn Jyoti multi crop planter (Fig. 4.64) has vertical cell type plastic rollers for seed metering and variable opening with agitator for fertilizer metering. The depth of seed placement is upto 3 cm and fertilizer is placed 5 cm to the side and below the seeds. The width of coverage is 70 cm, speed of operation 2 to 3 km/h and field capacity 0.10 to 0.12 ha/h.

Power tiller operated ridger planter (Fig. 4.65) developed at IIHR, Bangalore, has horizontal seed plates for seed metering (Figs. 4.53 & 4.54) and variable opening with agitator for fertilizer metering. The seed cut off and seed ejector is also provided on the seed metering plates. The seed cell size in seed plate could accommodate only one seed. The seed plate thickness for radish is 2.0 mm, okra 4.5 mm and French bean 5.0 mm. About 97% seeds are planted at the required spacing while the doubling and skipping of seed is less than 2%. The speed of operation is 2 km/h and field capacity 0.06 ha/h when the ridge spacing is 45 cm. The bottom width of ridge is 32 cm and the height is 14 cm.

Fig. 4.65 Animal drawn seed drill
IIHR, Bangalore

Fig. 4.66 Tractor operated seed cum fertilizer cum planter
PAU, Ludhiana

Fig. 4.67 Tractor operated seed fertilizer drill cum inclined (plate) planter
CIAE, Bhopal

Fig. 4.68 Tractor operated vertical disc multicrop planter
MPKV, Rahuri

| | |
|---|---|
|  Fig. 4.69 Tractor operated raised bed planter<br>CIAE, Bhopal |  Fig. 4.70 Tractor operated pneumatic planter<br>CIAE,Bhopal |
|  Fig. 4.71 Tractor operated pnematic seed planter with fertilizer applicator<br>www. monosem.com |  Fig. 4.72 Tractor operated pnematic seed planter with fertilizer applicator<br>www. monosem.com |

Tractor mounted seed fertilizer drill-cum-planter (Fig. 4.66) developed at PAU, Ludhiana, has inclined plates for seed metering for planter and flutted rollers for seed drilling. It can plant 6 rows and drill 10 rows with a width coverage 1.8 m. The field capacity is 0.30 ha/h when operating at a speed of 2.5 to 3.0 km/h.

Tractor mounted inclined plate planter with ridger attachment (Fig. 4.67) developed at CIAE, Bhopal has 1.8 m effective width of coverage. The field capacity is 0.44 ha/h when operating at 3.6 km/h. The depth of sowing is upto 7.5 cm.

Tractor mounted multicrop planter (Fig. 4.68) developed at MPKV, Rahuri has vertical plates with cells for seed metering. The working width is upto 2.4 m and field capacity 0.4 to 0.5 ha/h when operating at a speed of 2.7 to 3.2 km/h.

The commercial pneumatic planter (Fig. 4.74) developed at CIAE Bhopal is a six row planter suitable for tractor of 35 hp or more. It plants a single seed at predetermined seed and row spacing and is suitable for planting most of the vegetable seeds. It consists of main frame, aspirator blower, disc with holes for vacuum pick up and release by vacuum cut off of seeds, hoppers (Fig. 4.58), furrow openers, pto driven shaft and ground wheel etc. The field capacity is 0.5 to 1.0 ha/h when operating speed is 3 to 5 km/h and working width 2 to 3 m.

The 2-row vacuum seed planter is also provided with fertilizer applicator for granular fertilizer. The ridger planter cum fertilizer applicator for 2 and 4 rows are shown in (Figs. 4.71 & 4.72). The vacuum planter can singulate all sizes of vegetable seeds from the smallest like raddish, carrot, lettuce to large French bean and Lima bean size seeds.

**Potato planters**

Potato is one of the most important crop and its planting can be easily mechanized. Potatoes are generally planted on ridges with a spacing of 0.6 metres. Traditionally the ridges were made by animal drawn ridger and seed potatoes sown manually on ridges. Now potato planters have become very popular in Punjab and U.P. and the cultivation of potatoes is fully mechanized.

In semi automatic transplanters the manual feeding of seed potatoes is done to the metering system. In automatic planters, the seed potatotes are picked singularly by the metering system from the seed box and dropped into the ridge through furrow opener.

**Semi automatic potato planters**

Tractor drawn two row planter (Fig. 4.74) and four row planter (Figs. 4.74a and Fig. 4.74b) are have rotary cups /sectors (Fig. 4.73). The seeds are filled in the seed hopper. There is an opening at the bottom of seed hopper and a single layer of potatoes is available on the bottom trough. The operator sits on the seat and picks handfull of potatoes from the seed hopper by both hands and drops single potatoes in the cups in the rotary metering disc divided in sectors/cups. The movement of cups/sectors drop the seed potatoes in delivery pipe and further in furrows inside the ridge. The metering disc is rotated by ground wheel through power transmission system consisting of sprocket, chain and gears. The ground wheel (drive wheel) is provided with a mechanism to change its diameter. The seed/tuber spacing is varied by changing the diameter of power/drive wheel.

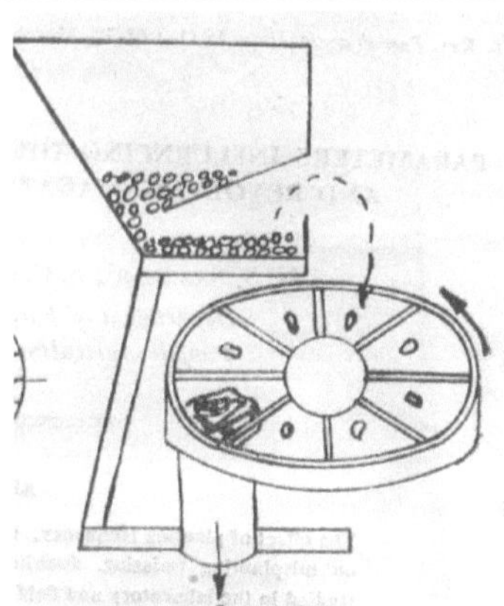

Fig. 4.73 Semi automatic disc type 3 row potato planter
S.C. Mandhar PAU, Ludhiana

Fig. 4.74 Semi automatic disc type 2 row potato planter
Guru Nanak Agriculture Implements Mfrs Jalandhar, http://www.indiamart.com/gurunanak-agriculture-implements/

Fig.4. 74 (a) Semi automatic disc type 4 row potato planter
http://www.moderninds.com/products.htm

Fig.4.74 (b) Semi automatic disc type 4 row potato planter woring in field
CPRI Jallandhar

The belt type semi automatic potato planter (Figs.4.76 and 4.76a) is generally 2 row planter. There are two openings (on left and right side) of the seed hopper. The seed potatoes fall under gravity into seed cups moving on endless belts (Fig. 4.75). The operator sits on the seats and regulates the feeding of individual seeds/tubers in each cup. The movement of cups on the belt makes these fall into seed pipe. The seed tubers then fall into the ridge made by ridgers. The seed spacing is varied by changing the diameter of ground drive wheel.

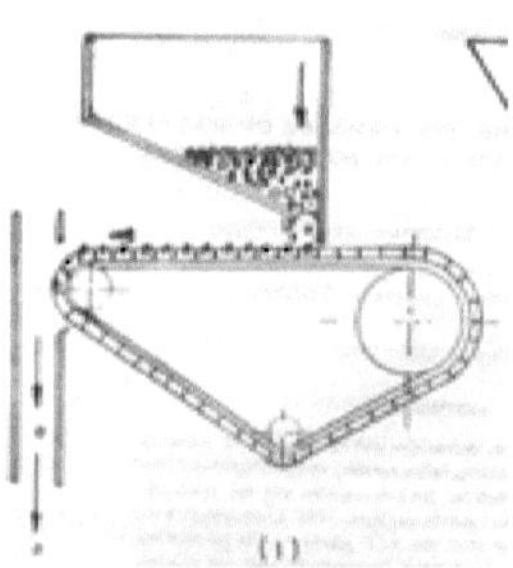

Fig. 4.75 Semi automatic belt type 2 row potato planter
S.C. Mandhar PAU, Ludhiana

Fig. 4.76 Semi automatic belt type 2 row potato planter
Droli Indudtries Regd, Moga, Punjab

Fig.4.76(a) Semi automatic belt type potato planter woring in field
CPRI Jallandhar

**Automatic potato planter**

The automatic potato planter (Fig. 4.77) has rotary picker wheels fixed on the sides of seed hopper. The picker wheel rotates with the rotation of the ground wheel. The cam operated picker arms pick the seed from the seed hopper and drops these into seed pipe. The seeds further fall in the furrow in the ridge.

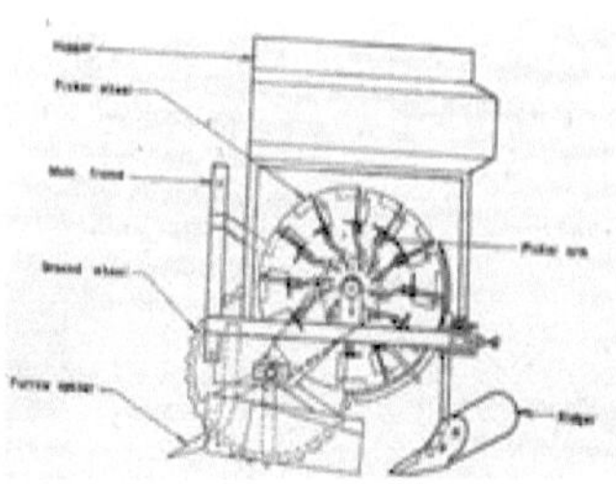

Fig. 4.77 Automatic finger type 2 row potato planter
B.C Srivastava, IARI, NewDelhi

Fig. 4.78 Automatic finger type 2 row potato planter
Droli Industries Regd, Majestic Road, Moga, Punjab

Fig. 4.78a Automatic potato planter
Manjit Singh and Sunil Gulati, CPRI, Jalandhar, Punjab

The cup type automatic potato planter (Figs.4.79 and 4.80) is provided with vertical bucket (cup) conveyor. The number of cup conveyors is equal to the number of rows to be planted by the planter. The conveyors pick up potatoes from the adjustable opening near the bottom of seed hopper. Individual seed tubers are picked by the cups on cup conveyor and then dropped in ridge through the seed pipe.

The seed spacing in most of the potato planters can be varied from 15 to 35 cm by changing the diameter of drive wheel. Two row potato planter needs about 35 hp tractor. The speed of operation of the semi automatic potato planter into the field is about 1 km/h. It needs frequent filling of the seed hoppers and seed bags are to be stored at both ends of the field. The field capacity of a tractor drawn 2 row semi automatic planter is about 0.1 ha/h. The field capacity of a tractor drawn 2 row automatic planter is about 0.4-0.5 ha/h while working at 4-5 km/h. Semi automatic planter and automatic planters are also available with fertilizer application systems.

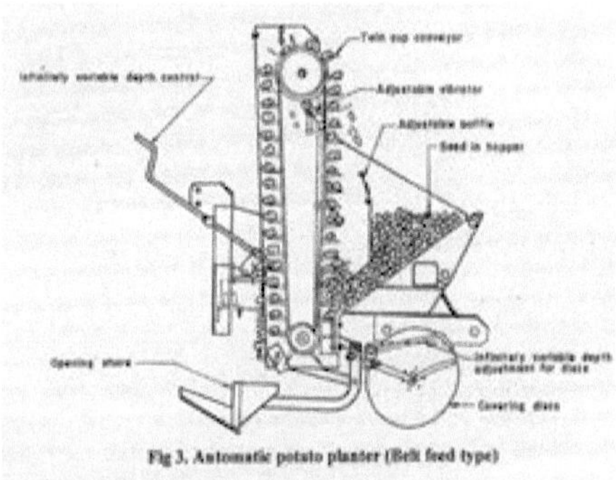


Fig. 4. 79 Automatic vertical bucket belt type potato planter
B.C Srivastava, IARI, NewDelhi

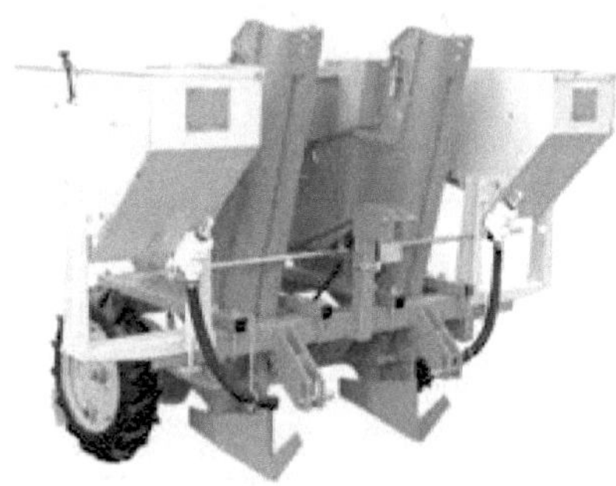

Fig. 4.80 Automatic vertical bucket belt type potato planter
Ganesh Agro Equipments , Vadpura, Gujarat, http://ganeshagro.tradeindia.com

# *Chapter 5*

# Nursery Raising

A large area is being covered under vegetables each season in the country. Vegetable seeds particularly of hybrid vegetables cost from Rs. 10,000 to Rs. 60,000 per kg and germination of these seeds is some times only 60 to 80%. Due to high price and low germination of seeds a large number of nurseries (about 6000) have been established in the country in recent years, about 1000 of these is around Bangalore. These nurseries raise the seedlings in protrays and sell them to farmers at Rs.0.60 to Rs.1.00 per seedlings depending on the crop and veriety. Farmers are able to purchase the exact number of healthy seedlings grown under greenhouse or nethouse and need not waste time, land and energy to raise their own nursery. These seedlings are also economical when compared to seedlings raised by farmers on their own fields.

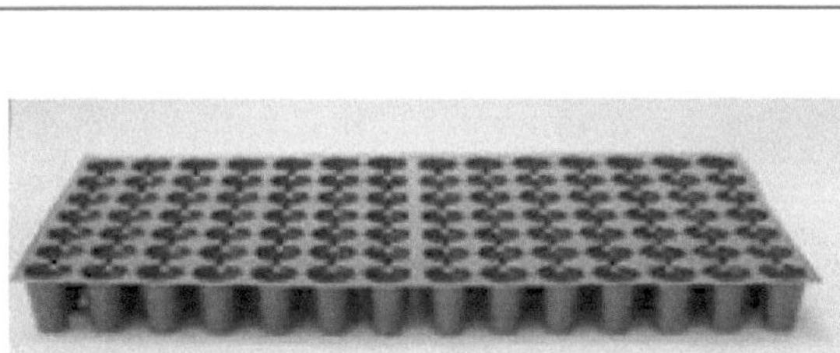

Fig. 5.1 Protrays used for raising seedlings
IIHR , Bangalore, India

Fig. 5.2 Pots with carrier for easy transport and mechanization
http://www.visserite.com

The protrays or pots (Figs. 5.1 & 5.2) for raising the nursery are generally kept in greenhouse or net house for filling with media, seeding and germination. The rooting media is sieved, sterilized, mixed and filled in the protrays or pots manually. Most of the nurseries in South India use composted and sterilized cocopeat as rooting media. Depressions are made in the protrays and individual seeds are sown in each cavity manually. The seeded protrays are topped with rooting media to cover the seeds. The protrays are regularly watered and it takes about one month to raise the seedlings suitable for transplantation. Ninety eight cell portrays are very common for chillies, capsicum, cabbage, tomato and brinjal etc. while 32 and 50 cell protrays are common for bitter gourd, bottle gourd and pumpkin etc. The dimension of the protray is about 27 cm in width and 53 cm in length. The depth of cavity is about 3.5 cm and top and bottom diameter is 3.0 and 2.2 cm respectively for 98 cell protray. About one kg cocopeat is required to fill a single protray.

Four labourers are required to sieve, sterilize, mix and seed 400 protrays (40,000 seedlings) per day. This is very slow and laborious process and the production capacity of most of the nurseries is limited due to non availability of that much labour.

## 5.1 Media Siever

A motorized rotary screen type media siever (Fig. 5.3) has been developed at the Indian Institute of Horticultural Research (IIHR), Bangalore. It is useful to sieve farm yard manure, vermicompost, cocopeat, sand and soil and removes all the particles bigger than 1.5 mm in the form of stones, clods, straw etc. The capacity of the machine is about 1 tonne/h and it is operated by 0.5 hp motor. One person is required to feed the media and one person is required to collect the sieved material. It can also be operated manually by one labourer.

Power Screener has a screen surface of over 1.5 square metre. It is portable and is operated by 3 hp electric motor or 7 hp diesel engine. It is easy to move around on site by hand as it is fitted on transport wheels (Fig. 5.4). Material is screened through the screen openings and is discharged directly onto an inclined conveyor belt that moves the screened material to a high discharge height. Power screeners combine a vibrating screen deck and an elevated conveyor, operating from a single power drive. These complete the time-consuming tasks of preparing and loading (or stockpiling) soil, sand and top dressing mixtures in one-step.

Fig. 5.3 Motorised media siever
IIHR , Bangalore, India

Fig. 5.4 Power screener
Royer Industries Inc.USA
www.royerind.com

## 5.2 Soil / Media Shredder

The soil shredder (Figs. 5.5 & 5.6) pulverize the rooting media, top soil etc to the required particle size and screen off the hard particles which can not be pulverized. The pulverized material is heaped with the help of inclined conveyor belt.

## 5.3 Screening Bucket

The screening bucket (Fig. 5.7) is a modified bucket of soil excavator. This is provided with the screen and rotary blades (operated by the hydraulic system) at the bottom. The media is picked in the bucket, pulverized and screened by the operation of rotary blades. The pulverized material passes through the screen and unpulverized hard items are left in the screening bucket. The unscreened media is discharged in a separate heap. The capacity of these machines are 10 to 20 tonnes per hour and are useful for big nurseries, golf courses etc.

## 5.4 Media Mixer

The mixture of different types of media like sand, soil, vermicompost, cocopeat, farm yard manure etc. are used as rooting media for raising the nursery (for seed sowing and rooting of cuttings). An electric motor operated batch type media mixer (Fig. 5.8) has been developed at IIHR Bangalore. Peddle type agitators cum conveyors have been provided on a shaft rotating at 60 rpm for thorough mixing. The mixer with 3 hp motor has a capacity of 1 ton per hour for mixing of sand, FYM, vermicompost, cocopeat etc. It has been observed that the rooting media with vermicompost + sand in equal proportion produces healthy seedlings in lesser number of days for vegetable crops in comparison to other media. The nurseries around Bangalore are using composted and sterilized cocopeat for raising vegetable nurseries.

Fig. 5.5 Soil shredder
Royer Industries Inc. USA
www.royerind.com

Fig. 5.6 Soil shredding by soil shredder
Royer Industries Inc.USA
www.royerind.com

Fig. 5.7 Screening bucket
REMU Ltd , Finland
http://www.remu.fi/

Fig. 5.8 Motorised media mixer
IIHR , Bangalore, India

## 5.5 Sterilization of Media

The commercially available baby boilers (Fig. 5.9) can be used to sterilize the media. The steam sterilization needs only one hour treatment at 80°C to 100°C in comparison to one week for chemical sterilization. The method of sterilizing the media is heaping the media on floor, and covering it fully covered by tarpaulin cloth from all sides. The steam from steam boiler is circulated through the media through perforated pipes placed on the floor (Fig. 5.10) for sterilization. The boiler can produce steam from 60 kg to 200 kg per hour at a working pressure of 0.5 atm. pressure and produces stream at around 180°C. In cocopeat production industry the cocopeat is sterilized by steam nozzles provided on the auger conveyor as a continuous operation.

Fig. 5.9 Portable steam boiler for media sterilization
http://www.visserite.com

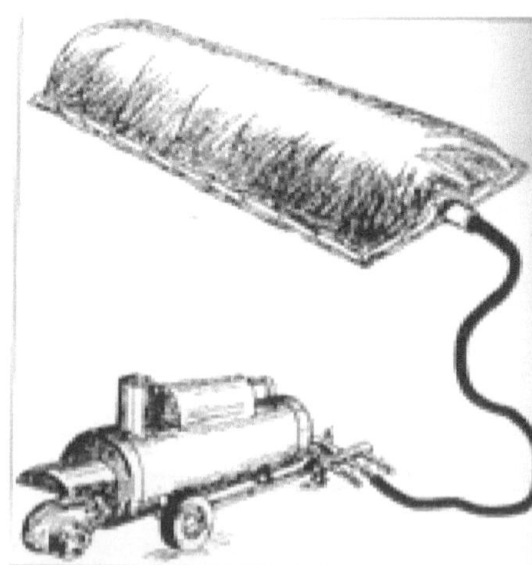

Fig. 5.10 Sterilizing the media
http://www.visserite.com

## 5.6 Filling the Protrays

In all the nurseries, the protrays or pots are filled manually (Fig. 5.11). Around 100 protrays can be filled manually in one hour. Commercially available media filling machine is shown in Figs. 5.12 & 5.13. The capacity of these machine varies from 200 to 1000 protrays per hour.

Fig. 5.11 Filling protrays manually
IIHR , Bangalore, India

Fig. 5.12 Protray filling machine
http://www.visserite.com

Fig. 5.13 Protray filling machine
http://www.blackmoreco.com/

## 5.7 Filling the Plastic Bags

Most of the nurseries still user plastic bags to raise nurseries for fruit, ornamental and forest plants. These can be easily filled by using the slant cut PVC pipe (Fig. 5.14).

## Pouch forming and Media Filling Machine

The commercially available pouch forming and filling machine (Fig. 5.15) can be modified and used to make the plastic bags and filling these with media. Upto 1000 bags can be filled in one hour. The size of the pouch can be varied. The pouches are made from continuous LDPE film.

Fig. 5.14 Bag filling pipe
IIHR, Bangalore, India

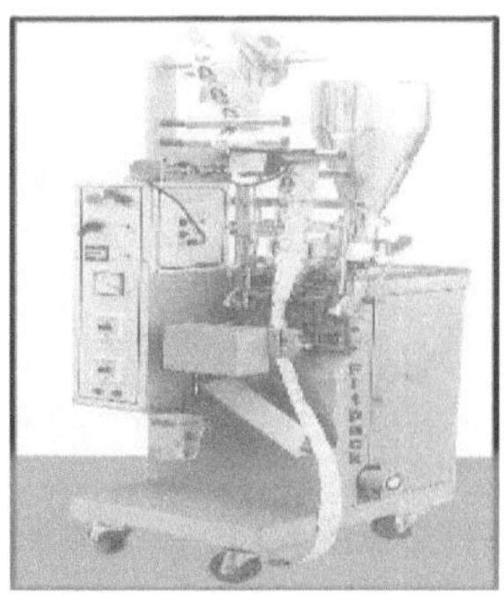

Fig. 5.15 Pouch forming and filling machine

## Media Sieving, Mixing and Bag Filling Machine

The machines shown in (Figs. 5.16 & 5.17) consist of media sieving, mixing, conveying and bag filling units. One person is required to feed media to the machine and two persons are required for filling the bags. Three persons can fill about 1000 bags per hr. The media has to be fed to the machine by buckets or bags in the ratio required for mixing. The sieving and mixing procedure is continuous. The mixed media is lifted by bucket conveyors for filling into plastic bags.

Fig. 5.16 Nursery bag filler
IIHR , Bangalore, India

Fig. 5.17 Nursery bag filler
IIHR, Bangalore, India

Pot Filling Machines

This machine consists of an automatic pot dispenser, media elevating unit and filling unit. The pots are dispensed on a rotating table at one point and get filled by media at another point (Fig. 5.18). These are then placed on the conveyor belt for planting of the seedling or cutting manually.

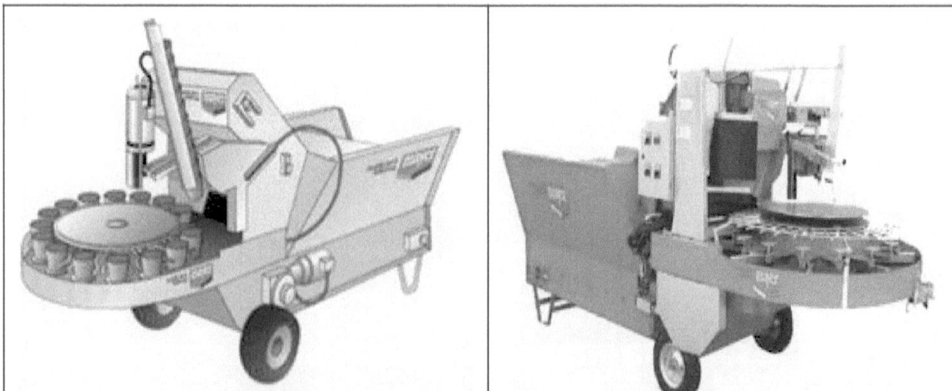

Fig. 5.18 Pot filling machines
http://www.visserite.com

## Sowing the Protrays

The filled protrays are dibbled to create depression in each cavity. The portrays are dibbled manually using fingers (Fig. 5.19) while in some nurseries dibble board is used for this purpose. The dibbled protrays are then seeded by sowing seed in each cavity. A seeder is required to sow the seed in each depression.

## 5.8 Dibblers

### Plate type dibbler /Dibble board

A tray type dibbler has been developed for small nurseries at IIHR (Fig. 5.20). It dibbles depressions/holes in standard 98 cell protrays, and can dibble 100 trays per hour. The dimension of the depressions is 12mm diameter and 15mm depth. It needs one person to operate it.

### Manually handled operated dibbler

This machine developed at IIHR consists of a dibbling board, and a spring loaded handle to move the dibbling board up and down (Fig. 5.21). A suitable base frame with sufficient working space is also provided to handle 4 trays. The dibbling board has dibbling pegs and each peg is fitted to match the centre of one of the cells of the protray. The protray filled with media is kept under the dibbling board. Small cavity is made in each cell by pressing the dibbler board on the protray by pressing the handle. The pegs have been modifed by making these conical for easy entry into media. Lesser force is required for dibbling and dibble board gets lifted automatically with spring.

Fig. 5.19 Manual dibbling by fingers
IIHR , Bangalore, India

Fig. 5.20 Dibble board
IIHR , Bangalore, India

Fig. 5.21 Handle operated dibbler
IIHR , Bangalore, India

### Manually operated rotary dibbler

A rotary type dibbler for medium scale nursery growers has been developed at IIHR (Fig. 5.22). It consists of a dibbler drum fitted with 98 conical pegs. These are arranged in 14 rows lengthwise and 7 rows along circumference. One movable trolley is provided with ball bearings as wheels for pushing the portray below the dibbler drum. The media filled portray is kept on the trolley and pushed below the dibbler drum. The pegs make 10 mm deep depressions for seed sowing in all the 98 cavities in portrays. The dibbler can dibble 600 protrays per hour. The rotary dibbler shown in (Fig. 5.23) is a part of continuous portray filling and dibbling machine, it dibbles upto 1000 trays per hour.

Fig.5.22 Manual rotary dibbler
IIHR , Bangalore, India

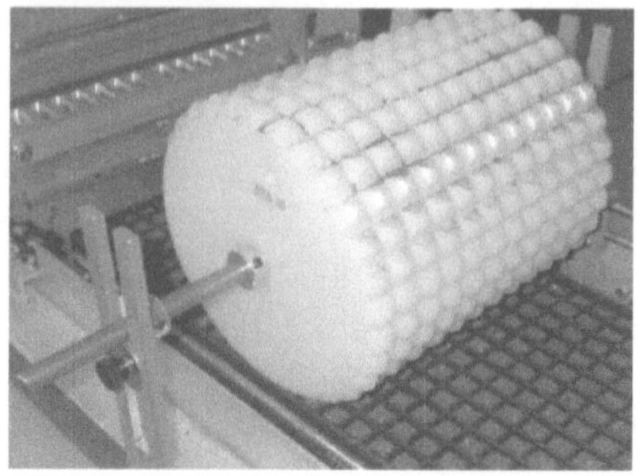

Fig. 5.23 Rotary dibbler
http://www.blackmoreco.com/

## 5.9 Seeders

Presently the seeds are sown manually in the dibbled protrays (Fig. 5.24). Three labours fill, dibble and seed about 300 – 400 protrays per day.

### a. Plate type seeder

Plate type seeder (Figs. 5.25 to 5.28) is a simple tool to sow seeds in the protrays. It consists of a tray with seed metering mechanism at the bottom. The seed metering mechanism consists of three plates with holes. While the top and bottom plates are fixed type, the centre plate is moving type. All the plates are drilled with holes for seed metering. The center plate has a thickness equal to seed thickness. The diameter of holes in the center plate is slightly bigger than the seed diameter. The seeds are filled in the tray and are brought into the holes of the top plate by oscillating the tray. The holes of centre plate are brought below the holes of the top plate. The individual seeds get metered into the holes of the centre plate. The seeder is now placed on dibbled protray. The centre plate is moved such that the seeds fall through the holes in the bottom plate in the dibbled protray.

### b. Vibratory hand seeder

The seeds are put into the seed holder which vibrates and singulates the seeds. The individual seeds are dropped into the dibbled protray (Fig. 5.29).

Fig. 5.24 Manual sowing
IIHR , Bangalore, India

Fig. 5.25 Tray type seeder
www.abcquickseeder.com

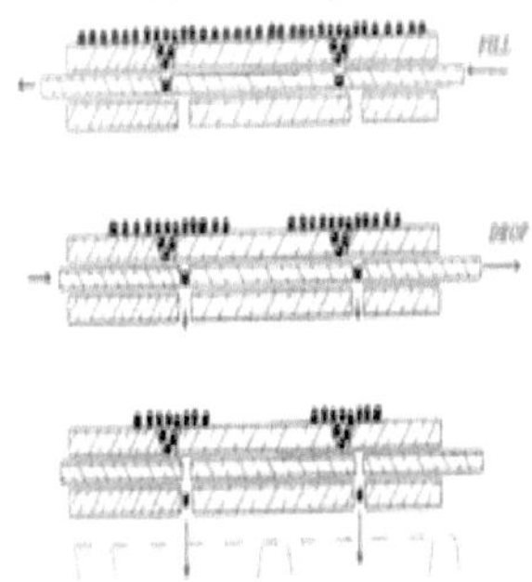


Fig. 5.26 Working of Tray type seeder

Fig. 5.27 The protray is brought below the seeder

Fig. 5.28 The seeding table
(Dibbler and tray seeder)

**c. Manually operated single row vacuum seeder**

This seeder (Figs. 5.30 & 5.31) consists of a tube which is fitted with nozzles to pick up and drop the seeds. The seeds are picked by creating vacuum in the seed tube and dropped by cutting of the vacuum. A hand operated valve is fitted on the seeder and the vacuum is created by vacuum pump. One row of seeds is dropped into the dibbled portray at one time.

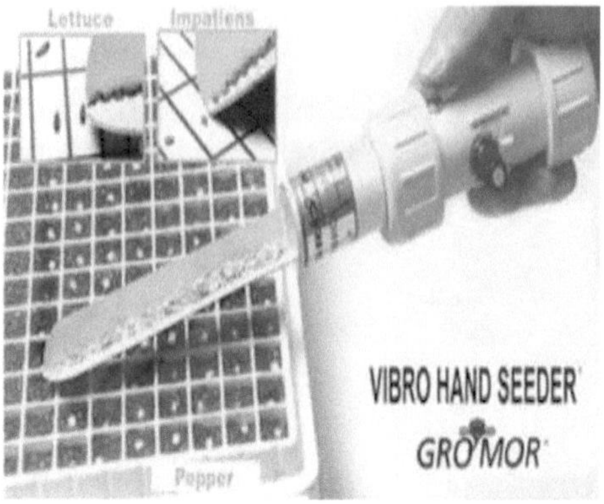


Fig. 5.29 Hand operated vibratory type seeder
http://www.gro-morent.com/wandseeder.htm

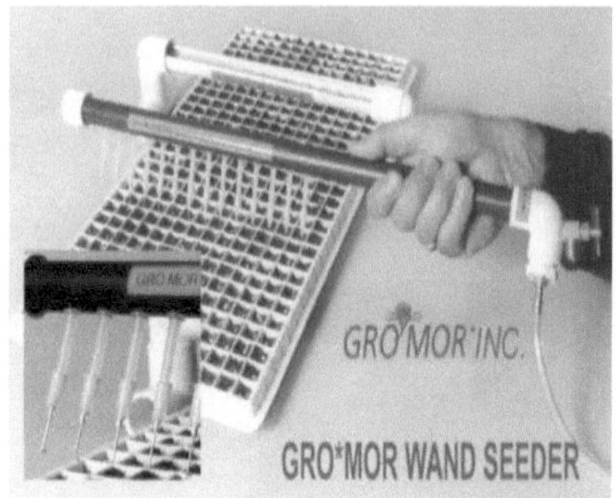


Fig. 5.30 Hand operated single row vacuum seeder.
http://www.gro-morent.com/wandseeder.htm

Fig. 5.31 Hand operated single row vacuum seeder.
http://www.blackmoreco.com

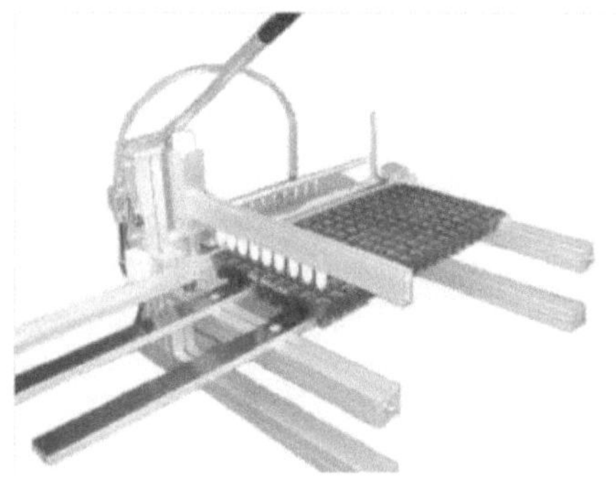

Fig. 5.32 Hand operated single row dibbler cum vacuum seeder.
http://www.kwautomation.com.au/

The dibbler cum seeder in (Fig. 5.32) is operating by pushing of the handle. It dibbles one row in portray and picks the seeds for one row for sowing at the same time. The manual rotation of the dibble/seeder bar by 90 degrees and operation of lever cuts the vacuum allowing the seeds to drop into the dibbled hole in the tray. Each seeder includes 6 sets of 20 needles as per the requirements of the crops.

**d. Tray type vacuum seeder**

The tray type vacuum seeder picks-up 98 seeds at one time using the suction of 0.5 kg/sq.cm is created by 1 hp vacuum pump has been developed at IIHR (Figs.5.33 & 5.34). The vacuum chamber is made out of acrylic sheet. The holes for picking the seeds have 0.3 mm diameter. The seed plate can also be made from aluminium. The holes should be drilled in a depression for easy picking of seeds. The seeds are put over the seed plate with holes. The seeder is oscillated slightly for picking of the seeds by each hole by vacuum. The excess seeds are removed and then the seeder is reversed over the dibbled protray. The vacuum is cut off and individual seeds fall in the dibbled protray. The seeder has to be light weight as it has to be oscillated atleast 2-3 time for seed picking and reversed 2 times for sowing one protray. It has a capacity of 50 protrays per hour.

Fig. 5.33 Tray type vacuum seeder
IIHR, Bangalore, India

Fig. 5.34 Tray type vacuum seeder
IIHR, Bangalore, India

Fig. 5.35 Manualy operated vacuum seeder
IIHR, Bangalore, India

Fig. 5.36 Single seed is picked by each nozzle of vacuum seeder
IIHR, Bangalore, India

### e. Manually operated vacuum seeder

The vacuum seeder consists of a vacuum chamber fitted with nozzles arranged to match the cells of the protray and a seed tray (Figs. 5.35 & 5.36). A suitable base frame with working space is provided to handle the trays. The vacuum is sourced in the nozzles through a vacuum pump of 0.75 hp and controlled by a valve. The vacuum chamber is moved by a handle to go up and down. The seed tray is kept under the nozzles. The vacuum chamber is brought down by operating the handle to touch the seed tray. Single seed is picked by each nozzle when the vacuum is sourced by opening the valve (Fig. 5.36). The vacuum chamber with seeds picked by nozzles is moved up. The seed plate is then removed and dibbled media protray is kept under it. The valve is closed to cut off vacuum and the seeds are released to the protray cavities. The capacity of dibbling and sowing is 200 trays per hour which is about 3.5 times higher than the manual operation. The manual operated vacuum seader has been developed by IIHR.

Fig. 5.37 Manual operated rotary dibbler cum vacuum seeder
IIHR, Bangalore, India

**f. Manual operated rotary dibbler cum vacuum seeder**

A manual operated rotary dibble cum vacuum seeder (Fig. 5.37) has been developed at IIHR. This consists of rotary dibbler system and the vacuum chamber fitted with nozzles for picking the seeds from the seed tray and dropping these into the dibbled protray. The capacity of the machine is upto 500 trays per hour.

**g. Step seeder**

This is a motorized version of the single row vacuum seeder. In some versions nozzles are provided for sowing two rows. The protray moves by one or two cavity/ steps (rows) and are dibbled by the reciprocating or rotary dibbler and are picked by the vacuum nozzles from vibrating seed tray and are then dropped into the dibbled holes. The process is repeated again. The seeder moves about 60 steps per minute. The nozzle assembly has to be changed if there are more number holes in one row of the protray (Figs. 5.38 to 5.41). The movement of protray is mechanically synchronized and the seeds are dropped into the dibbled protray automatically by gravity. The step seeder developed at IARI Delhi is shown in (Fig. 5.42).

Fig. 5.38 Step seeder
Visser International Trade & Engg. BV
http://www.visserite.com

Fig. 5.39 The vacuum nozzles and the vibrating seed tray of the motorized vacuum seeder.
Visser International Trade & Engg. BV
http://www.visserite.com

Fig. 5.40 Turbo seeder / step seeder
http://www.blackmoreco.com

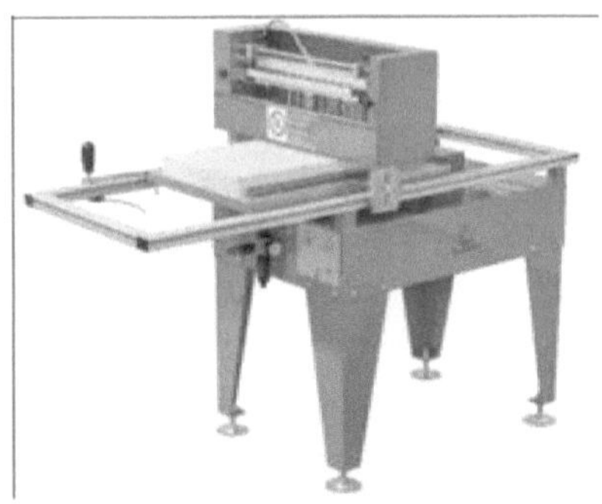

Fig. 5.41 Step seeder
http://www.urbinati.com

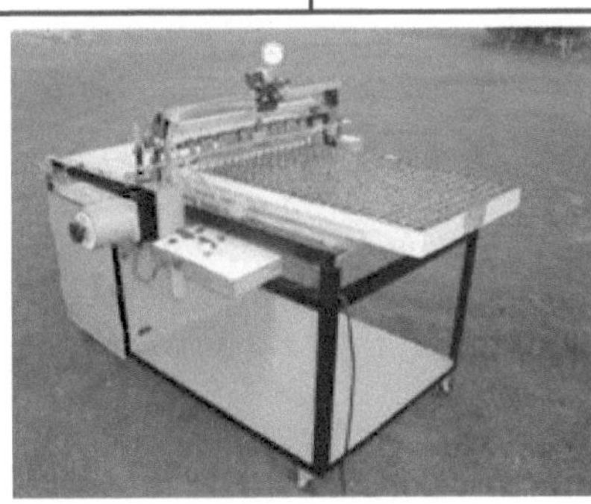

Fig. 5.42 Step seeder
IARI Delhi

**h. Automatic drum seeder**

This is a fully automatic seeding machine (Figs. 5.43 & 5.44) with a capacity of about 600 protrays per hour. The dibbled protrays are placed on moving conveyor. The seeds are stored in the seed plate. The holes in the rotating drum pick seeds by vacuum. Blow off nozzles are provided to remove the excess seeds and singulate the seeds. The seeds are dropped in the protray cell by cutting off the vacuum. The machine can sow 600 protrays per hour.

Fig. 5.43 Automatic vacuum drum seeder (600 protrays/ hr)
http://www.visserite.com

Fig. 5.44 Vacuum drum seeder
http://www.visserite.com

## 5.10 Protray Filling and Seeding Machine

The automatic protray filling, dibbling, seeding, media toping, and watering machines are also available commercially (Figs. 5.45 to 5.48). These have a capacity of 200-1000 protray per hour. Some of the machines are also fitted with portray unstacking/dispensing unit for supplying individual protrays to the protray filling and seeding machine. The protray stacking unit is also provided to stack the filled and seeded protrays at the end of the machine.

The machine shown in (Fig. 5.45) is suitable for big nurseries and has a capacity of filling 400 protrays or producing 39,200 seedlings per hour. The machine mixes the media and conveys it to the bucket conveyor. The bucket conveyor takes the media to the filling hopper. The empty trays are placed on continuous belt by the operator. Media is filled into the protrays continuously by filling hopper and is leveled by the rotary leveler. The excess media falls back into the mixing and conveying hopper. The machine is fitted with individual motors and gear boxes (3 nos–2 hp each) for media mixing, conveying, bucket conveyor and tray transport/ endless belt units.

Fig. 5.45 Media conveying, elevating, filling, dibbling and sowing machine
IIHR , Bangalore, India

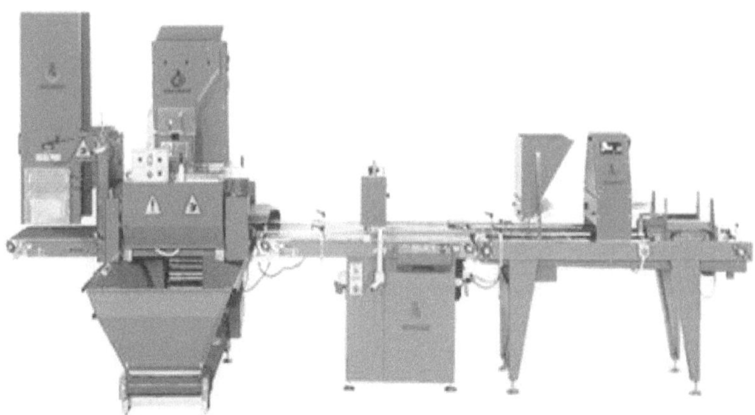

Fig. 5.46. Protray destacking, filling with soil elevator, rotating dibbling, nozzle seeder and media topping and watering unit (capacity 200 protrays per hour).
**URBINATI,** Italy http://www.urbinati.com

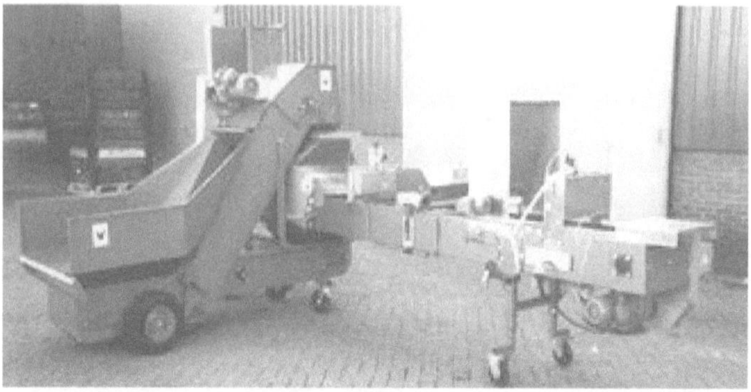

Fig. 5.47. Protray destacking, filling with soil elevator, rotating dibbling, nozzle seeder and media topping and watering unit (600 protrays/hr or 25 steps per minute capacity)

Fig. 5.48. High speed protray destacking, filling with soil elevator, rotating dibbling, rotary seed sowing drum with blowoff nozzle bar and media topping unit (1000 protrays/hr capacity)

The protrays filled by tray filling machine are conveyed to protray seeding machine by endless conveyor belt. Depression in each cell of the protray is made by rotary dibbler. The seeds are singulated by the rotary vacuum seeder with 0.3mm holes operating at 0.5 kg per sq.cm. pressure with one hp vacuum pump. The seeds are individually dropped in the depressions in cells. The Media is topped in seeded trays by the media topping hopper to fill the seeded depressions. The watering is done by the watering pipe fitted with seven nozzles. The machine is suitable for 98 cells protray of 53 x 25cm size. However, the modifications in the dibbler and vacuum seeder can be done for making it suitable for other protrays.

The commercially available machine may have reciprocatory dibbler and oscillating nozzles (like step seeder) for low capacity machine of upto 200 protrays per hour. The high capacity machines of upto 1000 protrays per hour capacity have rotary dibbler and rotary drum seeder.

## 5.11 Protray Transport System

A trolley for transport of media filled protrays and protrays with grown up seedlings has been developed at IIHR (Figs. 5.49 & 5.50). The capacity of the trolley is 80 media filled protrays or 16 grown up seedling protrays. Different types of commercially available trolleys for transporting seedling protrays and pots are shown in (Figs. 5.51 & 5.52). Special type of hooks are provided to connect the trolleys so that 12 to 20 trolleys can be connected and pulled at a time. The tractor pulled trollies are provided with tyres.

Fig. 5.49. Trolley for seeded protray.

Fig. 5.50. Trolley for seedling protrays
IIHR , Bangalore, India

Fig 5.50(a) Transport of seedlings in lorry for transplanting

Fig.5.51 For large capacity tractor pulled trolley.
http://www.visserite.com

Fig. 5.52 Trolleys for transporting pots.
http://www.visserite.com

## 5.12 De-Popping the Seedlings from Protray

The seedlings are to generally removed from the protray cells. These are generally pulled by hand. These can be easily removed by de-popper, which generally consists of pegs fitted on a board. The pegs are forced through the drainage holes under the protrays. The seedlings get de-popped and can be packed for sale. The seedlings raised in cocopeat do not need a de-popper as these are removed very easily by pulling, with a media ball intact. The nursery grower sells the seedlings along with protrays and the farmer returns the empty protrays to the nursery grower after the transplanting.

## Tray type de-popper

A tray type de-popper for standard 98 cell protray has been developed at IIHR (Fig. 5.53). This has a wooden board fitted with 25 mm length and 6 mm dia. pegs. The seedling portrays are kept on the pegs and the pegs are pushed through the drainage holes of the protray cells. The seedlings with media and root zone get detached from the cells and are then tied in bundles for sale to farmers. Upto 50 protrays can be de-popped per hour.

## Single row de-popper

Commercially available single row seedling de-popper is shown in Fig. 5.54. This consists of a handle fitted with pegs (equal to number of cells in one row of protray). The pegs are pushed through the drainage holes of the protray cells after inverting the protray. The seedlings get detached from the cavity.

Fig. 5.53. Manually operated seedling de-popper.
IIHR , Bangalore, India

Fig. 5.54 Single row seedling de-popper
GRO-MOR INC. USA
http://www.gro-morent.com

## Stand type mechanical seedling de-popper

The de-popper shown in Fig. 5.55 is a peddle operated tray type de-popper. The protray is held by the top plate fitted with pegs. The seedlings are de-popped by pressing the peddle (by foot) and pushing the pegs of the bottom plate in drainage holes of the cells of protray. The de-popper shown in Fig. 5.56 has hydraulic operated cylinder for pushing the bottom plate for de-popping.

Fig. 5.55 Stand type seedling de-popper. (foot operated)
The Blackmore Company, MI.U.S.A.
http://www.blackmoreco.com/www.plugpop.com

Fig. 5.56 Stand type seedling de-popper.
( hydaulic operated )
URBINATI, Italy http://www.urbinati.com

## 5.13 Machinery for Grafting

One of the important operations in a nurserys is grafting. The machines for grafting are available abroad and their working is shown in Figs. 5.57 to 5.60.

In the grafting machine shown in Fig. 5.57a the protrays of root stock are placed on stand on left of the operator. The protray of the root stock are kept on the right of the protray. The operator lifts the seedlings with media ball and cuts the scion and root stock by pneumatic operated blades at an angle of 65 degree (Fig. 5.57b). The scion and root stock are brought together by placing there in the grove, where the clip is provided at the joint in (Fig. 5.57c). The grafted seedling is shown in Fig. 5.57(d). The grafting machine is able to graft 120-200 plants per hour.

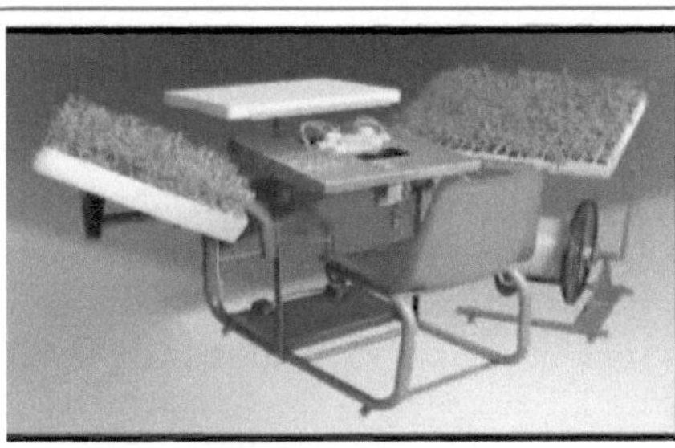

Fig. 5.57(a)Grafting roobt with rootstock and scion seedlings

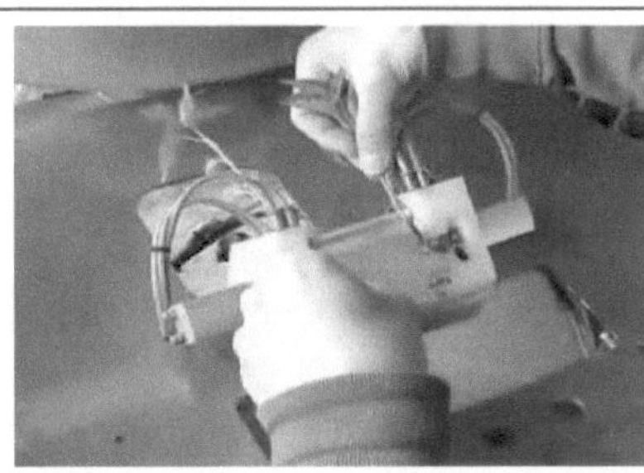

Fig. 5.57(b) Scion & rootstock is prepared

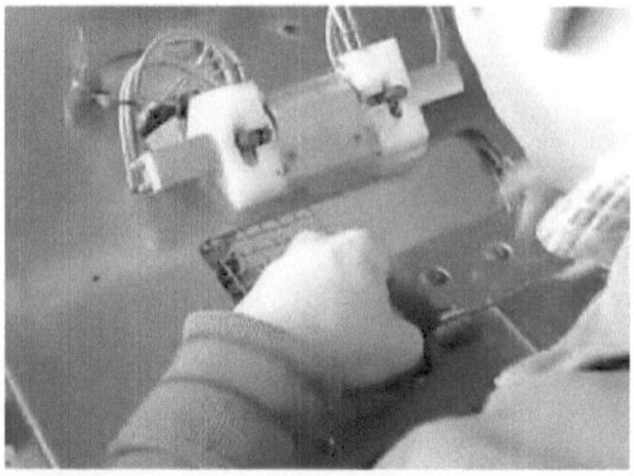

Fig. 5.57(c) Scion & rootstock is brought together and clip provided at joint

Fig. 5.57(d). Grafted seedling.
http://www.agrupamex.com

Fig. 5.57. Grafting machine, manual picking of the scion and rootstock seedling

Grafting machine
Helper Robotech

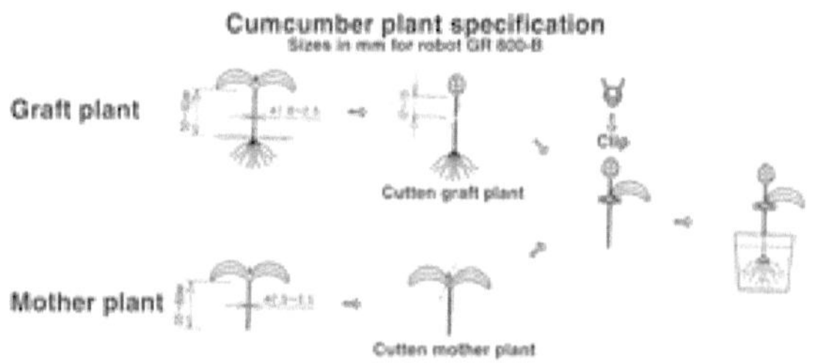



Steps of mechanical grafting
http://www.visserite.com

Fig. 5.58. Grafting machine, manual picking of the scion and rootstock seedling.

In the grafting machine shown in Fig. 5.58, the scion and rootstock are kept on table provided on both sides. One of the operator keep the scion at a required grove while the other operator keeps the rootstock seedlings at the grove provided on the other side. These are picked by the pneumatically operated holders. Both root stock and scion shoot are cut at a required angle by pneumatically operated sharp blade. The holders of scion and rootstock bring these closer and the clip is provided at the joint. The clips are singulated in the machine. The machine can graft about 650-900 plants per hour. The systematic operation steps are shown in Fig. 5.58(b).

Pin Method Grafting Robot– The grafting robot for single row (Fig. 5.59) takes one row of the seedlings from the protray, cuts the top and cuts grooves in the center of the root stock. The shoots of the mother plants are cut separately and inserted into the grooves in root stock plants. All the grafted plants are provided with clips. The process is repeated. It can graft upto 6000 seedlings per hour.

Pin Method Grafting

Robot Ideal System Co., Ltd.[Korea] & http://www.hort.purdue.edu/newcrop/history/lecture12/fig_12-18c.html

Scion supply & Seedling supply & cutting

Inserting

Grafting

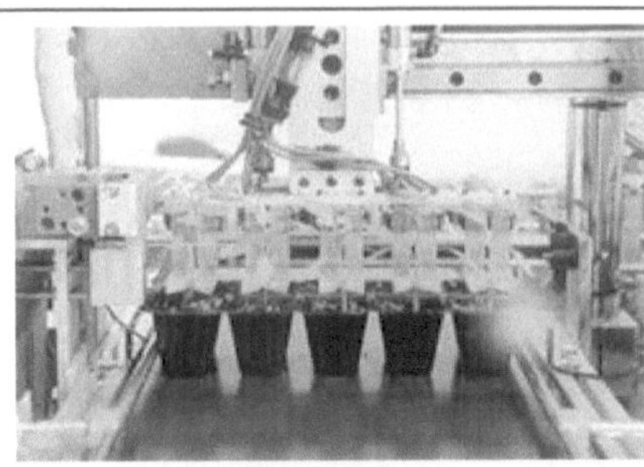

Completion

Fig. 5.59 Grafting robot for one row of the seedlings of the protray.

The grafting machine for grape shown in Fig. 5.60 is operated manually by a handle. It can cut the slot in rootstock and these are brought together with a plastic tape applied over the joint. The grafting tool shown in Fig. 5.61 is a hand tool and is supplied with dyes for slot and bud grafting. The necessary slots/cuts can be cut by the hand tool. The scion and rootstock are brought together and the plastic tape is applied over the joint manually.

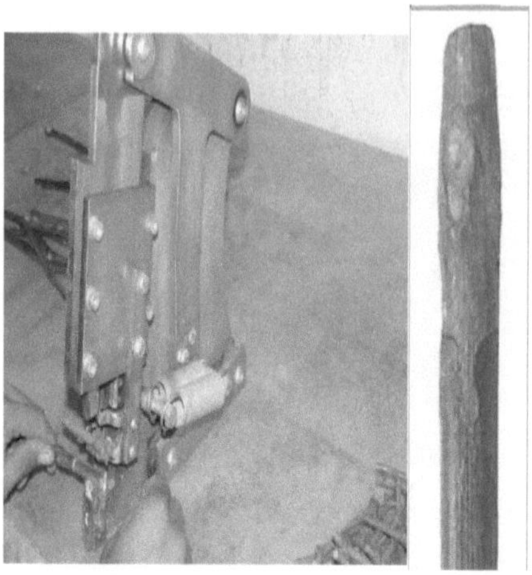

Fig. 5.60. Grafting machine for Grapes & a graft

NRC Grapes, Pune

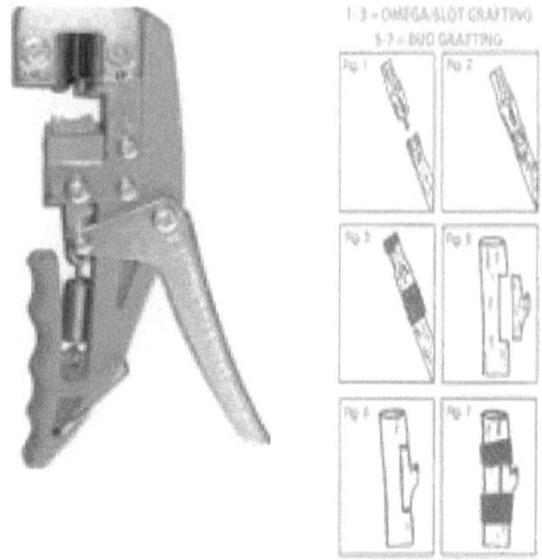

Fig. 5.61 Grafting tool with slot grafting and bud grafting tools

Falcon

## 5.14 Transplanter

The transplanter is used to transfer the seedlings from a protray with small size cavity to a protray with big size cavity or pots. The protrays with de-popped seedlings are conveyed to the transplanting machine on conveyer (Fig. 5.62). The fingers of the transplanter pick up the seedlings from protray and plant these into the row of pots. The protrays and pots move on two conveyors.

Fig. 5.62 Transplanter (protray to pots)

## 5.15 Irrigation

Generally watering in the nurseries in done manually by connecting the flower of watercan to the water hose (Fig. 5.63). The greenhouse however, has water outlets at several points and the hose pipe with flower is connected to the outlet. In the automatic irrigation system (Fig. 5.64) spray booms are used for watering the nursery. These move overhead and spray water on protrays for irrigation. The micro sprinklers and misters are also be used for watering the portrays.

Fig. 5.63 Manual watering of protrays
IIHR , Bangalore, India

Fig. 5.64 The atutomatic irrigation by using spray booms
http://www.visserite.com

## 5.16 Protray Washing Machine

After removing the seedlings for transplanting, the protrays are washed and sterilized for reuse with jet of hot water/steam through nozzles of a ptrotray washing machine. Protray washing machine is generally made of stainless steel (Fig. 5.64). A conveyor transports the trays and crates inside the machine. The nozzles from all sides spray jets of water to wash the trays and crates. The steam is also provided by electrical heaters or boilers for sterilizing the protrays at 90°C. The machine can wash/sterilize about 650 protrays or 30 crates per hour.

## 5.17 Protray De-Stacking Machine

For mechanized filling and seeding of protrays one person is required for de-stacking the protrays and placing these on the conveyor belt. The protray de-stacking machine (Fig. 5.66) separates individual protrays from the stack and places these on the conveyor belt of the filling machine

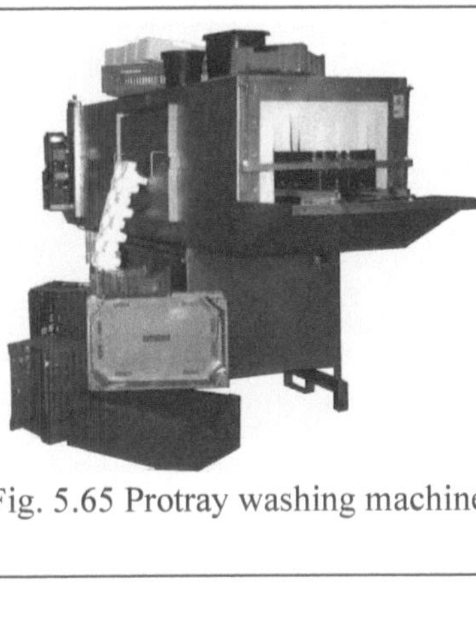

Fig. 5.65 Protray washing machine

Fig. 5.66 Protray de-stacking machine

# *Chapter 6*
# Hand Tools

A variety of hand tools are used for soil working, plant propagation, weeding, interculture and harvesting of horticultural crops. A soil working hand tool consists of a blade or working part, a handle and a joining member (eye) between working part and the handle. The working part comes in contact with soil or plant and is exposed to various mechanical stresses. For strength and abrasion resistance it is made of medium carbon, high carbon, alloy steel or tool steel. It is hardened and tampered for toughness to prevent fracture. The cutting edge is generally hardened to 350-450 HB (Brinell Hardness).

The handles are generally made from wood and are of round or slightly oval sections. The handle should be light, elastic and strong. The tools with small handles are for working in sitting or squatting position However, their working capacity is very low. Long handles are used for tool meant for working in standing position. There also need less energy. The operator gets easily fatigued if he has to work in bending, stooping or squatting position.

## 6.1 Spade (Indian) or *Powrah* or *Kaesi*

It is the most common and versatile hand tool for soil digging, irrigation, nursery bed preparation, earthing and ridge making. It consists of a thin sharpened blade (2.5 to 3.75 mm thick) set transversely on a wooden handle (Fig. 6.1) . The blade (digging head) has an eye for insertion of the handle. The shape and size of the blade varies from 180 to 230 mm in width and 215 to 305 mm length and 1.2 to 1.8 kg in weight for different areas (Fig. 6.1 ). The angle between blade and handle is about 80° in North India and about 30° in South India. The length of handle also varies from 45 cm in South India to 75 cm in North India, depending on height of the person and preferred posture of working. The blade of spade(Powrah) is made from medium carbon steel and forged to shape. The cutting edge is hardened to 350-450 HB.

| | | | |
|---|---|---|---|
|  West India Powrah |  Swan Neck Mamooty |  Swan Neck Mamooty |   Agri Powrah |
|  North India Powrah |  North India Powrah | |   East India Powrah |
|   West India Powrah |   North India Powrah / Kassi |   Tanged Powrah |   Mysore **Mamooty**tatasteel.com |

Fig. 6.1 Powrah (Indian spade) and powrah blades of different types

## 6.2 Hoe (Indian) or Kodali

It consists of a digging blade (4 to 7.5cm in width) with an eye through which wooden handle is inserted (Fig. 6.2) . The blade length is 15 to 40 cm and is sharpened at the digging end. The digging blade is slightly curved and forms an angle of 45 to 85° with the handle. The blade is made from medium carbon steel and forged to shape. The digging point is hardened to 350-450 HB. It is used for digging root crops and tubers besides general digging work. it weighs between 1 to 1.5 kgs for ease in working.

| | | |
|---|---|---|
|  K C Bhardwaj, M M Pandey, G Singh, CIAE Bhopal |  |  |

Fig. 6.2 Kodalis of differerent types

## 6.3 Pick Axe

The pickaxe is a double pointed tool with an oval shaped eye for fitting the wooden handle (Fig. 6.3). The pickaxe may have

- One chisel (wide) and one point end.
- Double point ends or
- Point and tee ends (beater).

The chisel point is used for digging normal soil and pointed end for digging hard or rocky soils. The overall length of the blade varies from 54 to 60 cm. It weighs 2.5 to 3.6 kg with hardness of 350-450 HB. It is used for general purpose digging.

| | | |
|---|---|---|
|  Miraj Steels | 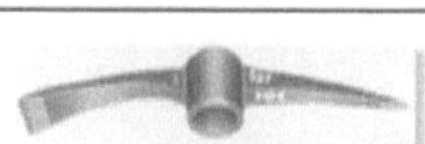 | Chisel (wide) and Point ends |
| |  | Point and tee ends |
| | 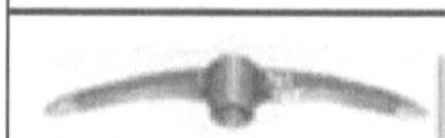 Fig. 6.3 | Point and tee ends tatasteel. com |

## 6.4 Digging Pronged Hoe

The digging pronged hoes (Fork Powrah) has 2, 3 or 4 prongs instead of a single blade (Fig. 6.4). This is used for digging hard and stony soils. The pronged hoe is also used like kodali and dragged in bending position for harvesting of tuber crops. The prong length is about 15 cm, top prong width 3.2cm, bottom width before sharpening (pointing) 1.0 cm and prong thickness about 6 mm.

The working width may vary from 8 cm to 24 cm depending upon the number of prongs. The handle length is 50 to 60 cm and handle diameter is about 3 cm.

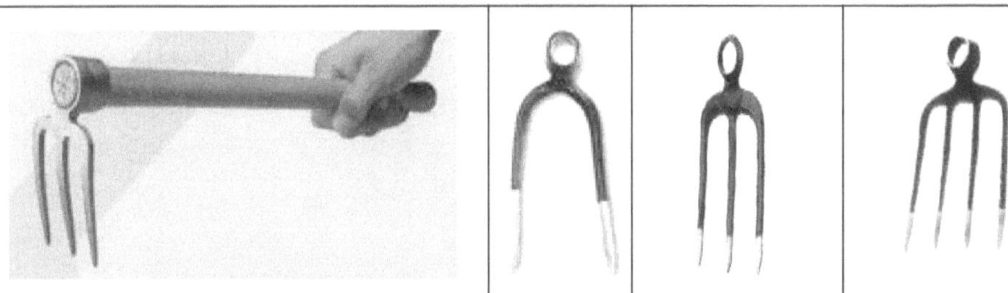

Dieter Schmid

Fig. 6.4 Digging pronged hoe and blades of different types

## 6.5 Mattock and Trenching Hoe

These two hand tools (Fig. 6.5) are double-ended and look similar. The mattock is a heavier implement in comparison to trenching hoe. One end of these implements consists of a flat rectangular blade similar to spade (Powrah) and the other end is a chisel shaped blade similar to the blade of pick axe or kodali. For mattock, the flat blade length is 18 cm. width 10 cm, overall length of blade about 40 cm. and it weight 2.3 kg. For trenching hoe , length and width of flat blade is 15 cm. overall length 33 cm. and weight is 0.7 kg. The mattock is generally used for digging and cutting of roots and stumps and the trenching hoe is used for digging or making field channels and drains for irrigation and drainage, earthing operations and breaking of clods.

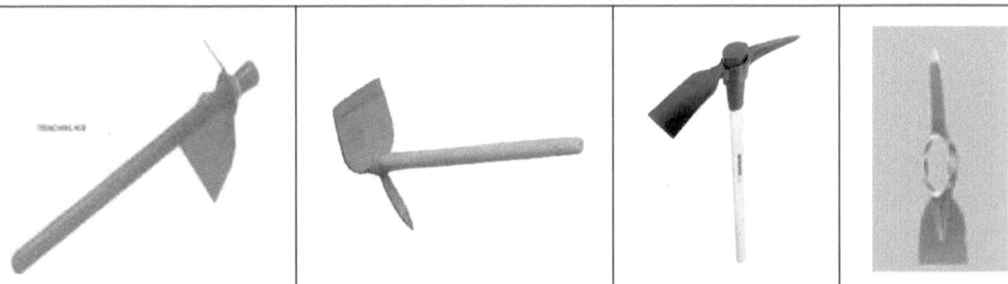

mudharallied.com | atcomaart.com | gardendecor.net | abmtools.com

Fig. 6.5. Trenching hoe (and mattock) and blades of different types

## 6.6 Shovel

A shovel consists of a digging blade, socket and wooden handle (Fig. 6.6). The digging blade is rectangular, round or heart shaped and has a small curvature along its length. Some of the blades are fitted with a foot rest for easy pushing in soil. The blade length is 20 to 30 cm, width 9 to 19 cm and thickness 2 to 3 mm. The blade is fabricated from medium carbon steel and the cutting edge is hardened to 350-450 HB.

A mild steel socket of 18 to 25 cm. in length is provided to the blade for fixing wooden or fiberglass handle of 3.0 to 3.5 cm diameter. The grip of D or T shape is provided on the wooden or fiberglass handle. For digging, the blade is pushed into soil and a soil mass is detached. The shovel is also used to throw the soil to small distance by swinging action of the arm. The shovel is used for digging of soil, making trenches and mixing of farmyard manure. The shovels with straight and heavy blades are known as Spades (Fig.6.7) and can be used for digging soil in standing position.

Square blade
Tata Steel

Round blade
Tata Steel

Fork blade
Tata Steel

Fig. 6.6 Shovel of different types

## 6.7 Fork

Forks are manually operated multipurpose hand tools with 3 or 4 prongs. The prongs are welded to the socket. A wooden or fiberglass handle usually with a "D" type grip is fitted in the socket.The prongs are either round or square forged to pointed end and are slightly curved for easy penetration and lifting of soil and other biomass. The prongs are fabricated by forging from medium carbon steel. The pointed ends are hardened to 350-425 HB. The mild steel socket is welded to fork for fitting the handle. The width of the fork varies from 18 to 23 cm., length of blades 27 to 31 cm., length of handle 70 to 120 cm. and total weight 1.8 to 2.3 kg. The forks are used for digging, weeding and handling of farmyard manure.

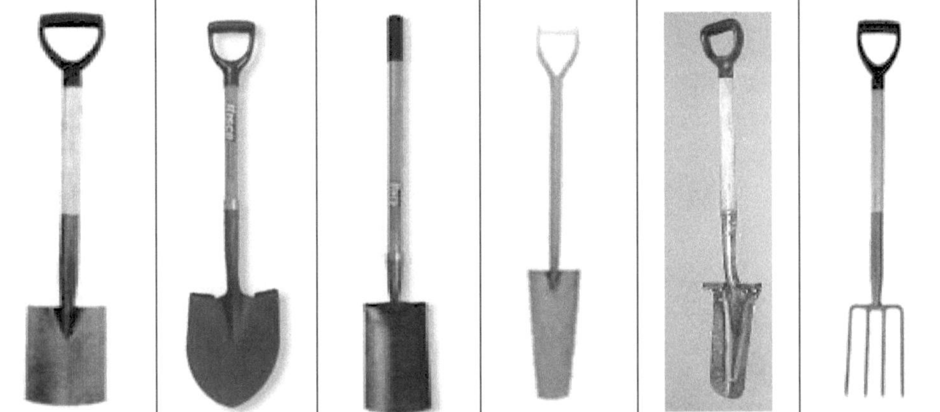

| Square blade Luthra Overseas | Round blade | Post Hole Spade | Drain Spade Luthra Overseas | Blade with foot rest | Fork blade Luthra Overseas |
|---|---|---|---|---|---|

Fig. 6.7. Spade of different types

## 6.8 Hoe – cum – rake

This is a multipurpose hand tool and consists of a flat blade (like spade/powrah) on one side and prongs on the other side (Fig. 6.8). A wooden or a fiberglass/plastic handle is fitted to the eye for operation. The flat blade is used for digging and rake side for weeding. The flat blade length is 12cm, width 6 cm and thickness at cutting edge is about 3mm. The number of prongs vary from 1 to 3 . The length of prong is about 10cm and these are pointed at the ends. The length of handle is about 60cm and thickness is about 3 cm. The Hoe cum rake is generally used for digging, earthing, weeding and root crop harvesting/digging.

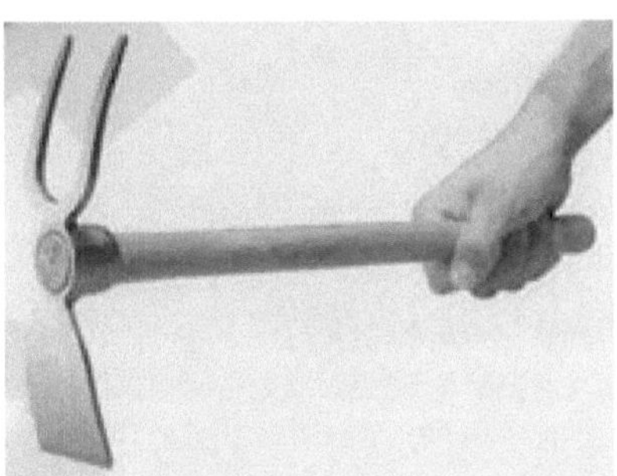

Dieter Schmid

Falcon

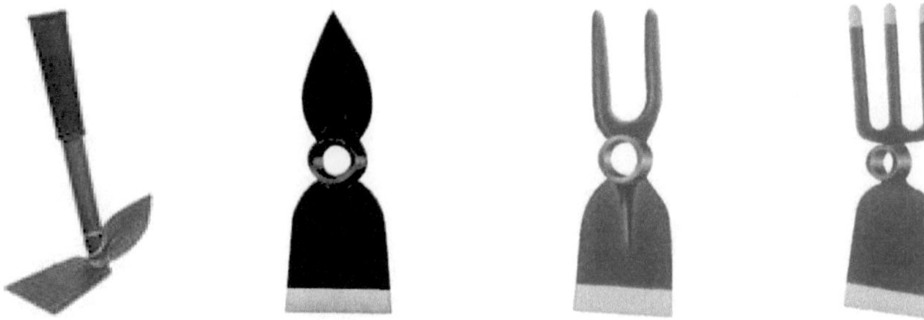

Fig. 6.8 Hoe cum rake and blades of different types

## 6.9 Transplanting trowel

The transplanting trowel consists of a blade and handle (Fig.6.9). The blade is slightly curved for easy handling/removal of soil. The width of blade varies from 6 to 10 cm; length 15 to 25cm and the thickness is 1.5 to 2.0 mm. The length of short handle is about 13 cm and the worker has to work in squatting or sitting posture. The length of handle for long handle trowel is about 75cm. The blade can be forked also. This is used for transplanting & uprooting of seedlings and bulbs and for earthing of vegetable and flower plants.

Standard Tata agrico | Fork RAMCO | Narrow Power Tools | Long and narrow blade www.progo-sports.com

Fig. 6.9 Transplanting Trowel

## 6.10 Hand shovel

This consists of a scoop shaped rectangular blade of mild steel sheet attached to a short handle (Fig. 6.10). The blade length is about 15cm, width about 13cm and total length about 40 cm. This is used for turning the soil in vegetable and flower beds and for filling the soil in pots.

http://gardendecors.net

montstar.com montstar.com

Fig. 6.10. Hand shovel of different types

## 6.11. Hand rake

It consists of a handle and a set of prongs bent at the tips (Fig 6.11). The prongs are welded to the socket for handle. The length of prongs is 20 to 30 cm and the length of handle is about 15 cm. The prongs and the socket for handle are fabricated from mild steel or spring steel. The working width varies from 30 to 45 cm. The tool is held in hand and dragged on soil surface for breaking. the crust for aeration in seedbed and lawns. It is also used for collection of trash

www.kisan76.com http://jspgardentools.com http://www.420hydro.com

Fig. 6.11. Hand rake of different types

## 6.12 Garden rake

The garden rake is similar to hand rake, but is provided with a long handle of about 1.5 meter length (Fig 6.12). For hand rake the operator has to work in squatting position and for garden rake in standing position. The working width varies from 30 to 45 cm and 8 to 18 spikes are provided. The pitch of spike is 25 to 31 mm. The spikes are fabricated from 1cm round rod or from 0.5 cm thick mild steel plate. The garden rake is used for leveling of beds, collection of uprooted weeds and aeration of soil.

http://jspgardentools.com | http://www.montstar.com | www.kisan76.com

Fig. 6.12. Garden rake of different types

## 6.13 Trowel and rake

These hand tools are used for soil working in pots and bonsai plants(Fig. 6.13).

Fig. 6.13 Trowels and rake

Falcon

## 6.14 Crow bar

Crow bars an fabricated in one piece to form an octagonal bar from high carbon steel (Fig. 6.14). There are manufactured in 4 sizes ie., 22, 25, 29 and 32 mm width across flats. The chisel end of the crow bar is forged to shape and heat-treated / hardened to 350 to 450 HB. The overall length is 1.65 meter and the weight is 5.1 kg for 22 mm crowbar and 10.5 kg for 32 mm crow bar. The 25 and 29 mm crow bars weigh 6.7 and 8.5 kg respectively. The crowbars are used for digging pits and holes for planting and fencing.

## 6.15 Auger

The auger is a manually operated hand tool for making holes or bores for planting and fencing (Fig. 6.15). The earth auger consists of a cutter head having semi circular blades connected to a pipe with the help of a clamp and flange. The connecting pipe is further connected to a handle for rotation. The cutter head is kept in moist soil and rotated manually by handle for penetration of head in soil. The cutter/head entraps the soil and is pulled for removal of soil. The soil is removed manually and again the head is inserted in the bore for cutting and removing the soil till the desired depth is reached. The cutter heads are available from 10 to 25cm diameter while the length of cutter head is about 25cm. The length and diameter of connecting pipe is 75cm and 1.5 cm while that of handle is 35 cm and 2.5 cm respectively.

## 6.16 Ridger

The ridger can be used to make ridges or furrows (Fig 6.17) for sowing seeds on there for earthing and weeding of crops sown on ridges. The working width may be upto 10 cm. The length of its handle is about 60 cm for working in sitting posture and 1.5 m for working in standing posture.

## 6.17 Dibbling stick

It is a simple manually operated device for creating a conical cavity in the soil for sowing seeds and seedlings (Fig 6.16). It consists of a long thin metallic cone fitted with a wooden or plastic handgrip. The dibbling stick is held at required spacing and the conical end is pressed into the seedbed to the required depth. The conical cavity is created in the soil and the seed or seedling is placed and soil is pressed manually around it. The conical end is made from mild steel sheet and fitted with wooden or plastic handle. Its overall length is 20 to 40cm.

<table>
<tr><td>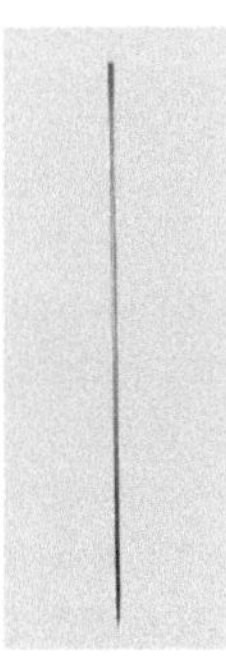<br>Ramco</td><td>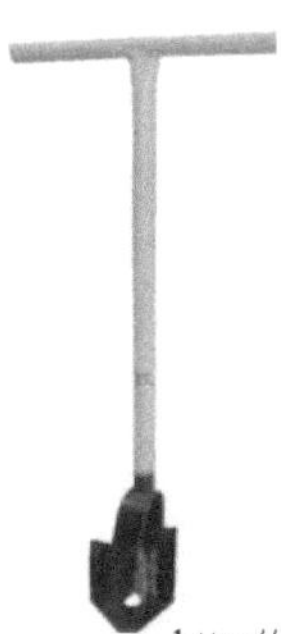<br>http://<br>jspgardentools.<br>com</td><td><br>WOLF-Garten</td><td><br>Wave Impex</td><td>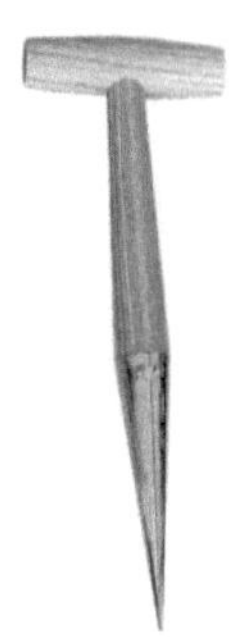<br>Ramco</td></tr>
<tr><td>Fig. 6.14<br>Crow bar</td><td>Fig. 6.15<br>Augar</td><td>Fig. 6.16 Ridger</td><td colspan="2">Fig. 6.17 Dibbling Stick</td></tr>
</table>

## 6.18 Bulb planter

This bulb planter consists of a spring loaded split cylinder and handle (Fig 6.18). with the spring fitted inside the handle. It is first pushed in soil to remove a plug of soil, the bulb is then held in the cylinder and released in the pit by squeezing the handle. Soil is also covered over the bulb. The overall length of bulb planter is about 45 cm and bulb planters with long handle are also available. One can work comfortably in standing posture as the handle length is about 75cm. The split cylinder is chromium plated with depth markings are marked on this. The handle can be of stainless steel or plastic.

| | | |
|---|---|---|
| 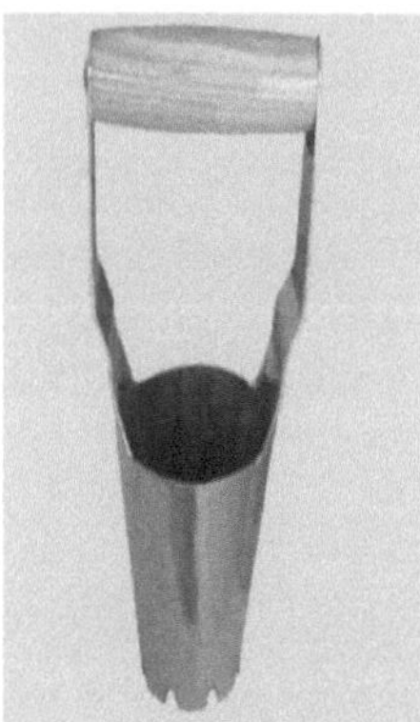 | 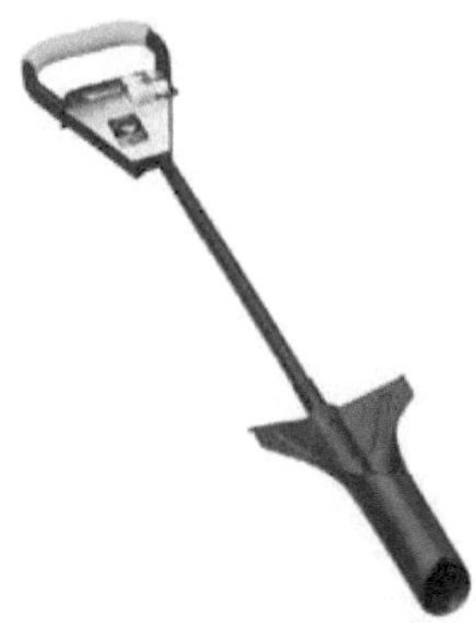 | 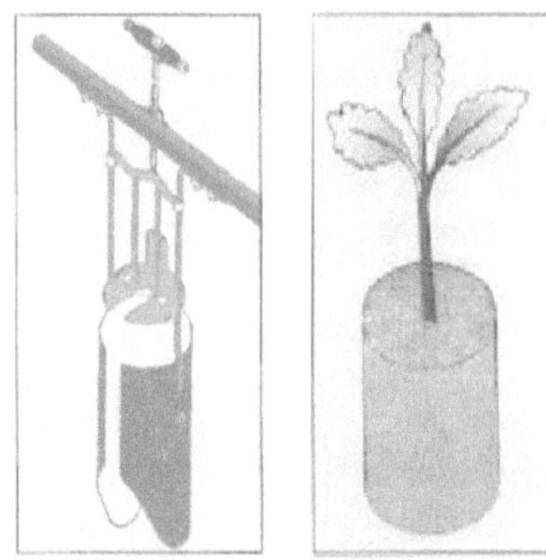 |
| Bulb planter<br>RAMCO | Long Handled Bulb Planter<br>greenfingers.com | Plant Replacer<br>K C Bhardwaj, M M andey, G Singh,<br>CIAE Bhopal |

Fig. 6.18 Bulb planters and plant replacer

## 6.19 Plant replacer

This is a manually operated hand tool for uprooting of saplings with lump of soil and root system for transplanting or grafting. It consists of an outer cylinder (auger shape at bottom), with a groove for enclosing the plant. The press plate with groove for enclosing the stem of the sapling is also provided with a handle. The cylinder diameter and length is 13 cm and 20 cm respectively and the press plate diameter is 12.5 cm. The outer cylinder is pressed in the soil (with sapling in the center) by the weight/force by the operator, press plate (with sapling at center) is pressed by handle to compact the topsoil around the sapling. After compaction a jerk is given to the cylinder and it is pulled out along with the seedling. The press plate is pushed to remove the sapling with soil lump. The cylinder is fabricated from 130mm diameter GI pipe and the handles from 40 mm diameter GI pipe "A" class. Other components are fabricated in mild steel and handle length is 56 cm. This is a handy tool for uprooting the saplings for planting or grafting.

## 6.20 Pit hole digger

The petrol engine operated pit hole diggers can be used for drilling the pits of 10 to 16cm diameter for planting the fruit crop saplings or erecting the pillars for purposes bower. These can be portable as shown in Fig. 6.19. The pit hole digger transported on pneumatic wheels is more convenient for field operation.

Pit hole digger
http://www.stihl.com.au

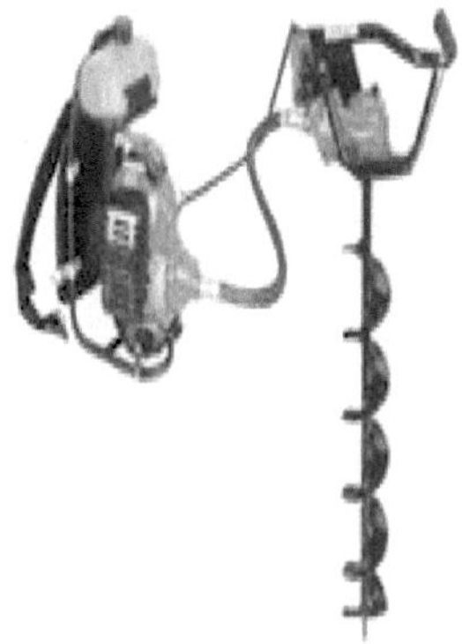

Pit hole digger with backpack engine
http://best-b2b.com

Pit hole digger
http://www.northwest rentals.net
http://www.groundhoginc.com

Fig. 6.19 Motorized pit hole diggers

# Chapter 7
# Transplanting

Transplanting of seedlings of fruit and vegetable crops is being done manually and is labour intensive. The pits for planting fruit crops are generally dug by labourers using pickaxe or spade and the seedlings are planted in these pits, manually. The pit digging capacity of 2 labourers is about 8 to 12 pits per day for 60X60X60 cm dimensions and it costs Rs. 40 to Rs. 60 per pit. Tractor operated pit hole digger can dig upto about 200 to 300 pits per day (of 60 cm diameter) and cost about Rs. 8 to 12 per pit. The self propelled mechanical digger or JCB/ tractor operated back hoe can dig upto 500 pits per day at a cost of Rs. 12 to 20 per pit. In USA and Malaysia, the hydraulically operated pit hole diggers with planting and pit filling system along with trailer for transporting the seedlings and transplanting the same are available. Their transplanting capacity is around 90 seedlings per day and the transplanting cost is around US $ 0.55 per seedling. The transplanter can be adopted under Indian conditions to increase the area under fruit cultivation.

Transplanting of vegetable seedlings is done manually in India and it takes about 20 to 25 mandays for transplanting most of the crops (except onion) and costs about Rs. 3000 to 4000 per hectare. The tractor drawn bare root transplanters for vegetable crops like cabbage, cauliflower, capsicum, chillies etc. are commercially used in Europe and USA. In India, 2 or 3 row transplanters are suitable with the transplanting capacity of around 50 seedlings per minute per row. It will take 3 to 5 hours to transplant one hectare with tractor operated transplanter at a cost of Rs. 1500 to 2000 per hectare.

A tractor drawn tow-row seedling transplanter (with seedling holders) for bare roots has been developed at CIAE, Bhopal and PAU, Ludhiana. These transplanters are available commercially.

These days nursery production is also grown commercially in protrays. Two or 3 row transplanter can also be adopted for plug type nursery to transplant most of the vegetable crop seedlings. The capacity of the plug type transplanter is 60 plants per minute per row and the establishment of crop stand is better.

A tractor drawn raised bed former cum seedling transplanter for plug type nursery has been developed at IIHR, Bangalore. The transplanter can plant 1 or 2 rows at a spacing of 120cm or 60cm respectively and plant spacing can be adjusted to 23, 30, 45, 60 or 90cm. The transplanter has a capacity of about 1 ha/day. It is a punch type transplanter and is available commercially

The adoption of mechanical transplanters for fruit and vegetable crops will increase the yield and area of cultivation under these crops in India.

Most of the fruit plants are grown by planting the seedlings or grafts in pits of size 30cm to 90cm width and 50cm to 100cm depth at spacing of 1.5m to 12m. The plants are planted in the pits and these are filled with mixture of FYM/fertilizer and soil and compacted. It needs around 4000 pits of 30cm width and 50cm depth for planting banana in one hectare and around 125 pits of 90cm width and 120cm depth for planting 1 hectare of mango. Pineapple is planted by digging trenches at a spacing of 60cm. Grapes are planted by digging 60cm wide and 60cm deep trenches at a plant spacing of 3m. Pits are also to be dug for raising bowers etc.

The pits and trenches are generally dug by manual labour using region specific hand tools based on the soil type. The hand tools generally used are crow bar, pick axe and spade etc. The top soil of the pit is generally mixed with FYM and insecticide and the seedling is planted manually in the pit and the soil is compacted around the seedling. This is a very labour intensive and cumbersome job. The digging of pits can be easily done by using tractor operated pit hole digger or the tractor operated excavator (JCB) and these equipments are available in India. The tractor operated oil palm seedling transplanters are being used in Malaysia and can be adapted under Indian conditions for transplanting. The vegetable seedlings are planted at the ridges prepared by animal drawn or tractor drawn implements. The bare root or plug type seedlings are transplanted manually at a row spacing of 15 for onion to 75cm for tomato and seedling spacing of 7.5 to 60cm. It needs 20 to 25 mandays for transplanting one hectare for most of the vegetable crops and needs 100-150 mandays for transplanting onions. The bare root and plug type seedling transplanters used in Korea, Japan, Europe and USA can be adapted under Indian conditions with suitable modifications.

## 7.1 Pit or Trench Digging for Fruit Crops

The tractor operated pit hole diggers have been developed by various tractor and farm machinery manufacturing companies (Fig.7.1). The augers of 30-60cm dia. are available for digging pits of different diameters. For digging pits of 60cm dia. and above, tractor of 50hp and above can be used. The pit hole diggers are operated by P.T.O. shaft and can dig 200-300 pits per day of 60cm dia. at a cost of Rs.8-12 per pit.

The tractor operated back hoe or excavators (JCB) (Fig. 7.2) are also available to dig the pits. These can dig the pits of dimension of 45cm and above. Upto 500 pits can be dug per day at a cost of Rs.8-12 per pit. These excavators can also be used for mixing FYM, with the top soil and refilling the pits for transplanting the seedlings. A deep trencher (Fig. 7.3) has been developed by Shiva Industries, Tumkur, Karnataka and is suitable for making trenches for planting pineapple or grape. This is able to make trenches in ploughed field at optimum moisture content. This can work at a speed of about 2 to3 km/hr. The rotary trencher (Fig. 7.4) available in developed countries can also be adopted for making trenches for planting of pineapple and grapes.

Fig. 7.1 Tractor operated pit hole digger- for digging pits for planting seedlings or fruit crops
Tirth Agro http://www.shaktimanagro.com

Fig. 7.2 Tractor operated back hoe and loader (mechanical excavator ) for digging pits and trenches for planting seedlings or fruit crops
http://www.bullindia.com

Fig. 7.3 Tractor operated deep trencher for making trenches for planting pineapple and grapes
(Shiva Industries, Tumkur)

Fig. 7.4 Rotary trencher in operation
http://www.forconstructionpros.com

## 7.2 Fruit Crop Seedling/Graft Transplanter

In Malaysia and Korea trailor type transplanters for 4-wheel driven tractors has been used for transplanting of palm seedlings (Figs 7.5 and 7.6). In addition to tractor driver one additional operator is required to operate the hydraulic controls of the machine. The seedlings are carried in the trailor/seedling bin. The conveying unit conveys the seedling during planting operation, near to the operator at the rear end of the trailor. The drilling mechanism prepares the pit for planting the seedling. The clamping mechanism holds the seedling, lowers it in the pit, covering mechanism fills the soil around the seedling and compacts the soil.

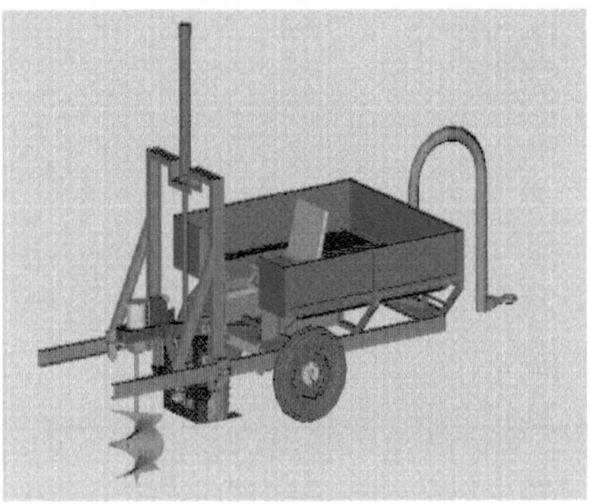

Fig. 7.5 Fruit crop seedling transplanter (oil palm seedling transplanter) (University Putra, Malaysia)

Fig. 7.6 Fruit crop seedling transplanter (American Society of Agricultural Engineers)

Fig. 7.7 Fruit crop seedling transplanter (with lifted auger) (University Putra, Malaysia)

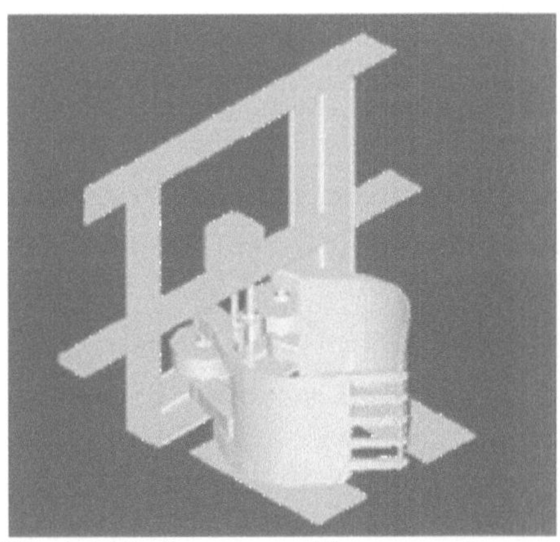

Fig. 7.8 Fruit crop seedling transplanter (Clamping – covering mechanism) (University Putra, Malaysia)

Fig. 7.9. Sequential operations of seedling, transplanting by fruit crop seedling transplanter

One double acting hydraulic cylinder is used for raising and lowing the auger bit, one double acting cylinder is used for tilting the auger bit and one bi-directional low speed high torque hydraulic motor is used for operating the auger bit (Fig. 7.7). One bi-directional hydraulic motor is used for operating the seedling conveyor, one double acting cylinder is used for horizontal movement of seedling placement unit and bi-directional motor is used for operating clamping-covering mechanism (Fig. 7.8). The machine can be used to transplant the fruit crop seedlings. It has a capacity of transplanting 99 seedlings per day and the cost of transplanting is US $ 0.55 per seedling. All the operations of the machine are shown sequentially in (Fig. 7.9).

## 7.3 Vegetable Crop Seedling Transplanter

Most of the vegetable seedlings are transplanted at a row spacing of 45-120 cm and seedling spacing of 30-90cm. The row spacing for onion is 15cm & seedling spacing is 7.5-10cm. The seedling population may vary from 9000 to 58000 per hectare for most of the vegetable crops except onion. It requires 20 to 25 mandays and costs about Rs.2,000 to 3,000 per hectare. The onion transplanting needs 80-100 mandays per hectare to transplant about 6,70,000 seedlings and costs Rs.8,000 to 10,000 per hectare. Bare root seedlings are traditionally transplanted on ridges or flat beds. The plug type seedlings grown in protrays have also become very popular in case of F1 hybrids. The following types of transplanters are commercially available in Europe, Japan, Korea and USA and can be easily adapted for vegetable seedling transplanting in India.

- Punch Planter
- Transplanter with manual feeding of bare root seedlings to seedling holders on planting wheels
- Tractor operated transplanter withmanual feeding of bare root seedlings to seedling holders on chain driven by traction wheel
- Transplanter with manual feeding of plug type seedlings to rotary cups
- Autmoatic self propelled/tractor operated seedling transplanter for plug type seedling.

**a. Punch planter**

The machine is provided with punch wheels to create pits/ depression on flat or raised bed by the planter (Figs. 7.10-7.20) and the seedlings are placed manually in the pits/depressions.

The main parts of the machine are -

- Frame
- Punch wheels and shaft
- Operator seats
- Pressing wheels
- Water tank
- Space for storing seedling protrays or seedling
- Space for storing empty seedling protrays or crates and
- Transport wheels

Fig 7.10 Single row punch transplanter with two operators, watering tank and racks for seedling protrays
http://www.kenncomfg.com

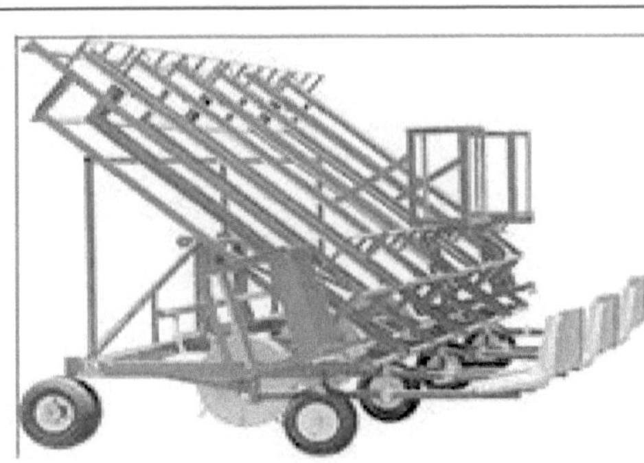

Fig. 7.11 Three row punch transplanter with one operator for each row, watering tank and racks for seedling protrays
http://www.irrigation-mart

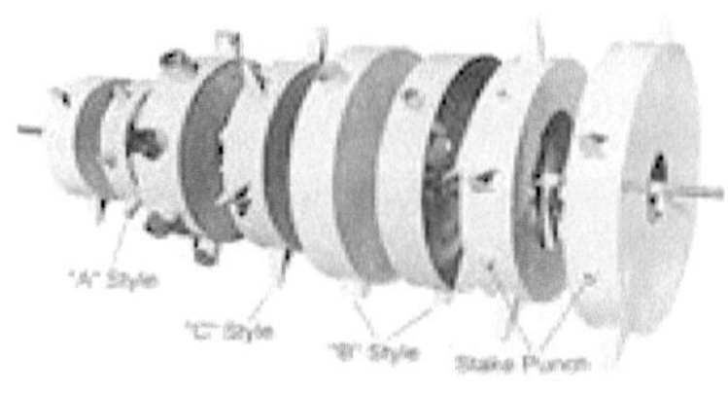


Fig. 7.12 Punch wheels of different diameter, punch shapes and spacing
http://www.irrigation-mart.com

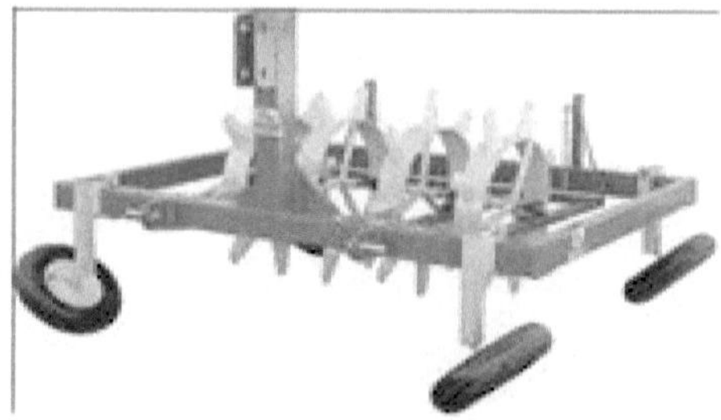

Fig.7.13. Four punch wheels fitted on a shaft and wheels for bed making
http://www.irrigation-mart.com

Fig. 7.14. Punch type transplanter with water tank for mulched field.

Fig. 7.15. Punch planter in operation in plastic mulched field
http://www.kenncomfg.com

Fig. 7.16. Punch planter in operation in plastic mulched field
http://www.kenncomfg.com

The frame supports all the components of the transplanter. Three point hitch is also provided to the frame (Figs. 7. 10 to 7.11). The punch wheels (Fig.7.12) are mounted on shaft and rotate when the transplanter is pulled by the tractor. The number of punch wheel is equal to number of rows to be transplanted by the transplanter. All the punch wheels can be mounted on one shaft (Fig. 7.13) or 2 to 4 punch wheels are mounted on individual shafts (Fig.7.14). The shape of the punch depends on the crop, type of soil and type of nursery (bare rooted or plug type) (Fig. 7.12). The circumferential distance in adjacent punches depends upon the spacing of seedlings in the row. Additional lugs/stakes/projections are provided on the punch wheel for smooth rotation in the field. There is generally 15 to 20% slip of the punch wheel in the field and the circumferential distance in adjacent punches is 15 to 20% less than the required plant spacing.

Fig.7 .17. Working of raised bed former cum two row punch transplanter with one operators for each row (with water tank).
IIHR, Bangalore

Fig. 7.18 Chilli crop raised by bed former cum two row punch transplanter
IIHR, Bangalore

Operator seat is provided for the operator to sit comfortable and take out the seedings from protray or bundle and place the individual seedings in the depression created in the soil/row by the punch wheel. One operator can place 50 to 60 plants in the row in a minute and generally one operated seat is provided for each punch wheel (Fig. 7.15-7.20). Two operator seats can also be provided behind a punch wheel for two labours to transplant one row to speed up the transplanting speed.

Pressing wheels are provided to press the soil around planted seedling. These wheels are in addition to transport wheels and also support the weight of transplanter in the field.

A water tank (Fig. 7.15, & 7.17) is provided to store water and provide life saving watering immediately after planting through the watering system. The water tank with a capacity upto 1000 liter can also be provided over the tractor. The regular watering of the transplanted crop can be done within 2 to 3 hours after transplanting.

The racks are provided on the machine to store the seedling protrays (Figs. 7.15 - 7.20). The racks are also provided to store the empty protrays. The operator can easily pick the seedling from protray, from the seat and can transfer the seedlings to the punched depressions. The bare root seedlings can be stored in crates on the platform. The transplanter is generally trailed type or semi trailed type implement for easy transport on road. These transport wheels also support the weight partially in the field and can be used to adjust the depth of transplanting.

Fig. 7.19. Chilli crop transplanting
IIHR, Bangalore

Fig. 7.20. Chilli crop transplanting
IIHR, Bangalore

Fig. 7.21. Chilli, Cabbage & Okra crop raised by IIHR Transplanter.

Single row punch transplanter with 2 operator seats and water tank is shown in Fig. 7.10. The raised bed former cum tansplanter developed at IIHR is shown in Fig. 7.17-7.20. Two ridger bottoms and leveller is fitted to form and level the bed of 1.2 m width and 15 cm height. Two punch wheels are fitted for planting seedlings on the bed. The transplanter operates at 1.2 to 1.8 km/hr based on the seedling spacing of 30-60 cm. The field capacity is 1.0-1.5 ha/day. The bold seeds like okra and beans can also be dibbled by the operators. The cilli, cabbage and okra crops raised by transplanter are shown in Fig 7.21. The transplanter can also plant potatoes and tubers of ornamental crops. Two row and three-row transplanter with one operator seat per row are shown in Figs. 7.11.

The punch transplanter shown in Fig. 7.22 is a bare root transplanter for onion. The platform is provided for storing the seedlings in crates. The punch wheels are very small and are known as peggers. Seedlings are planted on slightly raised beds approximately 1.2 m wide. Beds have 1.8 m width (center to center). Each bed have four rows spaced 35-45 cm apart and a seedling spacing of 12-15 cm within the row (Fig. 7.23). The spacing is determined by peg spacing on the pegger, the pegs are used to make holes in the bed surface (2-5 cm deep). Transplants are hand set manually in each hole. The crop transplanted by the onion transplanter is shown in Fig. 7.23a. The onion transplanter (Fig 7.24) is 6 row punch type transplanter with row spacing of 17 cm and seedling spacing of 10cm. This for transplanting on raised bed and operates at 0.4 km/h. The field capacity is 0.3 ha/day. The crop stand is shown in Fig 7.24a.

Fig. 7.23. Punch planter/pegger for bare root onion seedlings (for 3 beds) Onion Production Guide
http://www.ces.uga.edu/pubcd/B1198.htm#

Fig. 7.23a. Onion crop planted by pegger/ punch planter
Onion Production Guide
http://www.ces.uga.edu/pubcd/B1198.htm#

Fig. 7.24 Onion transplanter developed at IIHR, Bangalore

Fig. 7.24a Onion crop planted by IIHR transplanter

## b. Transplanter with manual feeding of seedlings to seedling holders on planting wheels

In these transplanters (Fig. 7.25 & 7.26) the planting wheel is provided with seedling grippers (in place of punch wheel). The operator places individual bare root or plug type seedling in grippers when these are in open position. These get closed to hold the seedling and release the seedling in soil at required depth during the rotation. The seedling holders are generally spring loaded and are opened by cam provided on the transplanter frame for placement of seedling and release of the seedling in soil. The number of rows transplanted by the transplanter is equal to the number of planting wheels and the row spacing is equal to the spacing between the planting wheels. The seedling spacing is approximately equal to the circumferential distance in seedling holders on the planting wheel. The compaction of soil around the seedling is done by planting cum compaction wheel.

Fig. 7.25. Transplanter with manual feeding of seedlings to holders fitted on planting wheels
Lee Shuknecht & Sons, Inc, NY 14058

Fig. 7.26. Manual feeding of seedling to seedling holders
Lee Shuknecht & Sons, Inc.4458 Ford Road Elba, NY 14058

Transplanter with manual feeding of seedlings to seedling grippers / holders on chain driven by traction wheel

The transplanter is tractor mounted or trailed type as per requirement. It consists of modules/units (Fig. 7.27) and one unit is used to transplant one row. Two or three row transplanter can be used for Indian tractors of upto 50hp (Fig. 7.32). Each unit consists of 1 or 2 seedling trays for storing the bare root seedlings (Figs. 7.27-7.29). The operators sit on the seat, pick up seedlings from the tray and place the same in rubber padded grippers / holders when these are in open position. These holders get closed to hold the seedling and move on endless chain (Fig. 7.30). The roots of the seedlings are pushed in the soil and released at the required depth by the holders (by opening) in soil as per the requirement. The holder chain may be operated by a separate pneumatic drive wheel/traction wheel (Fig. 7.25) or by the compaction wheel (Figs. 7.27 & 7.29). The soil is compacted around the seedlings by the compaction wheels. These are also watered if the water tanker is fitted on the transplanter. Upto 4000 seedlings can be transplanted by each unit in one hour. The minimum row to row spacing can be 46 or 23 cm based on inline or tandem arrangements of the units. The plant to plant spacing can be adjusted from 15 to 200 cm by changing the drive sprockets. If the seedling spacing is more, only one person is required to place the seedlings in one unit (Fig. 7.29). The modified holder for plug type seedlings and corms are also available (Fig. 7.31). Work on development of 2-row transplanter is going on at Central Institute of Agricultural Engineering, Bhopal (Fig. 7.32). The field capacity has been observed 0.1 hectare/ hr. The missing hills were 2.9 to 9.0% and the unplanted (lying down) seedlings were 4.5 to 13.4%. The transplanter has been tested for cauliflower, brinjal, tomato, cabbage and tomato at row spacing of 63 cm, plant spacing of 46 cm using 35hp tractor at forward speed of 1.20 km/hr and it costs around Rs.2500 to transplant one hectare. The commercially available 2 row transplanter is shown in Fig 7.29.

Fig.7.27. Single row bare root vegetable seedling transplanter with two operators, watering tank and compaction wheel as drive wheel
(Holland transplanter ,USA)

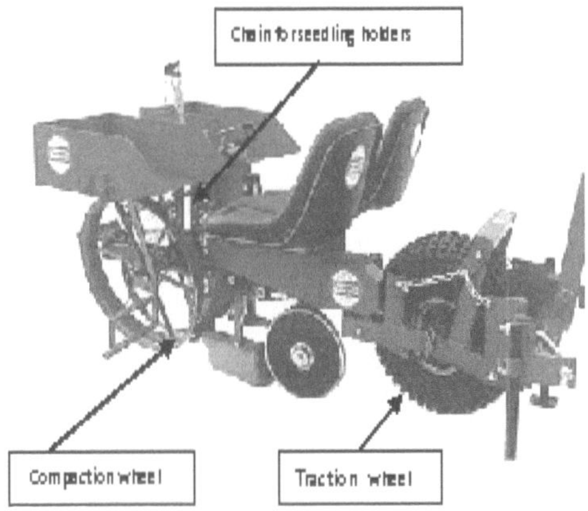


Fig. 7.28. Bare root seedling transplanter with separate compaction and pneumatic drivewheel
(Mechanical transplanters, USA)

Fig.7.29. Three row bare root seedling transplanter with one operator per row
(Mechanical transplanters, USA)

Fig. 7.30. Seedling holder chain driving mechanism
(Holland transplanter , USA)

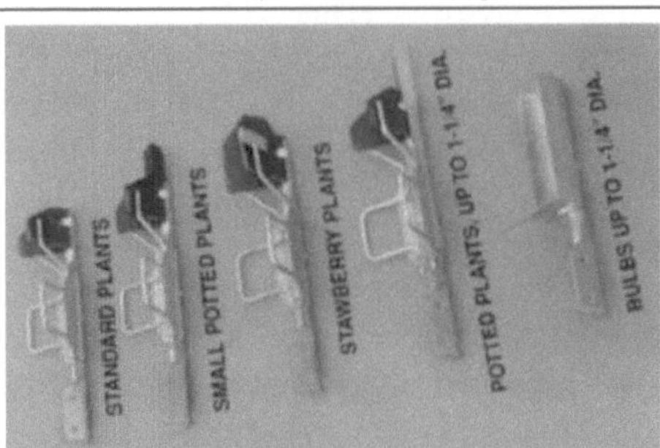


Fig.7.31. Seedling holders for different crops, plugs and corms
(Holland transplanter, USA)

Fig.7.32. Testing of seedling transplanter at Central Institute of Agricultural Engineering, Bhopal, India

**Transplanter with manual feeding of plug type seedlings to rotary cups**

These transplanters are either tractor operated (Fig. 7.35, 7.39,7.40 & 7.41) or self propelled (Fig. 7.42 & 7.43). The self-propelled transplanter (walk behind) is single row transplanter, while the tractor operated transplanter is 2 row (Fig.6.30 & 6.35) or multi row transplanter (Fig 7.39). The multi row transplanter have individual module/unit (Fig. 7.34, 7.36 & 7.41) for each row. The operator sits on the seat to pick up individual seedlings with hands from the hanging protrays (Fig. 7.40 & 7.41) and drops it into revolving cup. Then these are then transferred to the transfer mechanism and then to the furrow made in the soil. The soil is compacted around the seedlings after these are released in the furrow.

The rotary feed cups are rotated by a traction wheel (Fig. 7.33-7.35) or compaction wheel (Fig. 7.39 & 7.41) through the driving mechanism. Through rotation of the feed cups, the seedlings fall into the transfer mechanism. The transfer mechanism may have patented spike belt system to deliver the seedling to the packing wheels (Fig 7.41). The transfer mechanism can be spring loaded cup holder, which receives the seedling and transfers it to the soil by opening the pit (Fig 7.42 & 7.43). The seedlings can be dropped directly into the furrow by a transfer pipe, where the press wheels compacts the soil around it. The module/unit can have separate traction/drive wheel (Figs. 7.33-7.35) or the compaction wheel can also be used as drive wheel (Figs. 7.39 & 7.41). The seedlings are generally raised in conical plug/protrays (Figs 7.40 & 7.41) so that these have V-shaped root ball, for standing upright in furrow. The seedling trays are kept horizontally in self propelled (walk behind) transplanters (Figs. 7.42 & 7.43) and kept in hanging position for tractor operated transplanters (Figs. 7.40 & 7.41). The transplanter in Fig. 6.34 is for bare root seedlings. The minimum spacing between 2 modules or 2 rows can be 45 cm when the modules are fitted on one frame bar (Figs 7.35, 7.39 & 7.40). The spacing between rows can be reduced to 23 cm, when the modules are fitted in tandem in 2 rows. The spacing between seedlings is adjusted by adjusting the rotational speed of feed cups in relation to traction wheel. The seedlings are transferred to the furrow or ridge by the transfer mechanism or through gravity using the chute. The soil is compacted around the seedlings using press wheels. The details of rotary cups, drive mechanism, furrow opener and compaction wheels are shown in Figs. 7.37 & 7.38. The walk behind rotary cup transplanter with a holder and release linkage mechanism to plant the seedlings in soil is under testing at Central Institute of Agricultural Engineering, Bhopal Fig. 7.43.

Fig.7.33. Rotary cup seedling transplanter
(Holland transplanter, USA)

Fig.7.34. Rotary cup seedling transplanter with seedling protrays
(Mechanical transplanters, USA)

Fig. 7.35. Two Row Transplanter
http://www.opico.co.uk

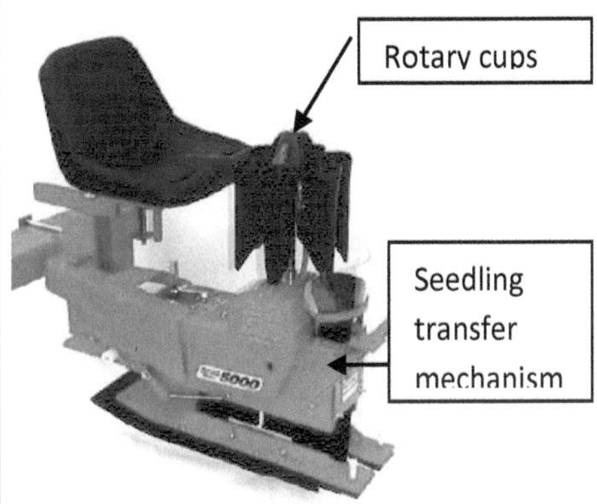


Fig.7.36. Module of single row rotary cup transplanter
http://www.opico.co.uk

Fig.7.37. Rotary cups and driving mechanism
(Holland transplanter, USA )

Fig.7.38. Furrow opener and compaction wheels
(Holland transplanter, USA)

Fig. 7.39. Four row rotary cup transplanter (bare root seedlings)
http://www.lannenplantsystems.com

Fig.7.40. Two row rotary cup seedling transplanter in operation
(Holland transplanter, USA)

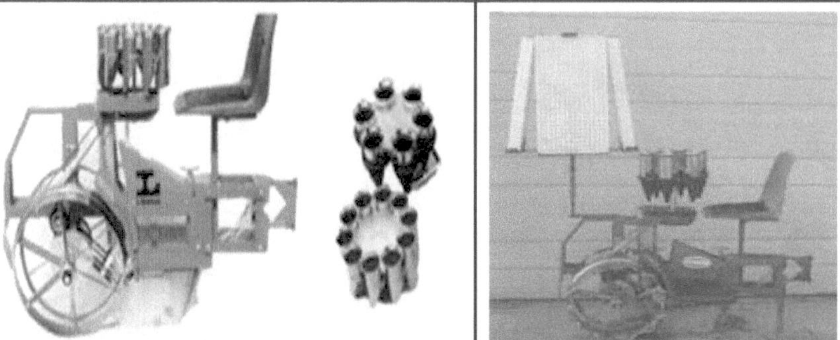

Fig. 7.41. Rotary cup transplanter using compaction wheel as drive wheel for rotating cups and a patented spike belt system to deliver the plant to the packing wheels. The spike belts do a better job of keeping the plants upright during soil packing
Piedmont Tobacco Equip. Co. 5870 Nc Hwy.268 Dobson, NC 27017
http:///www.marketfarm.com/lannenRTLG.com

Fig.7.42. Self propelled (walk behind) rotary cup transplanter. Seedlings are placed manually in the rotary cups which drops these in opening type holder cum planter mechanism
http://www.afftis.or.jp/nouki/ki3b3h.htm

Fig. 7.43. Testing of walk behind rotary cup transplanter
(Dae Dong Engineering Co., Ltd, South Korea) at Central Institute of Agricultural Engineering, Bhopal

## (e) Automatic transplanter for plug type seedlings

The walk behind automatic seedling transplanters are shown in Figs. 7.44 & 7.45, riding type in Figs. 7.47-7.52 and tractor operated seedling transplanters are shown in Figs. 7.53-7.55 for protray grown seedlings. The system is similar to Japanese rice transplanter. The seedlings are fed to the transplanter in protrays (Figs. 7.44-7.47& 7.53-7.55), individual row of seedlings are pushed out of protray cells (Fig. 7.46) and the seedlings are picked by the plug holders and kept on a horizontal conveyor belt. These are then transferred / held in vertical belts and transferred to two rubberized rotating discs. The seedlings held in the disc are released in the soil at the required depth in furrow made by the rotating discs. The soil is compacted around the seedlings by the compaction wheels. One walk behind seedling transplanter with rubber disc for transplanting the seedlings is shown in Fig. 7.57. The systematic operation of the transplanting mechanism is shown in Fig. 7.46. The operator feeds the seedling protray in to the system and removes the empty seedling protray from the system.

Fig. 7.44. Walk behind automatic transplanter
Visser International Trade & Engineering BV, S- Gravendeel, Holland
www. visseite.com

Fig.7.45. Walk behind automatic transplanter is in operation
www. visseite.com

The walk behind vegetable transplanter is 2 row transplanter, with row spacing of 45 to 100 cm., plant spacing of 5.8 to 52 cm and field capacity 0.14 to 0.2 ha/h. The tractor operated 8 row automatic transplanter(Figs. 7.53-7.55) has 1 ha/h capacity and needs 2 labourers for feeding seedling trays into machine and picking vacant trays out. The crop transplanted by tractor operated seedling transplanter is shown in Fig. 7.56. The tractor pulled vegetable planter (Fig. 7.53-7.56) (Minoru model VPTC-4) is specially developed for the European market because of the great demand for tractor pulled planters. The capacity is 240 plants per row per minute (around 45,000 plants per hour).

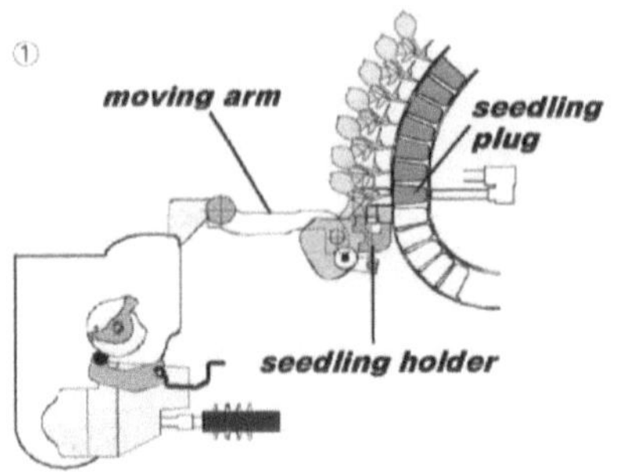


The moving arm which has the seeding holders at the top end raises to the uppermost position, and places the seeding holder in front of a seeding plug.

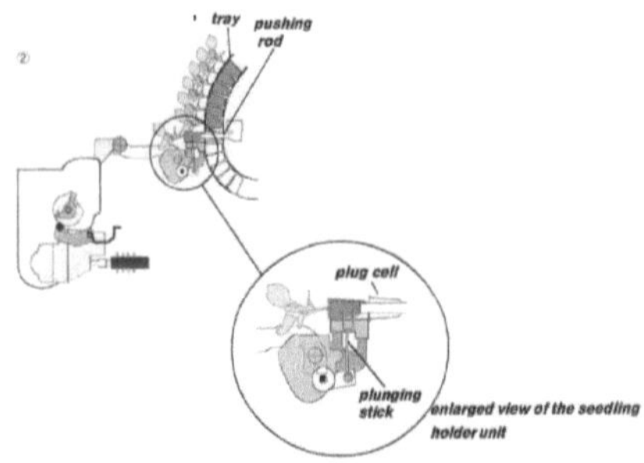


The pushing rod enters the plug cell from the bottom of the tray, pushing the seedling plugs out and into the slot of the seedling holder. Then the plunging stick fixes itself to the flank of the seedling plug on to the seedling holder.

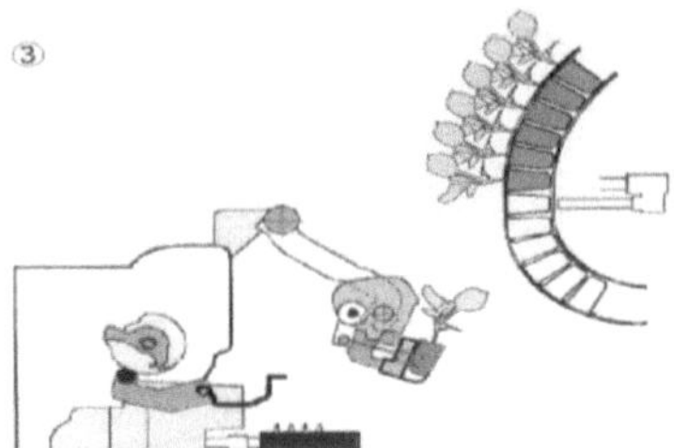


With the seedling plug inside, the seedling holder moves down, rotating at the end of the moving arm, thus changing the direction of the seedling plug.

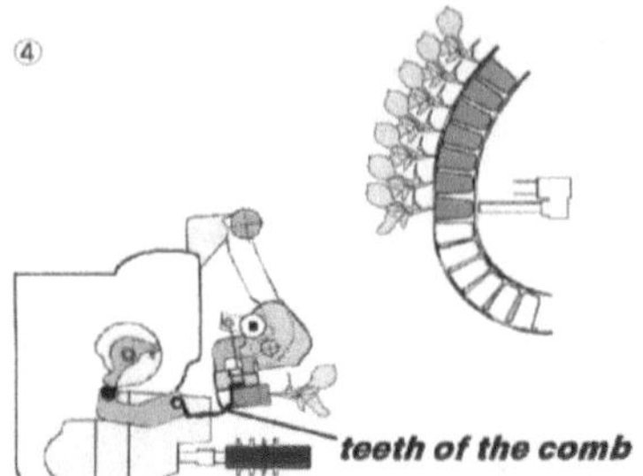


When the seedling holder moves down close to its lowermost position, the plunging stick comes off, and the teeth of the comb enter the seedling holder from behind, pushing the plugs out of the seedling holder

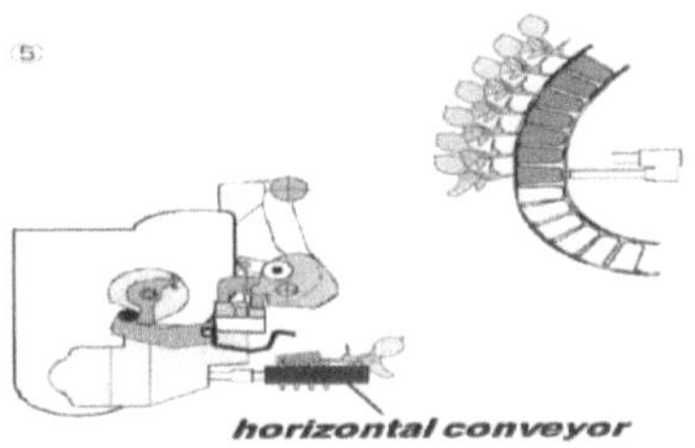


By bending itself forward , the teeth of the comb slap the seedling plugs down onto the horizontal conveyor belt.

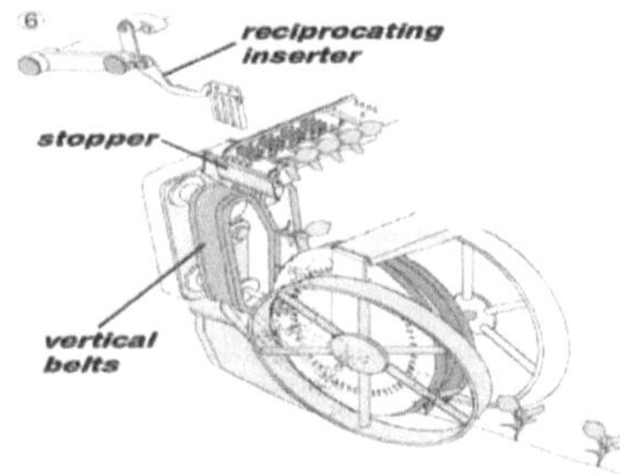


Placed on the horizontal conveyor belt, the seedling plugs are carried outwards and drop down at the of the horizontal conveyor belt where they are suspended temporarily by the stopper. The reciprocating inserter comes down onto the seeding plugs, thrusting the plugs further down to the merging interface of a pair of vertical belts, each one of which is driven oppositely.

MECHANISM OF THE MINORU TRANSPLANTER SYSTEM

Fig. 7.46. The systematic operational details of the automatic transplanting mechanism

www. visseite.com

Three models of riding type fully automatic transplanters have been developed at Institute of Agricultural machinery, Japan (Konosuke TSUGA,2000). Two prototypes are suitable for cell mold seedlings (Figs. 7.48-7.50) and one prototype is suitable for paper pulp mold cell pot seedlings (Figs. 7.51 & 7.52). These prototypes transplant 2 rows at a transplanting speed of 60 cells/row/min, with seedlings fed automatically. The transplanter works at a speed of 0.33 to 0.39 m/s (0.9 to 1.1 km/hrs) and transplant about 0.1 ha/h. The transplanters and the plug trays are manufactured commercially by different firms. The transplanting accuracy is above 97%.

The riding type machines handle mainly vegetables such as cabbage, Chinese cabbage, and lettuce with cell mold seedlings and pulp mold cell pot seedlings. The number of cells (pots) per tray is 128 and 200, standard trays can be used. Riding-type machine consisting of a 2-row planting device. The planting device is composed of the seedling feeding and transplanting sections. The vehicle has a 4 wheel-drive system and is well adapted to slopes. The standard adjustment range of the tread (distance between the right and left wheels) is 90 to 120 cm.This riding-type machine is suitable for relatively small fields. The models for cell mold seedlings (Fig.7.48 & 7.50) pull out seedlings one at a time from the tray using a seedling take-out claw of the holding type and discharge seedlings into the bill-type opener. The opener along with the claw, transplants the seedlings to the tamped/compacted surface of the ridge, and the tamping ring (compaction wheel) at the rear presses soil from both sides to cover the seedlings. The automatic-planting-depth control unit and the ridge unit allow stable transplanting with the use of hydraulic devices. The transplanter has 4-wheel-drive system and operated by a diesel or gasoline engine. The model for pulp mold cell pot seedlings holds seedlings that are raise in paper pot (pulp mold) and transplants them continuously by separating them into individual hills/ seedlings. It allows the transplanting of seedlings with a wide range of sizes over a long period of time and is suitable for transplanting regardless of the degree of growth of the individual seedlings.

Fig. 7.47. Riding type transplanter transplanting onions
www. visseite.com

Fig.7.48 Fully automatic transplanter Model PR2 for cell mold seedlings
Yammer Noki, Co., Ltd, Ogoori, Fukuoka http://ss.jircas.affrc.go.jp/engpage/jarq/34-

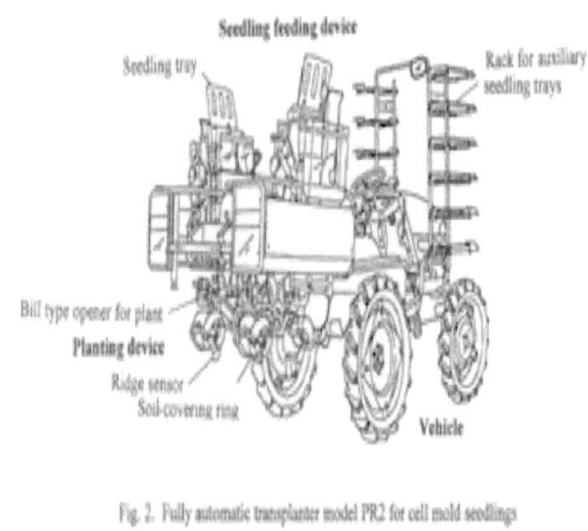


Fig. 7.49. Fully automatic transplanter Model PR2 for cell mold seedlings-detailed drawing
Yammer Noki, Co., Ltd, Ogoori,Fukuoka
http://ss.gircas.affrc.gu.jp/engpage/jarq/34-1

Fig.7.50. Fully automatic transplanter model SK20 for cell mold seedlings
Kubota, Co., Ltd., Takeda, Kumamoto http://ss.jircas.affrc.go.jp/engpage/jarq/34-1

Fig. 7.51. Fully automatic transplanter Model PVR 200 for pulp mold cell pot seedlings-
Iseki Noki, Co. Ltd., Kazuno Akita http://ss.gircas.affrc.gu.jp/engpage/jarq/3 4-1

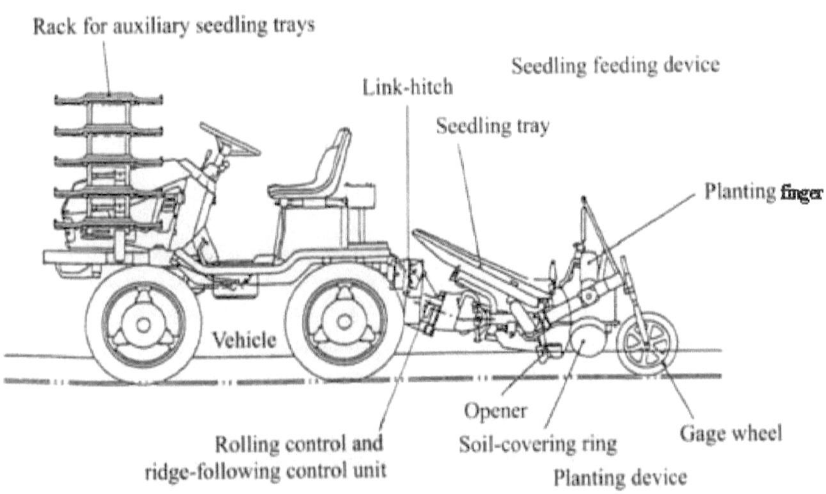


Fig. 7.52. Fully automatic transplanter Model PVR 200 for pulp mold cell pot seedlings
Iseki Noki, Co. Ltd., KazunoAkitahttp://ss.gircas.affrc.gu.jp/engpage/jarq/34-1

The test conditions for evaluation of all the 3 Japanese models (Figs. 7.48-7.52) included an adaptable ridge height of 0 to 25 cm, a hill space of 25 to 50 cm, and a ridge space of 45 to 60 cm, in which the model for cell mold seedlings (Figs. 7.48-7.50) allowed zigzag planting on 2 rows.

The seedling spacing, planting depth and soil covering can be easily adjusted and 60-70 seedlings can be sown per minute per row. The speed of two row transplanter is 1.2 to 1.4 km/h for row spacing of 60 cm and seedling spasing of 30 cm and field capacity is 0.1 ha/h. These can transplant seedlings at a slope of up to 10° with misplanting of up to 3%.

It was found that when 12 to 14 trays were mounted, it was possible to transplant 2,400 or more seedlings without requiring the operator to get off the vehicle.

Fig. 7.53. Tractor operated onion transplanter
www. visseite.com

Fig. 7.54 Tractor operated vegetable transplanter
www. visseite.com

Fig. 7.55. Tractor operated vegetable transplanter is in operation
www. visseite.com

Fig.7.56. Crop transplanted by tractor operated transplanter
www. visseite.com

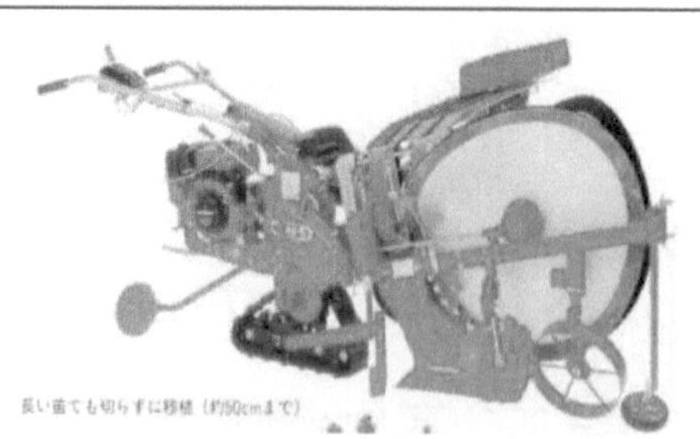

Fig. 7.57. Walk behind transplanter (with rubber wheels) to make furrow and release the plots (Welsh Onion Transplanter)
Journal of The Japanese Society Of Agricultural Machinery 1998 Vol. 60(4) http://www.afftis.or.jp/nouki/ki3b3i.htm

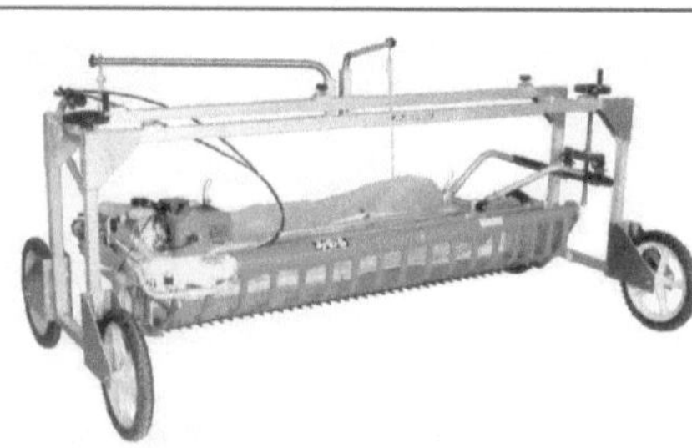

Fig.7.58. Mower for onion seedlings
(Visser International Trade & Engineering BV, Holland) www. visseite.com

## 7.4 Onion transplanter

Onions are grown in protray with a cavity of 1x1x1cm with 448 cavities and about 1500 protrays are to be raised in greenhouse for one hectare. These are to be moved to the height of 8 to 12 cm by mover (Fig. 7.58). The four row semi automatic transplanter with rotary cups (Figs 7.33 to 7.41) can be used for row spacing of 40 cm when the transplanting units are in line and 20 cm when the transplanting units are in tandem at plant spacing of 10-15 cm. The row spacing can be 24/27/30 cm and plant spacing can be 9.3/13.2 cm when planted with modified automatic transplanters. These can also be transplanted using punch planter /pegger (Figs. 7.23 & 7.24). The onion transplanter (Fig 7.24) is 6 row punch type transplanter with row spacing of 17 cm and seedling spacing of 10cm. This for transplanting on raised bed and operates at 0.4 km/h. The field capacity is 0.3 ha/day. The crop stand is shown in Fig 7.24a.

# *Chapter 8*

# Weeding, Mulching and Land Scape Maintenance

Weeding is generally done to uproot the unwanted plants and expose the basin of the soil to aeration. A different methods of Mechanical weeding is generally done by manual, animal, power tiller or tractor drawn implements. For chemical weeding sprayers are used to spray the chemicals to kill the weeds. Flame weeders use the heat energy to kill the weeds by burning. Different type of hoes of short, medium or long handles are used manually to dig out the weeds . Most of the soil working tools like spade, kodali and rakes are also used for weeding . The animal, power tiller and tractor operated weeders are also available.

## 8.1 Hand Tools

### Hand hoe (Khurpi)

*Khurpi* (hand hoe) is the most commonly used hand tool for weeding. It has a sharp hardened steel blade and a handle and this is used in squatting position. The *khurpi* has a handle for easy and safe operation. The design of khurpi blade and handle (Fig 8.1) vary from area to area and as per it use The overall length of khurpi is upto 35 cm, blade width varies from 6 to 15 cm and length of the blade varies from 15 to 25 cm. The weight of the khurpi is 300 to 700 gm. The wide blade khurpi is most commonly used in north India for weeding and digging. The field capacity for weeding is generally up to 0.025 hectare/day .

| | | | | |
|---|---|---|---|---|
| 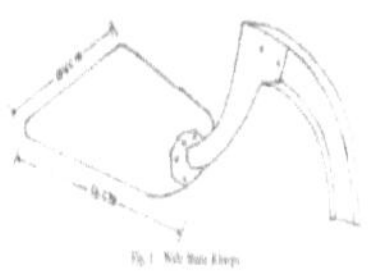 |  |  |  | 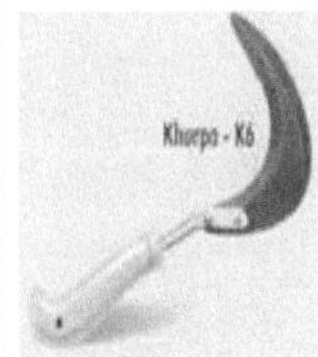 |
| CIAE | J.S.P. Lawn Mowers | Golden Industries | Tata agrico | Karimulla Group |

Fig. 8.1 Khurpi of different types

**Scraper**

This is similar to khurpi except that tha blade is in line of handle and is wider at the sharp end (Fig 8.2). This is mostly used for weeding in european countries.

| | |
|---|---|
| 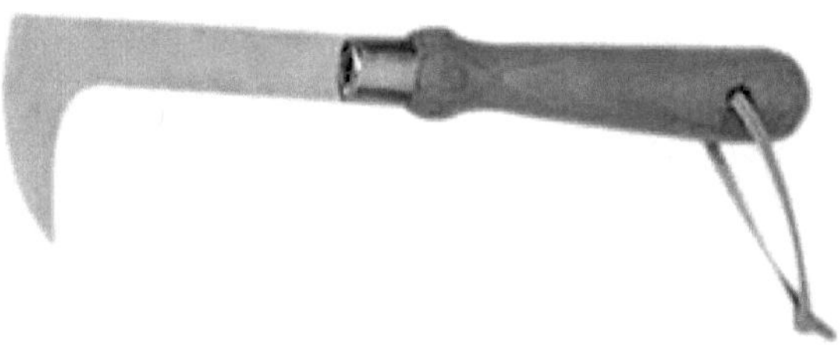 |  |
| RAMCO | Falcon |

Fig. 8.2 Scrapers of different types

**Weeding hook**

Weeding hook is used for loosening the soil and cutting and uprooting of the weeds (Fig. 8.3). The stem or root of weeds is generally held in the blade and the weed plant is uprooted by using the weeding hook as a lever. Its overall length is up to 31 cm, blade width 3.7 cm and thickness is 3mm. The weight of weeding hook is about 170 to 200 gm.

| | | | | |
|---|---|---|---|---|
|  |  |  |  |  |
| Draper | http:// jspgardentools.com | RAMCO | Fiskars | Garden Claw Ludhra Over-seas |

Fig. 8.3 Weeding hook of different types

**Sickle**

The sickle has a sharpened curved blade and a wooden handle (Fig. 8.4). The curved blade have serrations in some of the designs. It is generally used to cut the weeds above the ground. In some designs, the front portion of the blade is sharpened to use it as a khurpi. The front end of the blade may can been split to use it as a weeding hook.

| | | | |
|---|---|---|---|
|  | 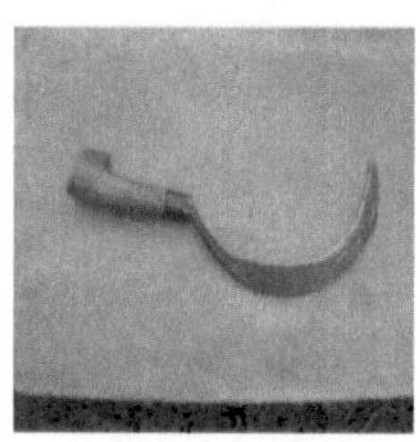 |  |  |
| IIHR | IIHR | abmtools.com | http://www.ergonica.net |

Fig. 8.4 Sickles of different types

**Scythe / Garden Sword**

Scythe has a hardened and sharpened steel blade of about 75 cm length and a wooden handle (Fig. 8.5). The front portion of scythe may be bent as shown in Fig. 8.5. The scythe is used to cut tall grasses and unwanted shrubs by hitting these with the blade. The leg guard should be worn for safety while working. The scythe should be kept in a leather cover during storage.

| | | | | |
|---|---|---|---|---|
| 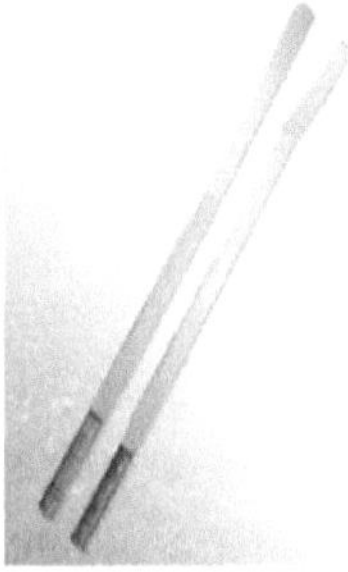 Mehar & Mukand Tools |  http://jspgardentools.com |  |  |  |

Fig. 8.5 Scythe of different types, scythe cover and leg guard

**Hand hoe (long handed)**

Hand hoes with about 1.5 to 2.0 meter handle are mostly used for weeding (Fig 8.6). The straight blade hoes are operated in the standing posture by pulling action. The blade pulverizes the top soil and cuts and uproots the weeds in between the rows of the crop. The weeds get buried under the soil and create a mulch.

The dutch hoe has a blade, sharpened at both edges and can be used by pulling and pushing in soil (Fig 8.6). The hand hoe with V blade creates a furrow in between the two rows of the crops and also does the earthling work for the crop in addition to weeding. Three tined hoe has 3 tines at the end of the handle. The tines uproot the weeds and pulverize the soil during the pulling action. This is the most widely used hand hoe for weeding and interculture of horticultural crop. The blade of multipurpose hoe generally has two ends. Of which one end is usually flat while the another is a triangular shape or has prongs for raking or collection of weeds.

| | | | |
|---|---|---|---|
|  Straight blade hand hoe http://kisan76.co.in |  Straight blade hand hoe (Dutch Hoe) Ludhra Overseas |  Straight blade hand hoe |  Straight blade hand hoe (Dutch Hoe) http://kisan76.co.in |

| | | | |
|---|---|---|---|
|  |  |  | 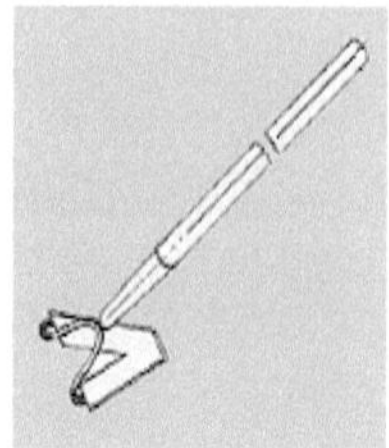 |
| Multi purpose hand hoe http://kisan76.co.in | Multi purpose hand hoe http://kisan76.co.in | Three tined hoe http://kisan76.co.in | V- blade hand hoe CIAE |

Fig. 8.6 Hand hoes (long handed) of different types

**Wheel hoe**

The wheel hoe is available in different designs and sizes (Fig.8.7). This generally consists of a wheel, frame, handle and soil working tool. The height of handle is suitable for easy operation of the hoe in a standing position. Generally the weeding/soil working is done by push and pull action of the hoe. One additional labour can also pull the hoe in front by using a rope. The soil working tool may have a straight blade, reversible shovel, a set of shovels or a miniature ridger etc. The wheel hoe can be used for weeding, interculture or ridging of the row crops. The maximum working width is 30cm and working speed 1 km/h

| | | | |
|---|---|---|---|
|  | 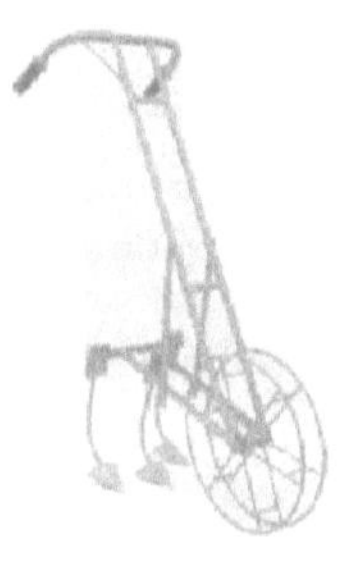 |  |  |
| Multi purpose wheel hoe http://www.marketfarm.com | Wheel hoe PAU | Wheel hoe http://jspgardentools.com | Wheel hoe CIAE |

Fig. 8.7 Wheel hoes of different types

**Hand cultivator**

This consists of a wooden handle and a set of prongs (Fig. 8.8). The number of prongs may vary from 3 to 7 and these are bent and sharpened at the working end. Generally this is used for weeding, interculture or soil mulching, while working in squatting position.

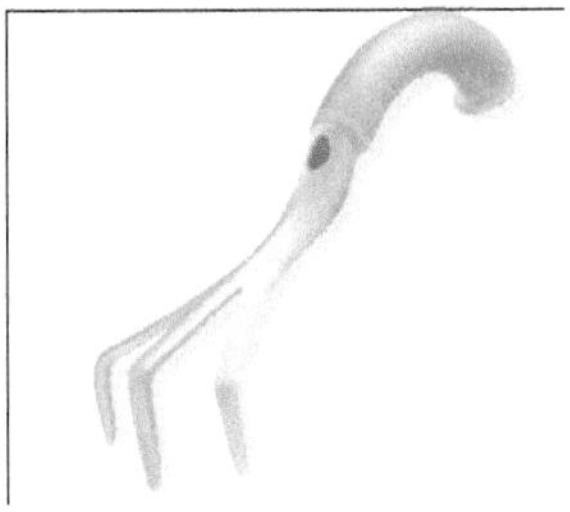

Zwicker Land Design

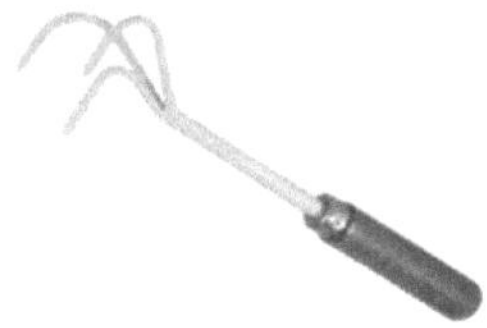

http://waveimpex.com

RAMCO

Fig. 8.8 Hand cultivators of different types

## 8.2 Powered Hand Tools

**Brush cutter**

Brush cutter consists of a light weight petrol engine, handle, clutch, drive mechanism, rotary blade and shield for rotary blade (Fig 8.9). The rotary blades rotate at 4000 to 5000 rpm. The brush cutter is used for cutting the tall grass and trimming the hedges.

http://www.stihl.com.au

IIHR

Fig. 8.9 Brush cutter

## 8.3 Powered Implements

**Rotavator/Rotary weeder / Self propelled weeder**

The power tillers and self propelled weeders (Fig. 8a-8d) of 1 to 8 hp are used to operate the rotavator for weeding. The engine operated rotary weeder (fig 8.10c) is to be pushed manually like wheel hoe for weeding the row crops. The small horse power tillers (Fig 8.10 e and f) are useful for basin cleaning of fruit crops. The higher hp rotavator is useful for weeding row crops. The engine operated row weeder (Fig 8.10g) is a engine operated (self propelled) weeder/cultivator and can be used to weed the row crops at a speed of 1 to 2 km/hr and has a field capacity of 0.06 to 0.10 ha/hr.

a. Power tiller with rotavator
http://www.vsttillers.com

b. Engine operated rotary weeder
http://www.weedes.jp

c. Engine operated rotary weeder
ciifad.cornell.edu/ sri/2004.html

d. Power tiller with rotavator
http://world.honda.com

e. Power tiller with rotavator
http://www.bolens.com http://www.mtdproducts.com

Fig. 8.10 Power tiller or self propelled weeders

## 8.4 Tractor Operated Weeders

### Cultivator

The tractor operated cultivator (Fig. 8.11) can be used for weeding or intercultivating the raised bed crops or widely spaced row crops. The shovel of the cultivator may be changed to V shape or sheep foot shovels. The arrangement of the showels is made to cover the required row spacing without damaging the crop. These implements can weed the crop upto 0.5 meter in height.

### Raised bed crop weeder

The raised bed crop weeder developed at Indian Institute of Horticultural Research is fitted with 2 ridger bottom for weeding in furrows. The shovels provided on times weed the area in between crop rows (Fig.8.12).

### Rotary Weeders (Tractor operated)

The tractor operated rotary weeder (Fig. 8.13) is developed at the National Research Centre for Soyabean, Indore for row crops. Fig 8.13a shows the flat blades fitted in front of rotary weeders. These blades uproot the weeds in the rows. The pto operated rotor drums with pegs are provided behind the blades for lifting and turning the weeds (Fig 8.13b). The crop shields are provided around the rotor drums, to avoid damage to the crops (Fig 8.13c). This implement is well suited for the crop is sown on raised beds or widely spaced crops (Fig 8.13d).

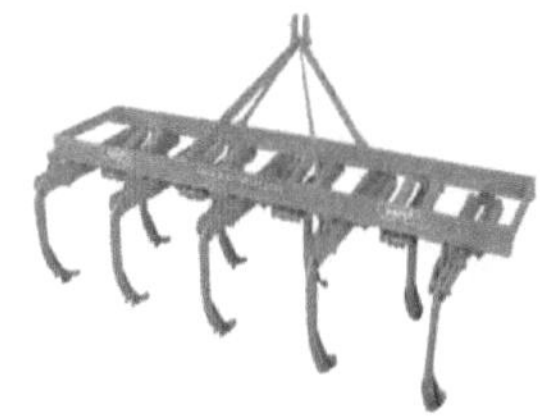

a. Cultivator

b. Cultivator with V shape shovels

eker-mak.com.tr

c. Arrangement of shovels for wide row

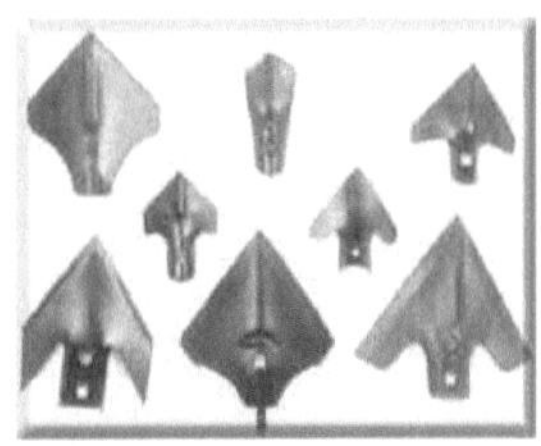

Fig. 8.11 Cultivators & different type of shovels and arrangements

Fig. 8.12 Raised crop weeder

IIHR Bangalore

Different versions of rotary weeders without crop shields are shown in (Fig. 8.14). The tractor operated weeders shown in (Fig. 8.15) are provided with rotary drum and crop shields. These weeders are suitable for weeding of spaces in between rows and do not weed the space in between plants. The field capacity of a 1.8 meter width weeder is 0.2 to 0.4 ha/h.

Blade harrow for row crop Weeding

b.Blade harrow with rotary blades for row crop weeding

c.Blade harrow with rotary blades and crop shield for row crop weeding

d.Blade harrow with rotary blades and crop shield for row crop weeding in operation

D.V.Singh,NRC for Soyabean, Indore

Fig. 8.13 Tractor operated rotary weeders

Fig. 8.14 Tractor operated rotary weeders

marketfarm.com

TNAU Coimbtore

http://www.gardenorganic.org.uk/assets/organicweeds/brush-weeder.jpg

Fig. 8.15 Tractor operated rotary weeders with crop shield

**Reigi weeders**

These are operated by the pto of the tractor (Fig. 8.16). These have horizontal disc fitted with vertical spring loaded pegs. The disc and pegs rotate horizontally in the space in between rows and uproot the weeds. The position of the disc can be controlled/oscilated by the operator sitting on reigi weeder/seat. In such weeders the space in between plants can also be weeded in addition to the space in between rows. All the components and operation of the reigi weeder are shown in Fig 8.16. The weeders uproot and mulch the weed plants thoroughly. These can work upto 1 to 2 km/hr and have field capacity of 0.1 to 0.4 ha/hr depending upto the size and type.

**Tractor mounted weeders cum basin cleaners**

The tractor rear mounted implements (Fig. 8.16a) are for weeding the inter row space and basin cleaning in the row. Generally it is cultivator with pulverizer roller for weeding the inter row space and a blade harrow or disc harrow or rotavator or a rotary harrow is mounted on both sides of the cultivator. The side implements are for basin cleaning and weeding the space in between the plants. Sensors are fitted on both sides in front of these implements to sense the plants or posts / pillars/ angle iron or other obstructions. The side implements retract from the obstructions or plants once these are sensed by the sensor and the space in between the plants is weeded without damaging the plants. The side brushes are provided in some cases for levelling the space in between the plants. The one side offset rotary harrows/rotavator/disc harrow with sensors are also available for lesser horsepower of tractors. The side rotary harrows can also be replaced with rotary brush for levelling the fields. In some of the cases only rotary harrows are provided on bothsides and no cultivator is provided for uprooting the weeds in between. The cultivators with adjustable width are also available for uprooting the weeds in between rows. These are generally tractor mounted. These are suitable for 50 hp and more hp tractors with cultivator, pulverizer roller and two rotary harrows.

| | | |
|---|---|---|
|  | 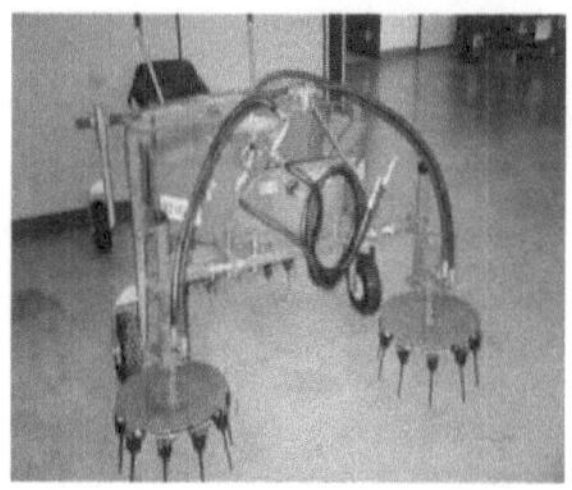 | 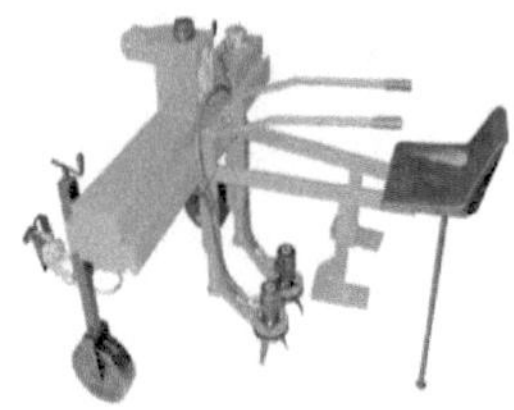 |
|  |  |  www.univerco.net/ en |

Fig. 8.16 Tractor operated rotary weeders (Reigi type)

| | | |
|---|---|---|
|  | 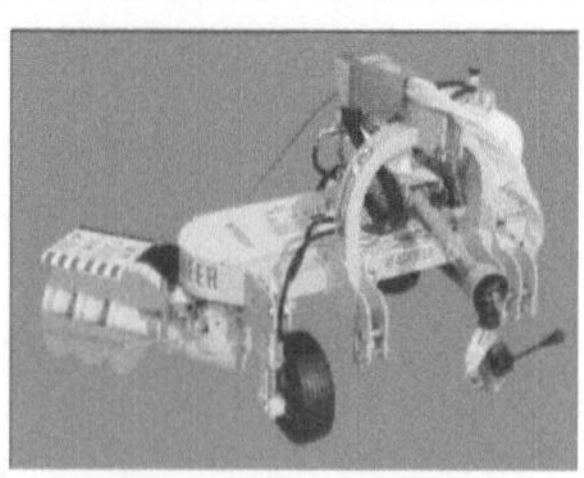 |  |
| Basin cleaner - Cutivator cum pulverizer roller with two blade harrows (fitted with plant sensor) | Basin cleaner – offset disc harrow for one row (fitted with plant sensor) | Basin cleaner – rotavator for one row (fitted with plant sensor) |

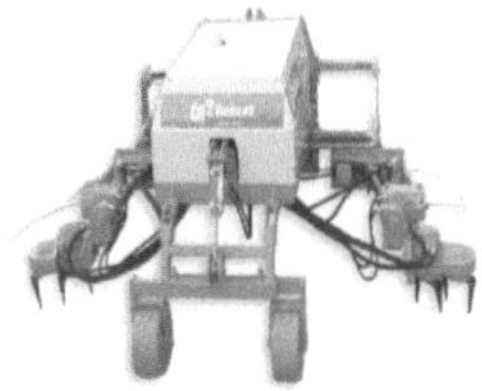

www. rinieri.com

Basin cleaner ( two row fitted with rotary harrows with plant sensors)

Basin cleaner (offset single row fitted with rotary harrow with plant sensor)

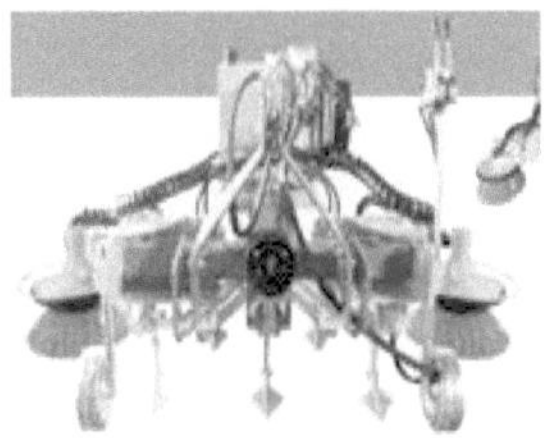

Basin cleaner ( with cultivator and rotary brushes for levelling the basin)

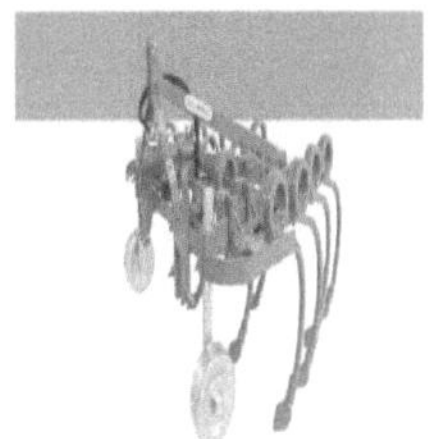

Variable width cultivator for basin cleaning

Fig. 8.16a Tractor operated weeder cum basin cleaner

## 8.5 Flame Weeders

Flame weeding is killing the weeds by heat, by bringing these in contact with flame or hot plate. The manual operated flame weeder consists of gas cylinder, tubes, regulator, handle, lighter, flame burner, crop shield and hot plate. The gas cylinder is mounted on the back of the operator and the flame or hot plate is brought close to the weeds to destroy these. The weeds get wilted or charred and the insects also get killed. These weeders in Figs 8.17a and 8.17b can weed in between rows and plants. The weeders in Figs 8.17c and 8.17d are provided with transport wheels and can weed in between rows only. Tractor operated flame weeders are shown in Fig 8.18. One burner with shield is provided for each row. The flame weeders are mostly suitable for organic farming and do weeding at faster speed than mechanical weeders at less cost.

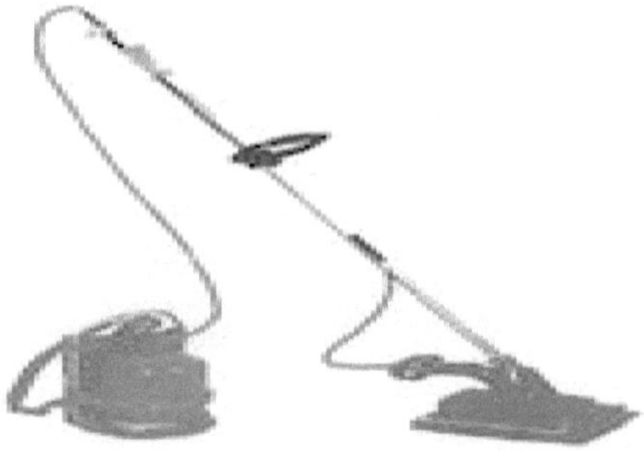

a)attra.ncat.org/attra-pub/flameweed veg.html

b)http://www.auri.org/proproj/flamewee.html

c) http://attra.ncat.org/attra-pub/flameweedveg.html

d)http://www.chemfree-weedcontrol.com/

Fig. 8.17 Manual operated flame weeder

http://www.flameengineering.com/Agricultural_Flamers.html

http://www.flameengineering.com/Agricultural_Flamers.html

Fig. 8.18 Tractor operated flame weeder

## 8.6 Lawn Mowers

Lawn mowers are generally used to trim the lawns or tall grass. These are of two types i) Reel type ii) Rotary blades. These are discussed in details in lawn maintenance.

## 8.7 Roto Slasher

This is a tractor operated implement (Fig 8.19). The rotor blades are provided with the safety shield and rotate horizontally through pto drive. The rotor blades The tail wheel may also be provided to control the height of the cut. The rotor cuts the grass and weeds and makes a mulch of these. The roto slasher is useful for controlling weeds in fruit crops and for mowing the lawns.

b. http://www. brownmanufacturing.us

c. IIHR

Fig. 8.19 Tractor operated roto slasher

### 8.8 Plastic Mulch Layer

Plastic mulch film of 0.03mm thickness is generally put on the raised bed (Fig 8.20). The seeds or seedlings are planted after then covering the bed with film. The use of plastic mulch reduces weed emergence and These water requirement for the crop because of less evaporation. The fertilizer use efficiency is increased and the crop yield is also increased.

The plastic mulch films are available in rolls of 1.2 meter width in India. Their length of is generally 100 meter and width of 1.1, 1.6 and 2.1 meter. The mulch layer is operated by power tiller or tractor. The film roll is kept on the steel rollers or held on the unwinding roller. It roll rotates and gets unwound when the film is pulled and laid on the raised bed by stretching wheels and closing discs. Some of the mulch layers also prepare the raised bed on which the seedlings are transplanted. The planting of seeds or transplanting of seedling is generally done by throught the holdes in the plastic mechanical punch/dibbling stick or gas heated punch. The drip irrigation line is also laid on the bed before laying the mulch film. The plastic mulch films are available in different colours and usually have a life of one cropping season. The plastic mulch remover (Fig. 8.21) scraps the film from the bed and the film can be folded for disposal. Some film removers have burner to burn the film for disposal. In India as the film laying and removing is done manually, the film can be used for 2 to 3 cropping seasons.

a. Tractor operated three bed mulch layer.
http://marketfarm.com/

b. Tractor operated single bed mulch layer.
http://marketfarm.com/

c. Tractor operated single bed mulch layer.
http://marketfarm.com/

d. Tractor operated single bed mulch layer with drip line laying.
Agribiz Corporation

e. Self propelled mulch layer.

f. Parts of mulch layer.
http://marketfarm.com/

g. Parts of mulch layer.
http://marketfarm.com/

h. Gas heated punch for mulch film
http://www.renaldo.org

i. Transplanter with gas heated punch
http://www.renaldo.org

j. Raised bed with plastic mulch
http://marketfarm.com/

k. Mulch remover
http://marketfarm.com/

l. Mulch Remover
http://marketfarm.com/

Fig. 8.20 Tractor operated / power tiller operated plastic mulch layer, transplanter & mulch Removers

## 8.9 Straw Mulch Layer

In cold countries, the seeds are sown in the fields by seed drill. The seed bed is covered with straw mulch, to avoid frost on the newly germinated seeds and to increase the soil temperature. The bales of straw are put into the machine, and cut into small pieces. These are then spread on the seed bed by aspirator/air blower (Fig 8.21 and 8.22). The straw mulch is then fixed on the seed bed by mulch fixer a tractor drawn disc harrow, with 180°C gang angle which pushes the straw in the soil. The demulcher is tractor operated rotor with brushes to remove the straw from the rows of plants (Fig. 8.21c).

a. Straw mulcher for Strawberry
http://www.marketfarm.com

b. Round Bale for Straw Mulching
http://www.marketfarm.com

c. Kasco Strawberry DeMulcher
http://www.marketfarm.com

Straw Mulcher with aspirator blower

Straw Mulcher with 9-15m throw

Mulch Fixer

Fig. 8.21 Tractor operated straw mulch layer and mulch remover
http://www.emeraldseedandsupply.com/

Straw Mulcher with trailor for transport of straw

Feeding Straw to Mulcher

Spreading the Straw

Fig. 8.22 Tractor operated aero straw mulch layer and mulch fixer

## 8.10 Tractor mounted weed remover cum windrower mulcher

These tractor rear mounted implements (Fig. 8.23) cuts/ shreds the grass / weeds and windrow these in two rows in line with plants. The working width can be 1.5 to 3.0 meter and the power requirement is 35-45 hp tractor.

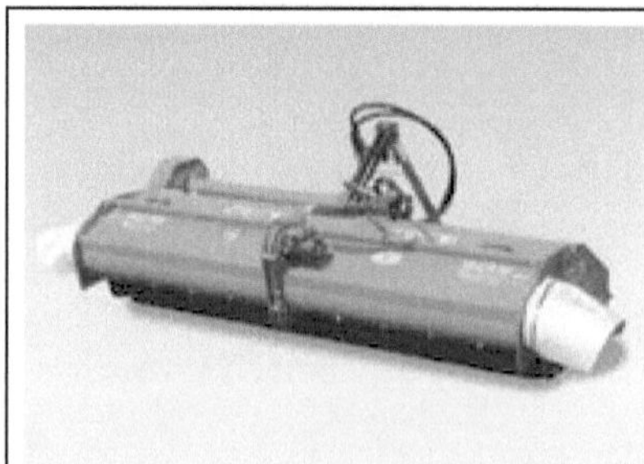

Fig. 8.23 Tractor mounted weed remover cum windrower mulcher

## 8.11 Shreddercum Mulcher

These tractor rear mounted machines (Fig.8.24) are used to shred the windrowed pruned vines. These can shred the dried vines with rotary hammers and mulch these in the field. These are operated by the tractor PTO and 40 -50 hp tractor is required to operated these. The working width of the require machine is 2 meters.

Fig. 8.24 Shredder for vines

a. Leaf mulcher

b. Shredding the leaves

c. Collecting the shredded leaves

Fig. 8.25 Leaf mulcher

www.leafmulcher.com

The dried leaves can also be cut in leaf mulcher (Fig 8.25) and spread over the seed bed or recently transplanted seedlings for avoiding frost damage.

## 8.12 Landscape Maintenance

The major operations in landscape maintenance are lawn maintenance and plant pruning. The irrigation of lawns and plants can also be done by providing sprinklers and drippers.

The lawn maintenance is the weekly operation and is very laborious. The lawn aeration is required once in a year. The lawn planting is required in beginning and may be required where the grass has not established properly. The pruning of hedges, shrubs and tees is done at different intervals. Most of these operations can be mechanized to different levels depending upon the area under landscape. The following are main operations in landscape maintenance.

- Lawn maintenance
- Weeding of flower beds or plant basins
- Lawn Aeration
- Pruning
- Plant protection
- Irrigation

**Lawn planting**

The lawn planting and maintenance is discussed in this chapater as other items of land scape maintenance has been discussed in different chapaters

**Lawn planting**

Preparing the ground for lawn planting: Generally the lawns are planted by planting the grass cuttings. To prepare the bed for the lawn, the ground is ploughed up to 15 cm. depth (Kalbag 2012). If the existing soil is of not of good quality, then it is better to replace it with good fertile soil. For good drainage it is better to put about 3 cm. thick layer of brickbats or coarse gravely soil at the bottom. Over this give a layer of soil and farmyard manure, thoroughly mixed, in proportion of 3 parts soil and 2 parts farmyard manure. All stones larger than 2 cm. are to be removed from this mixture. The uppermost layer for lawn surface should be made up by mixing thoroughly, equals parts of soil and farmyard manure. Stones larger than ½ cm. from the soil should be removed before mixing.

Using a rake levels the soil surface. By raking not only leveling is done, but, large stones are also removed. Large stones left near the soil surface are capable of damaging the lawn mower blades. After this, drench the site thoroughly with water. This will help the soil-manure settle well, without leaving large air pockets. When the soil surface turns drier the site is compressed by using a light lawn roller. If the soil is very wet, then it will stick to the roller. Unless rolling is done, the soil starts settling down unevenly, thus spoiling the lawn level.

Lawn planting using grass cuttings : It is advisable to water the bed regularly for a few days before the lawn is actually planted. In this period the weeds start sprouting. The weeds should be removed regularly. At the time of planting, the soil should be moist. Clumps of lawn grass cuttings should be planted at about 5 cm. distance. . Water immediately, and then every day for about two weeks. Water just enough to get the roots and soil moist. The perfect lawn can be prepared in 30 to 40 days.

Lawn planting using grass cuttings : After the site is comparatively weed-free, sod / tufts of lawn grass can be planted. The sod must be fresh with moist roots, weed and brown spot free. Water the lawn immediately and the daily till it is established.

**Sowing grass seeds:** Spread FYM and starter fertilizer and gently work it into the soil, before sowing the seed. Then, use either a handheld or regular spreader to sow the seeds for small lawns. For larger lawns self propelled or tractor drawn split seeder is to be used. This makes the grove in soil and sows the seed in the grove. Don't try to sow them with just your hands, as it will result in uneven coverage. After sowing, water until the soil is moist to about 2 to 3 cm. deep. Don't overwater, as seeds may got washed and collect into clumps. Water the lawn twice a day, just enough to keep the soil moist, for several weeks. As seedlings thicken, water the lawn once a day and then a couple of times a week. The grass should be ready to mow in about six weeks or so.

## Lawn Maintenance

The major lawn operations are edging, mowing and blowing and area required every week. The lawn aeration is done once in a year and lawn planting as per the requirement/need. The following are the main equipments for lawn maintenance:

### Edging

Edging is a vertical cut to clearly separate the lawn from a walk way or other paved surface, such as a concrete side walk or asphalt path. The half moon edger (Fig.-8.26) is pressed by foot at the edge of the lawn to remove the spread of grass on to the foot way / pathway. The power operated edger (Fig.-8.27 and 8.28) has a vertical rotating blade, with a petrol engine (1 to 2 hp) or battery or electrical motor. The blade of the edger is usually having 20 to 25 cm diameter and made from the cast or hardened steel. The stick edger (Fig.-8.27) is generally fitted with 25 cc/ 1.3 hp petrol engine and weighs about 6 kilogram and walk behind edger (Fig.- 8 .28) is generally fitted with 50 cc/ 2.5 hp petrol engine and weighs about 20to 25 kilogram. The electrical and battery operated are suitable for small lawns. Use of edger defines a clear separation between the lawn and the walk way, imparting a finished appearance that is neater than can be achieved by moving over the border of the lawn and walk way.

### Lawn Mowing

Lawn mower is for trimming the grass or lawn. In large lawns, the ride on lawn mowers, Lawn tractors or turf terracer with working width of 1.2 to 1.8 meter are used first. The lawn mower with 0.5 to 0.9 meter width is used to mow the area where turf terracer or lawn tractor cannot reach. The lawn trimmer or brushes cutters are having working width of 0.3 to 0.4 meter are used to mow near the walls and the places where other mowers could not reach. The details of different lawn mowers are as given below: The lawn mower can be provided to with grass catcher or it can be mulcher type to cut the grass in small pieces and spread those in lawn.

| | | |
|---|---|---|
| <br>Helen Yemm | 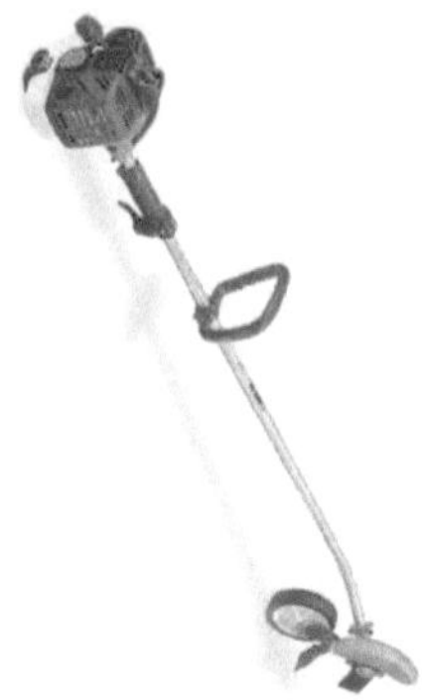<br>Tanaka Power Equipment |  |
| <br>Fig. 8.26 Lawn Edger | <br>Fig. 8.27 Stick Edger (engine operated) | <br>Fig. 8.28 Walk-behind Edger (engine operated) |

Lawn tractor: It is a regular tractor (from 25 to 35 hp) and a rotary lawn mower is fitted underneath this (Fig. 8.29). The rotary lawn mower has horizontal rotary blades to cut the grass. Usually it has 2 rotary blades. The working width is usually upto 1.8 meter, the operation speed is 5 to 10 km/hr. and the field capacity is 0.6 to 1.5 ha/hr.

http://www.toro.com

http://www.deere.com

Fig. 8.29 Lawn Tractor

Turf terracer: Turf terracer is rotary type mower with working width of 1.3 to 1.5 meter and fitted with 24-25 hp petrol engine. These can be ride on (Fig.-8.30), seat fitted on the machine) and having zero turning radius. In some of the models ride on platform is provided at the back of the machine. The operator can ride on the platform (Fig. 8.31) and need not walk. The ride on turf terracer work at the speed of 5 to 10 km/hr and have field capacity of 0.5 to 1.2 ha/hr. The walk behind turf terracer (Fig. 8.32) work at speed of 2 to 3 km/hr and field capacity is 0.2 to 0.4 km/hr.

Fig. 8.30 Ride on Turf Terracer

http://www.exmark.com

Fig. 8.31 Ride Platform Turf Terracer
http://www.exmark.com

Fig. 8.32 . Walk behind Turf Terracer
http://www.exmark.com

**Lawn Mower (Powered)**: These have working width of about 0.6 meter and are walk behind type. These can be rotary type or reel/cylinder type. In reel or cylinder type the grass is pushed on the fixed blade by the helical cylinder and gets trimmed. The rotary mowers have only powered rotary blade and rotated by engine or electric motor and driven by power source / self propelled type or pushed manually.

In most of the models the rear wheels are given power from engine and these are self propelled type. The reel type lawn mowers have electic motor or petro engine to rotate the reel and these are pushed manually. The power source is usually a petrol engine (2 to 7 hp) or electric or battery operated motor. The battery operated mowers are fitted with rechargeable battery of 12 or 24 volt. The field capacity can be 0.10 to 0.15 ha/hr.

http://www.hondasuperstore.info/

Fig. 8.33 Walk behind Powered Lawn Mower

**Lawn Mower (Push Type):** This is a reel type lawn mower . The reel is operated by the rotation of wheels while pushing the mower. The working width is 0.3 to 0.5 meter. The grass catcher is provided in front of the reel/cylinder. This type of lawn mower is suitable for small lawns in houses.

**Animal / Tractor Pulled Lawn Mower :** Animal drawn lawn mower is suitable for the areas where engine operated lawn movers are prohibited to control the pollution, as Taj Mahal, Agra. The gang type lawn mower can be pulled by animals or tractor.

http://www.jspgardentools.com

Fig. 8.34 Walk behind , push type, electric powered lawn mower

http://www.blackanddecker.com/

http://www.fromtherumbleseat.com

Fig. 8.34 a Walk behind , electric powered lawn mower

Fig. 8.34b Walk behind chordless or battery operated lawn mower

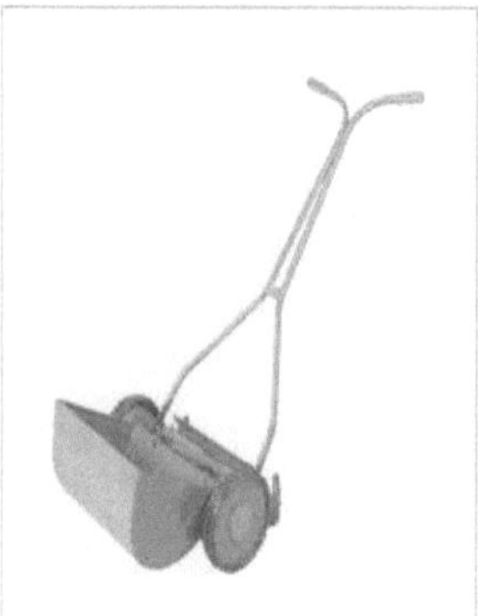

goodwillagrotech.com

Fig. 8.35. Manual reel rotation and push type lawn mower

http://www.greenplanet.in/lawn-mowers.htm

Fig. 8.36. Tractor / animal pulled, gang type, reel lawn mower

Fig. 8.36a. Animal pulled, reel rotation type lawn mower
http://www.7is7.com/otto/travel/photos/20051019/tajmahal_lawnmower.html

Fig. 8.37 Electric motor powered reel rotation and push type lawn mower
http://www.jspgardentools.com

http://www.greenplanet.in

Fig. 8.38 Petro engine powered reel rotation and push type lawn mower

Lawn Trimmer or Brush cutter: These have rotary blade or nylon line (1.2 mm) thick and this rotate at about 10,000 rpm in horizontal plane with help of 25 to 35 cc two or four stroke engine . The working width is 0.3 to 0.4 meter. The nylon line or rope type are fitted with spool to extend the line after its wear and tear. The lawn trimmer can also be electric or battery operated.

In some cases the roller is operated on the mowed lawn to given the smooth finish and to remove the impressions of the wheels of the lawn mower.

http://www.hondasielpower.com

IIHR, Bangalore

Fig. 8.39 Lawn Trimmer or Brush cutter ( petrol engine operated)

Fig. 8.40. L awn Trimmer or Brush cutter with nylon line ( petrol engine operated)

#29242 contains 3 of these replacement spools of dual line trimmer line

Fig. 8.41. Spool of Nylon Line
http.//www.mowersdirect.com

Fig. 8.41. Brush cutter with nylon line (electrical operated)
http://www.blackanddecker.com/

Fig. 8.42. Brush cutter with nylon line (electrical operated)
http://www.blackanddecker.com

Fig. 8.43. Brush cutter with nylon line , chordless ( battery operated)
http://reviews.costco.com

Blower: Blower is operated by a petrol engine or electric motor it can be hand held or back pack type. It is used to blow the cutting of grass and leaves fallen on foot path to the lawn.

**Lawn Aeration**

An aerator pokes holes into the soil to alleviate compaction and to allow air, fertilizer and moisture into the soil closer to the roots of the grass. Aeration encourages stronger and deeper development of the root system. This also helps microbes to decompose the grass cuttings and leaves to turn into humus and usable by the grass plants. It can be lawn aerator sandals (Fig. 8.45), manually operated push type (Fig. 8.46), or tractor pulled type (Fig. 8.47 & Fig. 8.48 ). It consists of a cylinder fixed with pegs. The water tank and roller are also fitted with some aerators for watering and finishing/pressing of lawns.

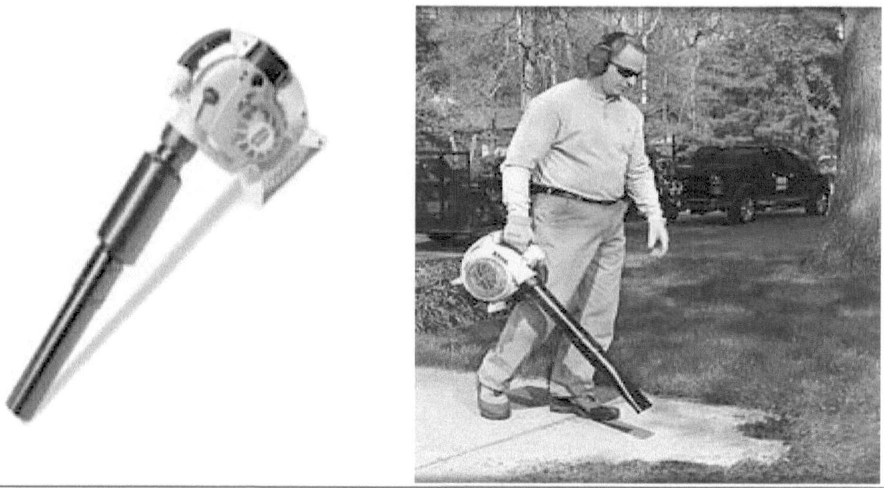

Fig. 8.44 Lawn blower

Fig. 8.44 a Lawn blower (back pack)

Fig. 8.45 Lawn Sandal Aerator

Fig. 8.46 Manually pushed pulled roller lawn aerator

Fig. 8.47 Tractor roller lawn aerator

## Lawn/Split Seeder

A slit seeder makes slits / grooves in the soil and drops seeds into the slits. The machine has number of rotating blades/pegs fitted on roller to cut narrow grooves. A seed hopper mounted on the machine is filled with grass seed and this is dispersed into the groove at a rate that is determined and adjusted as per requirement. In some of the machines the seed is first dispersed in rows and then a pegged roller cut the groove in soil and mixes the seeds in the soil. The lawn or slit seeder is used to sow fresh lawn. This is also used to sow seeds in patches with poor growth in the established lawn.

# *Chapter 9*
# Pruning

Propagation of the mango is done by grafting. This needs cutting of scion from mother stock by secature or pruning knife and grafting this on the rootstock using pruning and grafting knife. The propagation of rose needs budding of eye from mother stock to rootstock and this needs tools like secateur and budding and grafting knife. The pruning of grapes is a routine operation and needs tools like secateur or pruners. The renovation of old mango orchards needs heavy pruning of branches by axe, saw or pruners. The dead wood in lime plants is to be cut regularly. Secateaurs are required to harvest the rose flowers. The hedges are to be pruned by hedge pruner or shear. The lawn grass has to be mowed or cut regularly. All these tools have a blade (or a pair of blades) and a handle (or set of handles). The blade is generally made from medium or high carbon steel and sharpened and hardened at cutting edges. The blade should be wear / abrasion resistant and tough to absorb the impact forces. It should also resist corrosion as well as have anti-rust properties. The handle may be of wood or plastic covered metallic handle. The leverage is such that single or both hands can easily operate the hand tool.

The hand tools used for cutting, pruning and plant propagation are described below.

## 9.1 Secateur

### Hand operated secateur

This is the most common tool for pruning grapes roses and other crops. This is also used to harvest the roses. Pruning secateur is generally of 2 types anvil type (Fig.9.1) and shear type (Figs .9.2, 9.3 and 9.4).

The main parts of the secateur are blades, handles, center pin, spring and lock. The branch or twig to be cut is placed in the center of blades and cut by the shear action, when handles are closed by pressing. In anvil type secateur one of the blades is modified as an anvil and the twig or branch is placed on this and pressing of the other blade shears it. The blades are made from high carbon steel, tool steel or alloy steel by forging or casting, sharpened at the cutting edge and hardened to 460-510 HB (Brinnel Hardness). The spare blades are also available for replacement and sharpening stone is also available (Fig .9.5 and 9.6). The blades are replaced when these get worn. Handles are made from aluminum or mild steel and in some cases a cover of plastic is provided on the arm of the handle. The cutting capacity of the secateur is generally up to 20mm and the size (total length) is 15, 17.5, 20, 22.5 or 25 cm. The weight is about 180-200 gram. Left hander's secateurs are also available in the market (Fig .9.3). The spare blade sets, spring sets and sharpening stones are required for repair or easy maintenance. The leather holder is for easy carrying of the secateur by hanging in belt (Fig .9.5). The secateur with spray arrangement for the wound is also available (Fig .9.4). Compression sprayer of one litre capacity synchronizes the spraying with cutting. The pruning of grapes and rose etc. need continuous working with secateur and is very tiring operation. The battery operated (Fig 9.7), pneumatic operated (Fig. 9.8) and hydraulic secateurs (Fig.9.9) have been developed for easy pruning.

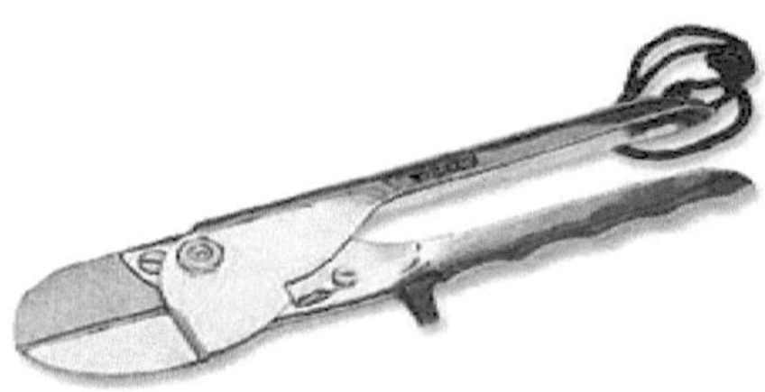

Fig. 9.1 Pruning Secateur (Anvil type)
Kisan

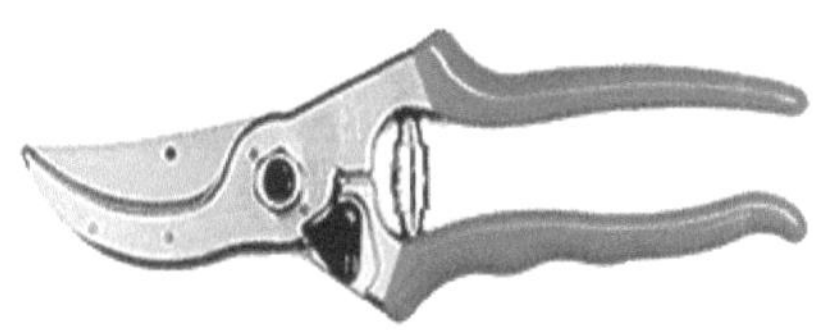

Fig. 9.2 Pruning Secateur (Shear type)
Felco

Fig. 9.3 Pruning Secateur (Left-hander's )
Felco

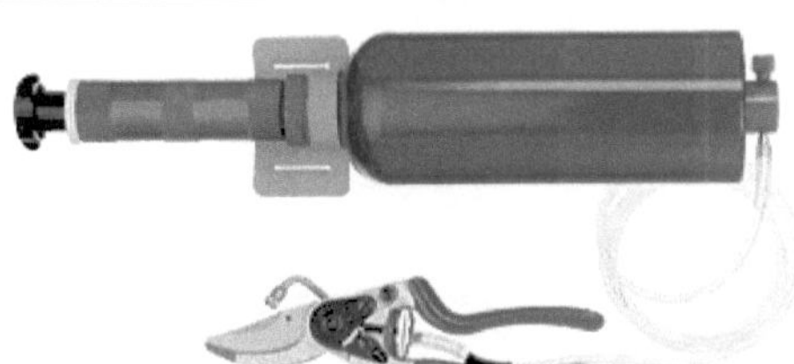

Fig. 9.4 Secateur with spray system for wound
(Felco 19)Felco

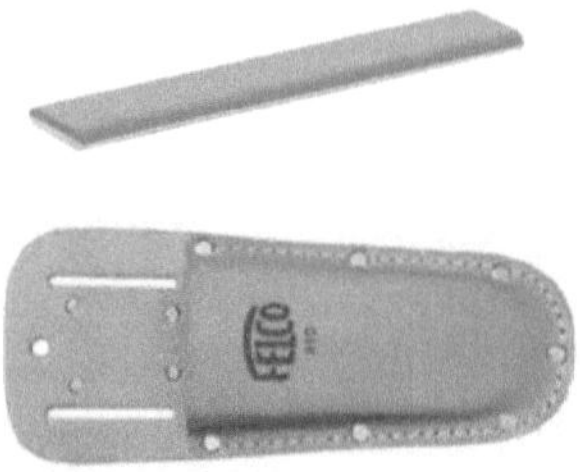


Fig. 9.5 Sharpening Stone Hand Pruner Holster, Leather
Felco

Fig. 9.6 Blade Spring Pack
Felco

**Battery operated secateurs**

The battery-operated secateurs (Fig .9.7) is having cutting capacity of 30 mm and weighs about 820 gm. The battery pack is supported in belt and weighs about 1.8 Kg and contains battery and operating circuit. The Li-ion battery is having 44.4 volts, 150-watt average power, 4 ampere hour and one day working capacity. It takes about 5 hours to charge the battery. The cutting head responds immediately to the movement of trigger. The battery needs replacement after 600 chargings.

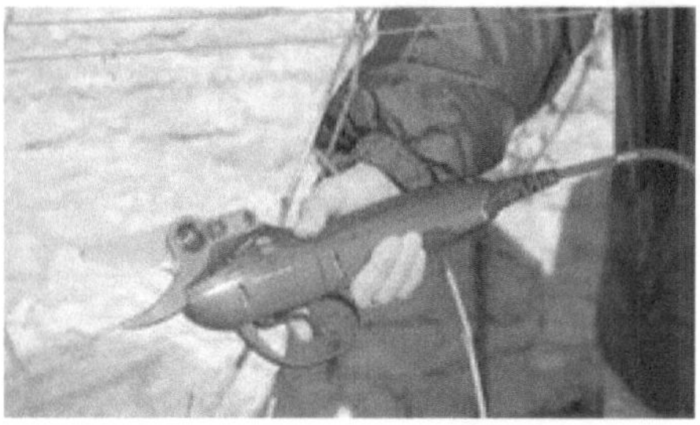

a. Battery operated Secateur
http://www.felco800.com

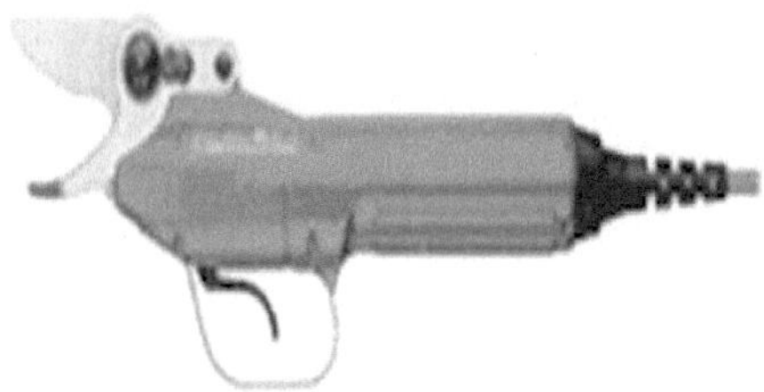

b. Battery operated Secateur
Felco

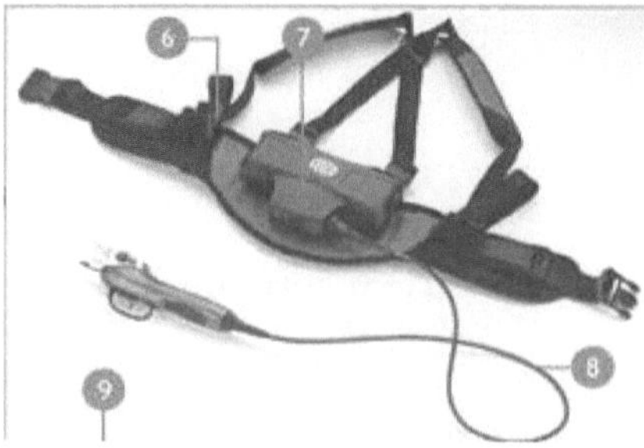


c. Battery operated Secateur with battery and belt

d. Battery operated Secateur (operating mechanism)

Fig. 9.7 Battery operated Secateur

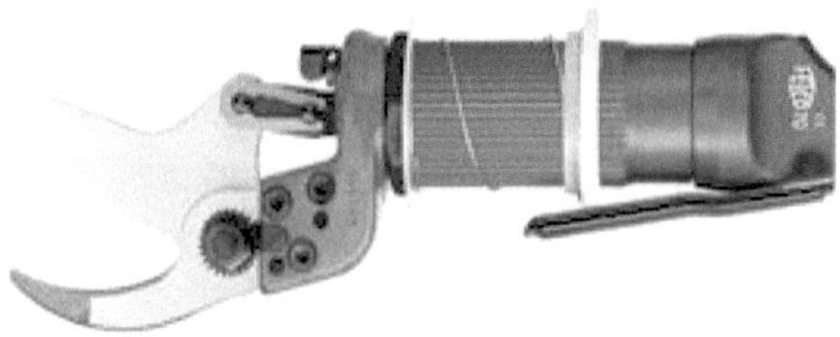

a. Pneumatic Secateur

Felco

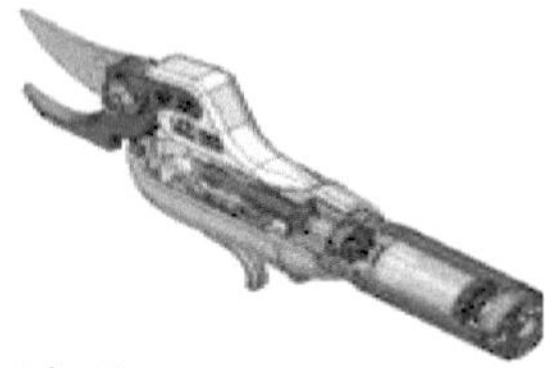

Portable Air Compressor

Paterlini, Italy

Portable Air Compressor

Paterlini, Italy

c. Pneumatic Secateur with extension rod Felco

Fig. 9.8 Pneumatic Secateur

**Pneumatic secatuer**

The pneumatic secatuer (Fig. 9.8) is operated with portable air compressor of 7 kg/ sq. cm. pressure capacity. A double acting piston is provided for easy operation of the secateur. The length of basic secateur is about 26 cm and weight about 685 gm. The extension rod of 60 cm or 120 cm or 150 cm can be fitted for easy reach for pruning at different heights. The compressor for the secateurs can also be operated by the pto shaft of the tractor and upto 7 secateurs can be operated by the tractor.

**Hydraulic pruner**

The hydraulic secateur (Fig.9.9) or pruner is operated by the hydraulic pump fitted with a diesel or petrol engine or operated by pto of the tractor. These are operated by double action hydraulic cylinder. These secateurs can be fitted with extension rod of upto 1.2 mtr length. These secateurs are also known as hydraulic pruners because of higher capacity of cutting. These can prun upto 50 mm thick branches.

Long reach hydraulic pruner

Afron , Isreal

a. Hydraulic pruner

b. Long reach hydraulic pruner

c. Palm & parrot cutting heads

d. Portable hydraulic pump

Fig.9.9 Hydraulic pruner

## 9.2 Axe

The axe (Fig .9.10) consists of a cutting edge and a handle (fitted through eye). The head is made from high carbon steel and cutting edge is hardened to 550-650 HB. The axes are available in various shapes and sizes with weight of cutting edge

1 to 2.7 kg, length of head 15 to 27 cm and width of cutting edge 6.5 to 12 cm. The axes are used for pruning mango and other trees and clearing of bushes.

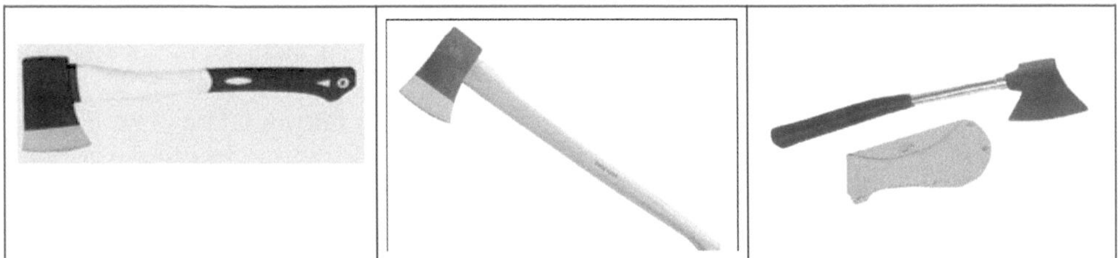

Ramco | Rapid Tools | Omson House

Fig. 9.10 Axe of different types

## 9.3 Billhook

Billhook consists of a curved blade fitted to a wooden or plastic handle (Fig.9.11). The billhooks are available with single and double cutting edges. The blade and tang are made from 3 mm thick high carbon steel and cutting edge is hardened to 400-480 HB. The total length of billhook is up to 75 cm and blade length varies from 30 to 60 cm. Some of the billhooks are provided with curved blade to hold the branch and long handle for easy reach. These are also known as slashing knives.

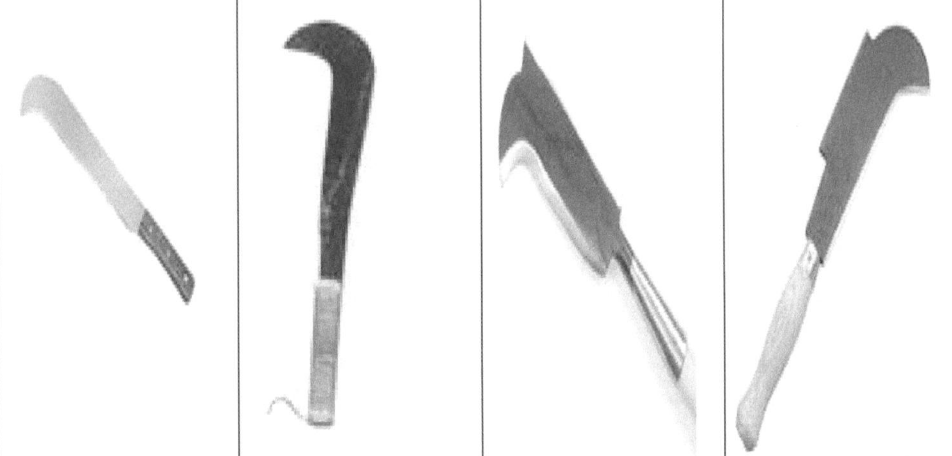

Abm tools | Subhash Agro | The Woodsmiths Store | Buxton's Ltd

Fig. 9.11 Bill hooks of different types

## 9.4 Budding, pruning and grafting knives

Most of the horticultural plants are propagated vegetatively, either through cuttings, layering, grafting and budding. These operations need the cutting of parts of the plants like bark, stem, scion, stock and buds etc. The stock and scion are given proper shape by giving the diagonal cuts, v-grooved cut, edge cut, T incision on root stock, lifting the bark of T-incision for placing the bud etc. Specialized knives (Fig .9.12) are generally used to perform these operations. These knives are generally foldable and resemble household knife. The blade is made from high carbon steel, tool steel, alloy steel or stainless steel and hardened to 450-510 HB. The outer part (handle and case) is made from the horn, plastic or fine quality wood and the internal fittings from brass. A spring steel strip is provided inside case to lock the blade in the operating position. A nail mark is provided in the blade to pull the blade out of handle.

Pruning knife is provided with one blade, which is slightly curved at the end. One edge of the blade is sharpened. This is used for cutting the stems for preparing the cuttings and the defoliating the cuttings.

a. Pruning knife
Kisan

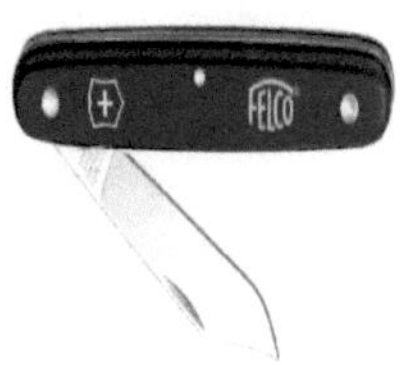

b. Budding knife
Felco

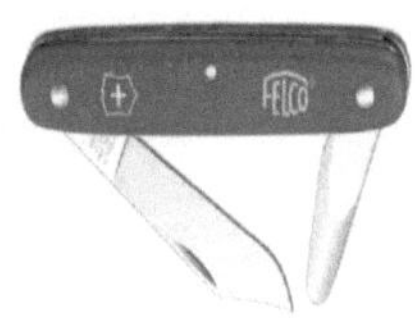

c. Grafting knife
Felco

d. Budding & grafting knife
Falcon

Fig. 9.12 Budding Pruning and Grafting knives

Grafting knife is used for cutting the scion and stock and giving them the proper shape by slant cutting, V- groove cutting etc. for (a) splice or whip grafting (b) tongue grafting (c) cleft grafting (d) Veneer grafting (e) bark grafting (f) approach grafting or (g) root grafting. The grafting knife has only one blade and the cutting edge is sharpened all along the length and the other edge is blunt. The grafting blade is also used for defoliating.

The budding knife has two blades (a) blade and (b) budder. The long blade is sharpened on one edge and is used for preparing buds and creating "T" opening or slot on the bark of the mother branch for insertion of the bud. The budder is sharpened at the end and is used for raising the bark of the slot for insertion of bud.

The budding and grafting knife has 3 blades and can be used for all the operations like cutting of stems, defoliation, cutting grooves in scion and stocks, preparing the buds, cutting T- grooves and lifting the bark of T- grooves for insertion of buds.

## 9.5 Tree Pruner

The tree pruner is a manually operated pruning tool for cutting the branches of orchard trees or plants which are beyond the reach of human hands. This may consist of a curved blade fitted at the end of a long pole (Fig 9.13b). This may consist of a curved anvil, spring actuated cutting blade, links for actuating the blade. The anvil and blade are fitted on the end of a long wooden or metallic aluminum pole. The cutting blade is actuated by pulling the rope by hand or through the lever/ handle (Fig 9.13a, b and c). Branch or stem is brought under the hook/anvil and cutting is done by pulling the rope or pressing the handle. The stem or branch is cut due to shearing action of the cutting blade. When the rope or handle is held loose the blade returns to its original position due to action of spring. The cutting blade is generally made by forging (425-500 HB) and sharpening the edge by grinding. These pruners have the cutting capacity of 20 mm diameter.

The two handed pruner are also available (Fig 9.13d and e) These are similar to pruning secateur, but have long handle and heavy duty cutting blade (6 to 8 mm thick) and anvil. The blade length is 15 to 25 cm and overall length is up to 1 meter. The cutting capacity is up to 60 mm diameter. These are also known as loppers and both hands are used to operate these.

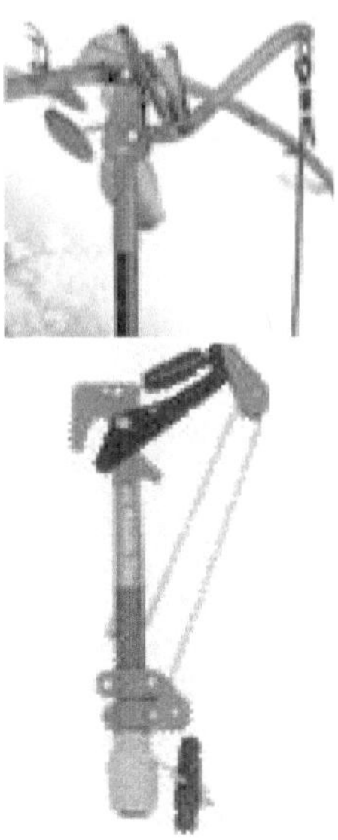

a. Pruner Mehar & Mukand Tools. Falcon

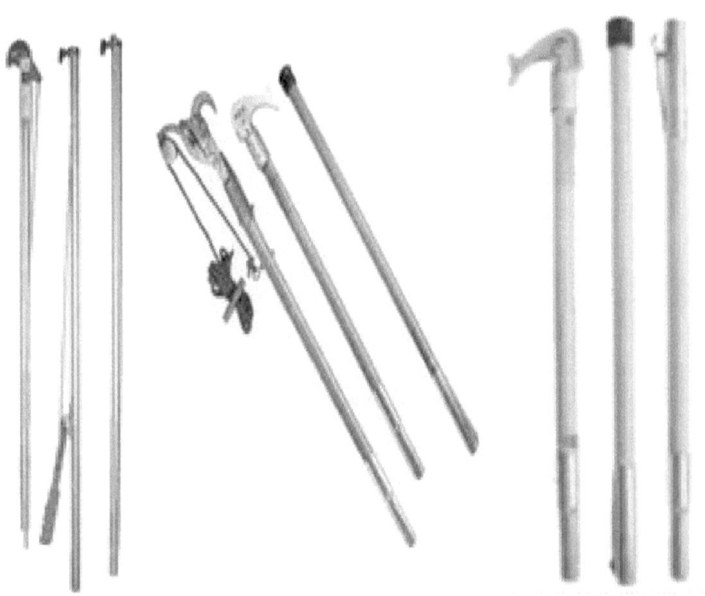

b. Easy reach pruners

Jameson

c. Snapper long reach cut & hold

d. Pruner -two handed Felco

e. - Pruner Heavy duty, Felco

Fig. 9.13 Tree pruner of different types

## 9.6 Pruning Saw

The pruning saw (Fig.9.14) is used for cutting and trimming the branches which are beyond the capacity of secateur, tree pruner or lopper. The main parts of the saw are handle and blade. The handle is generally of wood and riveted to the blade. In folding saw the handle is hinged to the blade. The handle is of steel pipe for bow saw. The blade length may vary from 30 cm to 45 cm, and width is 4 to 5 cm. The blade is 1.6 mm thick, made of tool steel and the cutting teeth are sharpened and hardened to 45-48 HRC. The blade has 4 to 5 mm pitch to avoid clogging, while cutting green branches. Some of the imported saws are available with safety sheath and belt loop for easy carrying. In pruner with saw a long handle is provided for saw to cut branches at a distance. The pruner operated with rope and lever is also available to do pruning (Fig 9.15).

| | | | | |
|---|---|---|---|---|
|  a.Garden Saw All Tools Direct, Draper | 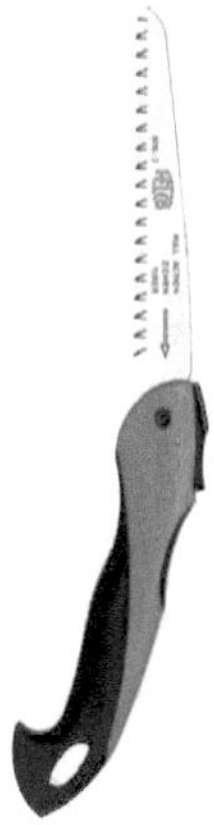 b. Pruning Saw Felco | 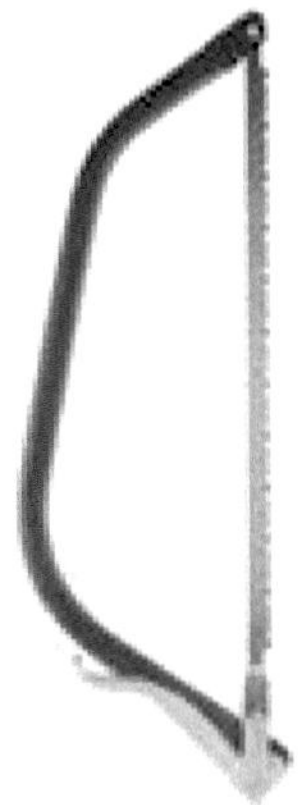 c. Bow Saw Fiskars |  d. Folding Saw Felco |  e. Pruning Saw with safety sheath and belt loop. Felco |

Fig. 9.14 Pruning saw of different types

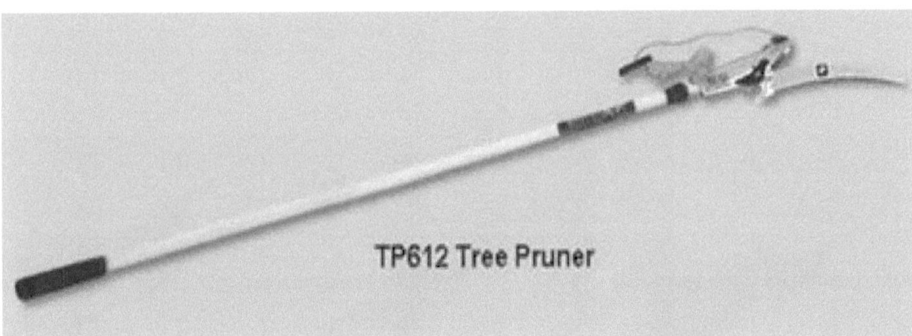


Fig. 9.15 Pruner with saw

Hoover Fence Co.

## 9.7 Hedge Shear

The hedge sheer (Fig 9.16) is used for pruning and trimming of hedge and giving them the desired shape. This is also used for cutting of shrubs and removing of haphazard growth in gardens and lawns. This consists of a pair of pivoted blades of high carbon steel, tool steel or alloy steel, fitted with wooden handles. The blade length is 150 or 200 or 250 or 300 mm and the thickness is 8 mm. The blades are hardened to 420-470 HB and sharpened to obtain bevel angle just less than 90degree. The handle has 27-30 degree angle with blade and the maximum opening angle of blades is 95 degree. Some of the hedge shears are provided with pruning notch near the pivot of blades for cutting of thick twigs.

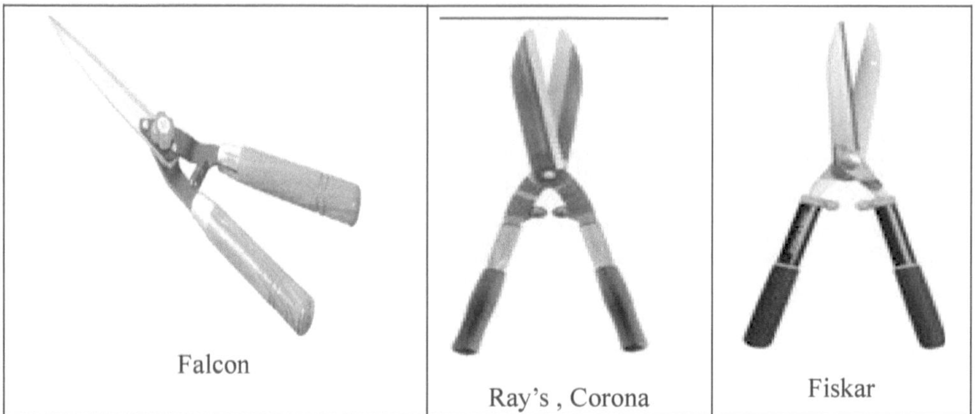


Fig. 9.16 Hedge shears of different types

## 9.8 Grass shear

The grass shear is used for trimming of the grass in lawns. These are generally of 2 types (Fig. 9.17). In one type the action of blades is horizontal and the action of spring loaded handle is vertical. The other type consists of 2 blades, joined to a "U" shaped spring steel handle. The length of blades is 150, 175 or 200 mm, thickness is 2-3 mm and overall length of the grass shear is about 300 mm. The cutting blade is made from high carbon steel or alloy steel and hardened to 425-450 HB.

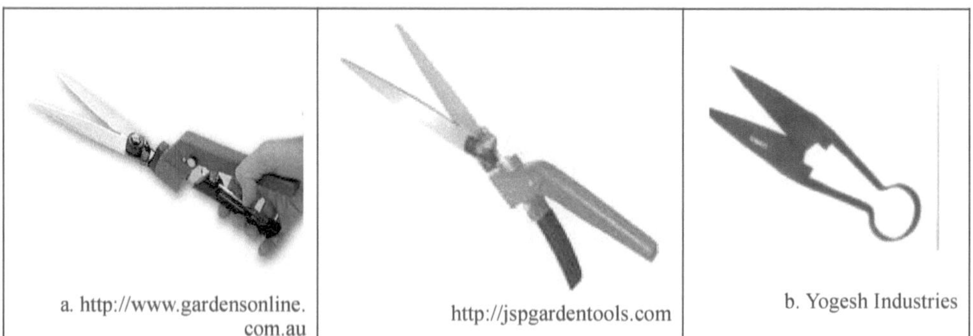


Fig. 9.17 Grass shears of different types

## 9.9 Pruners and Scissors

Single hand held, fruit pruners are used for grape trimming and general pruning (Fig. 9.18). These are also used for harvesting kino. Vine scissors with long nose and serrated edge are used for delicate pruning jobs. Cut and hold scissors are used for harvesting roses. The scissors of 100-200 mm length are used for plucking the flowers.

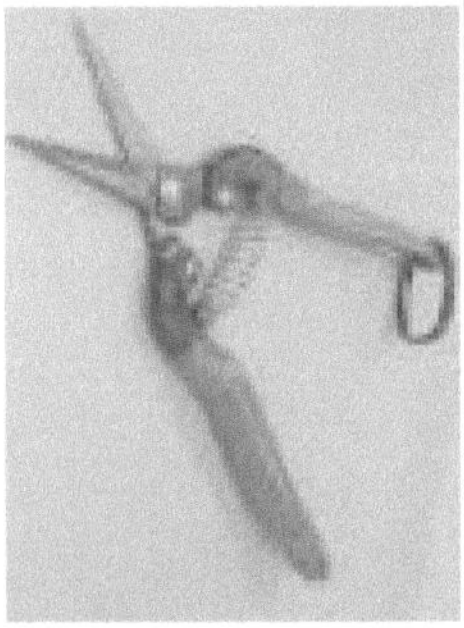

a. Fruit Pruner Pointed
queenswood.com

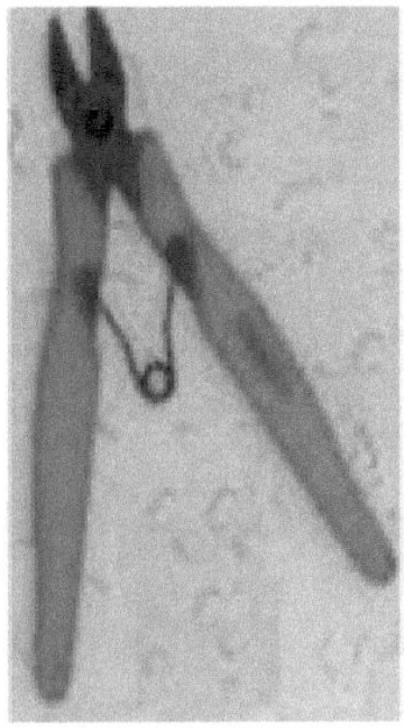

b. Kino harvester
CIPHET Ludhiana

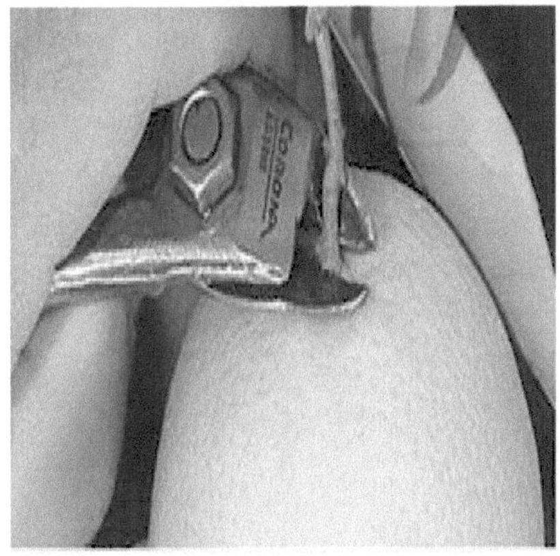

b1. Kino /orange harvester
Corona Clipper

c. Grape Trimming Pruner
Draper

d. Vine Scissors (long nose)
D.T. Brown

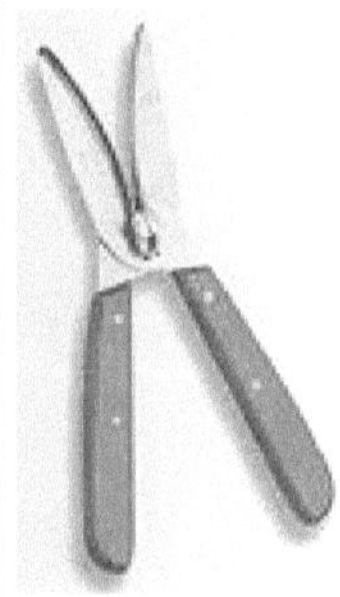

e. Cut and hold Scissors
Felco

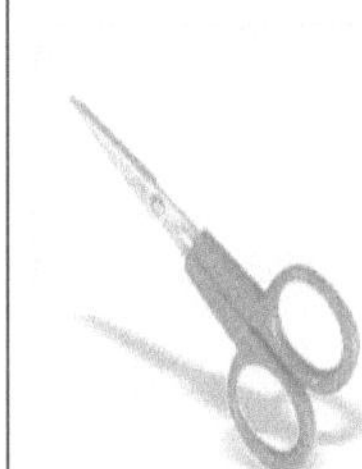

f. Scissors
Falcon

Fig. 9.18 Pruner and Scissors

## 9.10 Chain Saw

Chain saw (Fig. 9.19 a,b and c) is used to prune thick branches (upto 40 cm diameter). The Chain saw is either electrical operated, petrol engine operated, compressed air or hydraulic operated. The saw length is 30,35,40,45 or 50 cm. The power source is 0.9 to 4.4. kilowatts electric motor or 25 to 75cc petrol engine. The weight varies from 3 to 7 kg for light to heavy duty saw. The chain speed is 7 to 27 meter per second and the operator has to be very careful while operating these. The chain saw fitted with telescopic pole is known as pole saw pruner (Fig 9.19 d,e. and f.). The engine is close to the operator while the chain head is at the distal end up to 3.8 meter away from the operator. The guide bar length is 30 to 35 cm and the total weight of the pole pruner is 5 to 7 kg. Back pack battery operated pole chain saw pruner is having 44.4 volt Lithium ion battery. The batteries of 4.4 to 17.6 ampere hours are available and weigh 2 kg to 6.6 kg. The energy stored is 210 to 844

Watt hour and can be recharged at least 600 times. The chain saw can work 1 to 3 days after charging. The hydraulic operated chain saw are also available, these can be operated by hydraulic pump or tractor hydraulic system.

a. Petrol Engine Operated Chain Saw
STIHL Pty Ltd Australia, http://www.stihl.com.au

b. Electrical Operated Chain Saw
http://www.mafell.com

c. Power Chain Saws

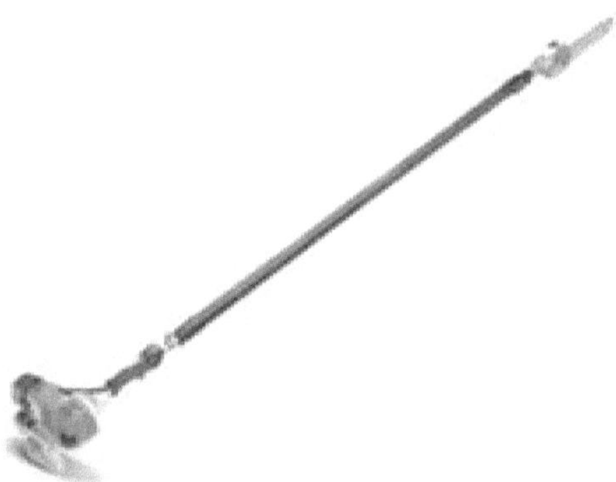

d Engine operated Chain Saw / Pole pruner
http://www.stihlusa.com STIHL , USA

e. Hydraulic Operated Pole Chain Saw Pruner
Afron , Israel

f. Back Pack Battery Operated Pole Chain Saw Pruner
http://www.pellenc.com.au

Fig. 19.19 Power chain saw

## 9.11 Hedge Trimmer

Hedge Trimmer (Fig. 9.20) is used to cut the excessive growth of hedges and to shape the hedges. The cutting takes place by reciprocating blade of cutter bar. This is operated by 22 to 25 cc petrol engine or 18 volt 1.5 Ah batteries. The blade length is 43 to 70 cm. and cutter bar strokes are 25 to 39 per second and blade pitch is 35mm. The weight of hedge trimmer is about 5.4 kg. Some of the hedge trimmers are fitted with the pole (up to 2.4 meter length) and blade angle adjustment is provided for shaping the tall hedges (Fig. 9.20 d and f).

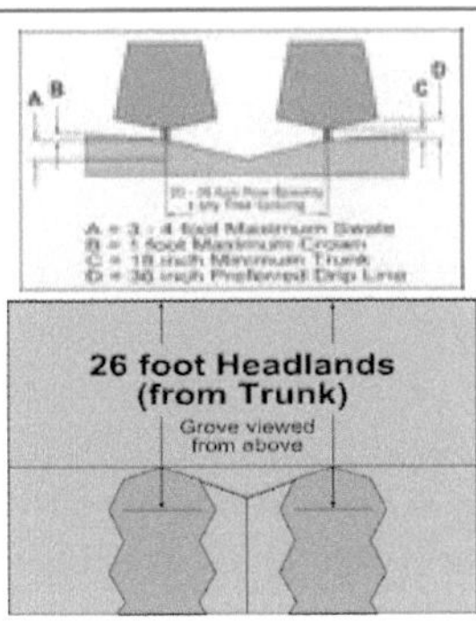


a. Pruned Citrus Tree
Korvan Ind.USA

b. Engine Operated Pruner With Cutter Bar
mcconnel.com/history.htm

c.Tree Pruner (Hedging Machine)
www.lal.ufl.edu/history/ crecfacultyaccomplish.htm

d. Tree pruner with 2 set of blades fitted on 2 bars
globalcitrusresources.com

e. Tree pruner with 2 set of blades fitted on 2 bars
globalcitrusresources.com

f. Tree pruner with 2 set of blades fitted on 2 frame
globalcitrusresources.com

**Fig 9.21 Tree Pruner**

## 9.12 Powered Tree Pruner

The citrus plants are pruned to the shape in California, U.S.A (Fig.9.21 a) for ease in mechanical harvesting of the crop. The mango plants are pruned to cubical shape in Australia for ease in mechanical operations. The first tractor mounted pruner in the year 1959 (Fig.9.21. b) consisted of a petrol engine operated cutter bar fitted on the tractor. The height of cutter bar and engine could be adjusted to adjust the height of pruning. The tree pruner (Fig.9.21c) consisted of one set of serrated circular blades fitted in a line. This machine was fitted on tractor trailer. The blades were operated by an engine. The height and angle of the blades could be adjusted in the machine. The tree pruner (Fig.9.21 d and e) consists of 2 set of blades fitted on 2 bars. The pruner is mounted in front of tractor and the bar angles are adjustable. The tractor moves in between the rows and prune the plants to the required shape on both sides. The tree pruner (Fig.9.21 f) has two set of blades and one set of these circular blades can prune the top of the tree at the required height.

## 9.13 Grape Pruner

The grape plants require pruning for fruiting. The tractor mounted grape pruner (Fig.9.22 a) consists of vine lifter in the form of rotary rods. The vines are lifted by the lifter and pruned by the rotary serrated blade. He grape pruner (Fig.9.20 b to f ) are mounted to the side of the tractor. The full row of grapes / trellis is taken inside opening. The pruner moves along the trellis. The rotary blades on sides and top (Fig. 9.20 c) cut the unwanted wines. The pruner move at a speed of 4 to 5 km per hour and several acres are pruned in a day.

a. Grape Pruner
http://picturethis.pnl.gov/picturet.nsf/by+id/PNLM-3RJVRR?opendocument

b. Grape Pruner
www.boston.net.au/ services.html

c. Arrangements of Blades in a Grape Pruner for Pruning around Trellies
binger.fr.tree_pruner.htm

d. Grape Pruner in Operation
silkwoodwines.com.au
http://www.pellenc.com.au

e.Grape Pruner in Operation
http://www.napierwebdesign.co.nz/hydralada
http://www.pellenc.com.au

f.Grape Pruner in Operation
http://www.pellenc.com.au

Fig. 9.22 Grape pruners

## 9.14 Ladders and Elevating Work Platforms

Ladders are very useful in pruning the tall trees. Folding ladders, self propelled elevating work platforms (Figs. 9.23a and b), tractor mounted platforms and tractor trailer mounted platforms (Fig.9.23c and d) are also available for pruning , spraying and harvesting tall trees. The working height of most of the platforms is upto 6 metres .Some of these have hydraulic system for operation of hydraulic operated pruners (Fig. 9.23b).

a. Self Propelled Platform Working in Crop
http://www.edp.com.au/powate.html

b. Self Propelled Platform with Operator
IIHR Bangalore

c. Tractor Mounted Elevating Work Platform

afron.com

d. Tractor Trailer Mounted Elevating Work Platform

IIHR Bangalore

Fig. 9.23 Elevating Work Platforms

# *Chapter 10*
# Spraying

Pest control is one of the important operation for improving the productivity and quality of the horticultural crops. Different kinds of sprayers and dusters are used for control of diseases and pests. The most commonly used sprayers for annual crops are knapsack sprayers and for perennial crops rocker and foot sprayers. The power operated sprayers are also becoming popular for spraying on perennial crops. The air assisted sprayers are becoming popular in Maharashtra for spraying of grapes and in UP for spraying mango. These can spray the annual and perennial crops very efficiently.

Sprayers can be classified as High Volume (HV) Sprayers (more than 400 liter per hectare (l/ha) to Low Volume (LV) Sprayers (5-400 l/ha) and Ultra Low Volume (ULV) Sprayers (less than 5 l/ha) capacities. The HV sprayers are manually operated knap sack sprayer, rocker and foot sprayers and power or tractor operated hydraulic sprayers. The HV sprayers are suitable for spraying insecticides, fungicides and herbicides. The LV sprayers are generally motorized air assisted knap sack sprayer, air assisted sprayer, air craft sprayers. The LV sprayers are suitable for spraying insecticides and fungicides. Motorized knapsack sprayer, air assisted sprayers and air craft sprayers and spinning disc (low rpm) sprayers are general types of LV sprayers.

The ULV sprayers are generally spinning disc (high rpm) or air craft sprayers and are suitable for spraying insecticides.

## 10.1 High Volume Sprayers

The High Volume (HV) sprayers can be either manually operated sprayers or power operated hydraulic sprayers. The manually operated HV sprayers for horticultural crops are: a)knap sack sprayer; b) rocker sprayer and c) foot sprayer. The single barrel or double barrel bucker sprayers are hardly used for spraying horticultural crops.

## a. Manually operated sprayers

The knapsack sprayer (Fig. 10.1) is the most commonly used sprayer for annual crops. This is a low pressure (100-200 kpa) manually operated sprayer with approximately 15 litre tank.

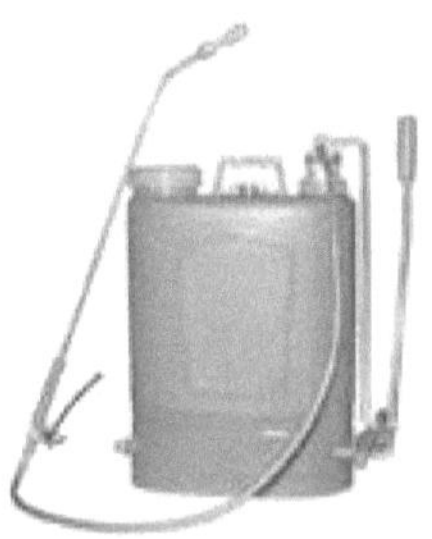

skybirdagro.com

Hymatic Agro

Fig. 10.1 Knapsack sprayer manual operated

Fig. 10.2 Knapsack sprayer with boom

Fig. 10.3 Hood / Shield for spraying pesticide

aspee.com

The parts are strainer, lid, cylinder, piston, ball valve, pressure chamber, agitator, delivery hose (1.8 meter length), lance, cut off valve, nozzle, operating handle and straps. The sprayer can be left hand or right hand operated. The tank can be of brass or UV resistant high density polyethylene for resistance to corrosion by chemicals. The piston is made from polyamide plastic. The pressure chamber, cut off valve, lance and nozzle are generally of brass and delivery hose of low density polyethylene pipe. Some of the recent sprayers have plastic pressure chamber. The weight of empty sprayer is about 3.5 to 5.0 kg. It is suitable for annuals and perennials not taller than 1.5 meter and the penetration of spray is up to 0.5 meter. Generally one nozzle is used with one sprayer, however more nozzles can be provided on lance to cover more than one row (Fig.10.2). Two knapsack sprayers left and right hand can be integrated and a number of nozzles can be fitted on lance to cover more rows to increase the spraying capacity (Fig. 10.2). The hood on nozzle (Fig. 10.3) can be provided for targeted spray for control of weeds. About 625 to 750 litres of solution is sprayed per ha (l/ha) and the field capacity is about 0.2 hectare per day (ha/day) per labour.

Hand compression sprayer is also used for spraying small areas. In these sprayers the tank is provided with barrel and plunger for compressing the air. The tank is filled to ½ of the capacity and the air is pumped inside to produce air pressure of about 4 atmosphere. The sprayer is then put on shoulders as backpack (Fig. 10.4) and spray is done. The air pressure in the tank goes on reducing and air is again pressurized for spraying. The throw of spray and particle size reduces as the pressure in the tank reduces.

Rocker sprayer and Foot sprayer are high pressure manually operated sprayers and work at a pressure of about 200-400 kpa. These are generally provided with adjustable nozzle. These are suitable for annual and perennial crops of upto 5 meter height. Lance extension can be used to reach higher heights. Most of the perennial fruits are generally sprayed using these sprayers. The lance is also tied to bamboo to reach he higher heights. It takes about 750-1000 litre to spray one hectare of perennial crops and about 625 to 750 l/ha for spraying annuals. Three persons (one for rocking and 2 for two lances) can spray 0.4 ha/day.

**Lances and nozzles.** The hand operated sprayer are provided with lances and nozzles. The lances may have trigger type or wheel type cut off valve. The nozzles may be non-adjustable or adjustable for volume of spray. The extension rods are available for tall trees and booms are available for spraying with more than one nozzle to increase the field capacity.

**Power sprayers.** Motorised high pressure knapsack sprayer (fitted with 22.5 cc petrol engine) has come recently in the market. The sprayer operates at about 1400 kpa and is suitable for spraying annuals and perennials of upto 5 meter height. The field capacity is upto 0.6 ha/day. The power sprayers are high pressure (500-900 kpa) and operated by a portable petrol engine or power tiller or tractors or self propelled. The main parts of a power sprayer are prime mover, pump, solution tank, strainer, inlet hose, pressure regulator and bye-pass, control valves, distributor, bye-pass hose, outlet hose, spray guns or spray boom fitted with nozzles.

Portable sprayers can be trolly mounted (Fig. 10.9 and Fig. 10.10). These are generally used for spraying perennial crops. The adjustable spray gun can spray upto 10 meter height and can cover most of the tree crops. For spraying lower portion of trees, the adjustable spray gun gives fine spray while for spraying the top of the trees the particle size is increased. Upto four spray guns can be operated by the sprayer. For power tiller operated sprayer the pump is mounted on the body of the power tiller and is directly operated by a belt from the engine (Fig. 10.10 & 10.14). The tank for insecticide is kept in the trailor. The labourers can also go around the tree while standing in the trailer. A ladder can also be fixed in the trailer for the labour to climb and spray the top of the trees. The booms can be provided for spraying the crops on the trellis or under bower system. The capacity is about 2 ha/day for power tiller operated sprayer while using two to four spray guns.

| | | |
|---|---|---|
|  |  |  |
| Fig. 10.4 Providing vertical booms for spraying perennials | Fig. 10.5 Self propelled boom sprayer<br>PAU Ludhiana | Fig. 10. 6 . Providing boom for crop on trellis |

In tractor operated sprayer the spray pump is operated with the pto shaft The sprayer tank of 400-700 liter capacity is generally 3 point linkage mounted and sprayer tank of larger capacity are trailed type. The boom length varies from 4 meter to 12 meter and it is foldable for easy transport Spray lances (vertical drop) with double swivel nozzles can be fitted for spraying under the leaves.

| | | |
|---|---|---|
|  |  |  |
| Fig. 10.7 High clearance sprayer<br>PAU Ludhiana | Fig. 10.8 High clearance sprayer<br>http:// www.agapeq.com/ postemerge.htm | Fig. 10.9 High clearance sprayer<br>http://gkmachine.com<br>http://vimeo.com/8459694 |

The nozzles are usually spaced at 30 to 50 cm interval. The nozzles can be provided to vertical lances to spray tall trees (Fig. 10.5). The boom can be modified for spraying the crop on trellis (Fig. 10.6). The boom sprayer can be made self propelled (Fig. 10.5) or high clearance sprayer (Fig. 10.7 to 10.9). The self propelled sprayer (Fig. 10.5) has the shade for the operator and the high clearance sprayers are have 0.9 to 1.2 metre ground clearance and are suitable for tomato and bushy crops like jasmine.

Fig. 10.10 Boom sprayer with continuous long hood to avoid drift
http://www.star-texas.com/RGV%20photos.htm

Fig. 10.11 Hoods for targeting the row crop
http://msucares.com/crops/cotton/sprayer.html

http://www.star-texas.com / melon%20hooded%20spraying.jpg

Fig. 10.12 Hoods for spraying weedicide in rows

http://omaha-ne.americanlisted.com

Fig. 10.13 Hoods for spraying crop rows

Fig. 10.14 Tunnel sprayer for complete tree
http://www.lipco.com/english/products/TUNNEL_Sprayers/tunnel_sprayers.htm

Fig. 10.15 Hood for spray of complete tree

Fig. 10.16 Tunnel sprayer for crop on trellis
http://www.lipco.com/english/products/TUNNEL_Sprayers/tunnel_sprayers.htm

Fig. 10.17 Tunnel sprayer for complete tree
http://www.spray.com

Hoods can be provided to avoid the drift while spraying (Fig. 10.10), to target the crop (Fig. 10.11 and 10.13), to control weeds in rows (Fig. 10.12). The tunnel sprayer sprays a complete tree in one go (Fig. 10.14 to 10.17). These can spray the plants upto 2 meter height and the crop on trellis. These can be tractor operated or self propelled with a working speed of 4 to 5 km/h.

## 10.2 Low Volume Sprayers

Low Volume Sprayers are generally air assisted sprayers and spray from 5-400 l/ha. These sprayers have a high capacity blower and the air stream is directed to the spray target. The chemical is sprayed in the air stream by the nozzles of a power sprayer. The air stream further atomizes the spray particles and carries these to the target. The air stream must have adequate velocity and volume for best possible spray distribution.

aspee.com

Fig. 10.18 Knap sack mist blower

Hymatic Agro

aspee.com

aspee.com

Fig. 10.19 Shoulder mounted air assisted turbo sprayer for tall trees

The power operated knapsack mist blower (Figs. 10.18) is generally fitted with about 75 cc petrol start and kerosene run engine. The spray tank capacity is about 15 litre. The blower blows an air stream of 20 $m^3$/min (0.33 $m^3$/sec) at a speed of 125 m/sec (440 km/h). The spray solution drips under gravity into the nozzle, where it is atomized. These sprayers have horizontal spray reach of about 15 m and vertical reach of about 10 meter. The air assisted shoulder mounted sprayer (Fig. 10.19) is used for spraying very tall trees like rubber and can also be used for spraying coconut tree.

Fig. 10.20 Tractor operated air assisted sprayer

Fig. 10.21 Spraying annuals with air assisted sprayer

Fig.10. 21a Tractor mounted air assisted sprayer for grape orchards and medium height trees
(spraying from below)

Fig. 10.21b Tractor mounted air assisted sprayer for grape orchards, medium and tall trees
(spraying from below and top)

Fig. 10.22 Tractor operated trailed type air assisted sprayer
aspee.com

Fig. 10.23 Tractor operated trailed type air assisted sprayer
http://www.star-texas.com/large%20 airblast%20 application.jpg

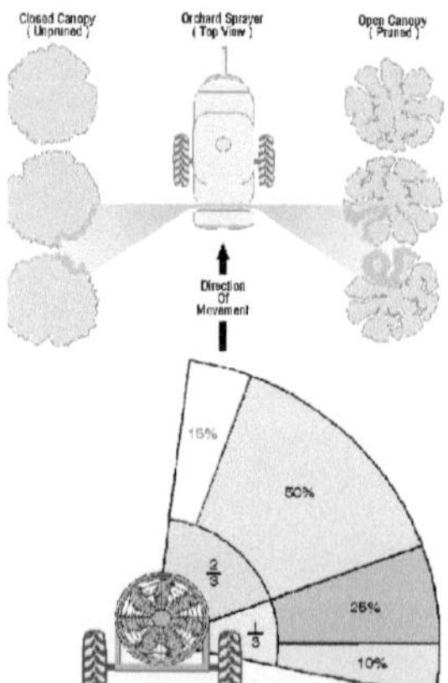


Fig. 10.24 Pruning improves the spray penetration and adjustable spray quantity as per canopy

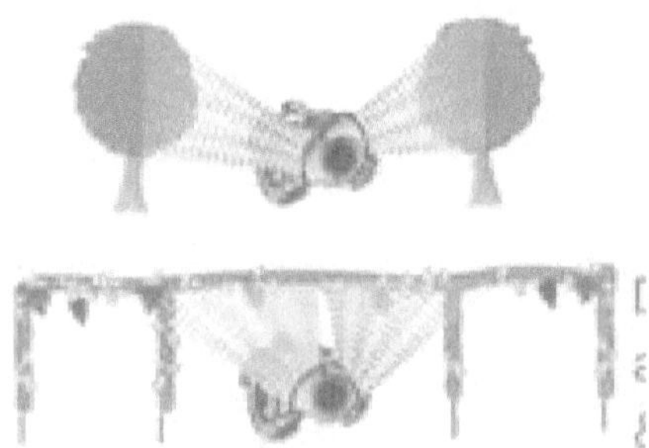

Fig. 10.25 Spraying row tree crops and under the bower

Fig. 10.26 Spraying annuals with air assisted tycoon sprayer

Fig. 10.27 Spraying with air assisted tycoon upto 30 m coverage

Fig. 10.28 Spraying perennials with adjustable nozzle of air assisted (Tycoon sprayer

aspee.com

Fig. 10.29 Spraying perennials with air assisted (Tycoon) sprayer

CIAE

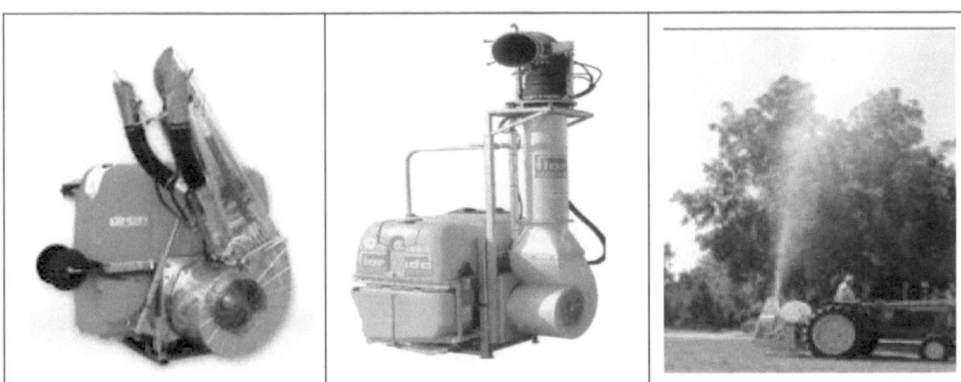

Fig 10.29a Tractor mounted air assisted sprayer for tall trees and field crops, with nozzle control from tractor operator seat

In tractor operated air-carrier or air blast sprayer, the spray solution is atomized either by pressure nozzles or rotary atomizers by a high velocity air stream. The atomized liquid is then carried to the target by air stream. The tractor operated sprayer is fitted with a blower with air flow rate of 2.5 to 30 $m^3/s$ and air speed of 125 to 240 km/h. The concentration of the chemical in low volume sprayers is three times than HV sprayers. The sprayers with tank capacity of about 500 liter can be tractor mounted (Figs. 10.20, 21, 25 and 26), while the sprayers with higher capacity tanks are generally trailed type (Fig. 10.22 & 23). The air stream outlet can be designed for spraying of annuals, perennials, very tall crops and green house crops.

Fig. 10.30 Air assisted boom sprayer

Gursukh Agro Works, Ludhiana

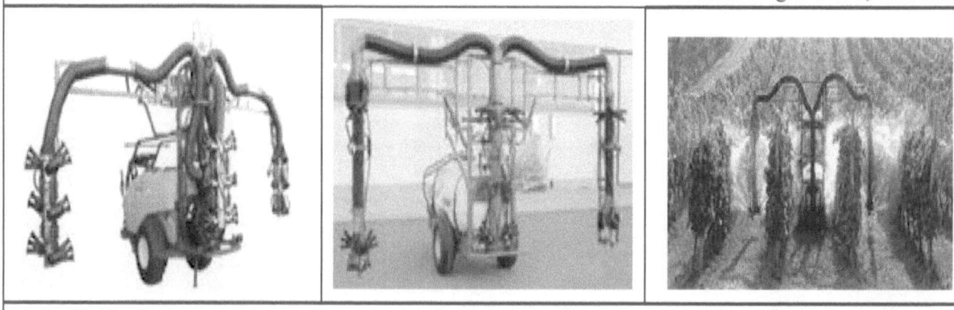

Fig 10.30a. Tractor mounted air assisted sprayer for grape orchards for 2 row (spraying from below and top)

The nozzles can be fitted properly to spray the whole canopy (Fig. 10.23 & 24). For spraying annuals, the sprayer can spray upto a distance of 30 m (Fig. 10.26 & 27). For spraying the perennials, the sprayer can spray upto a height of 15 m. For covering fruit crop trees, the air assisted tycoon type sprayer is provided with a swivel outlet (Fig. 10.38) and the outlet can be adjusted to different angles to spray the whole canopy.

Spraying the annuals on one side

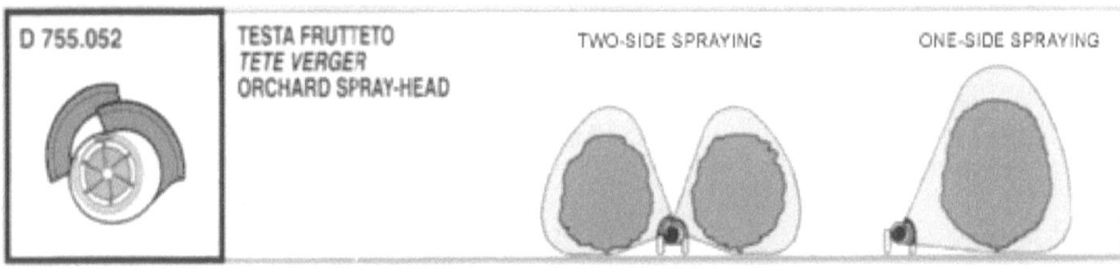


Spraying  on both or single side

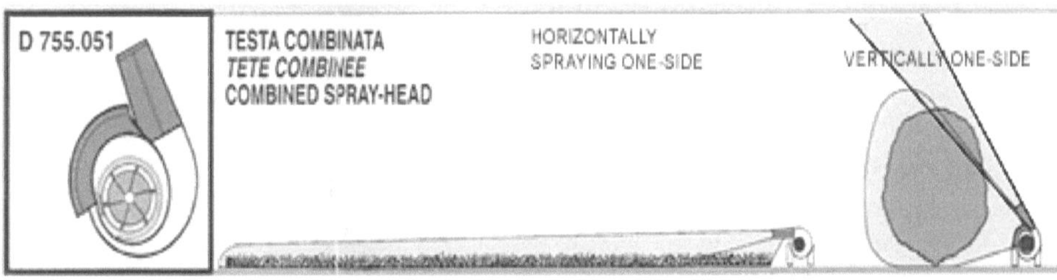


Spraying with two outlets to cover more area or height

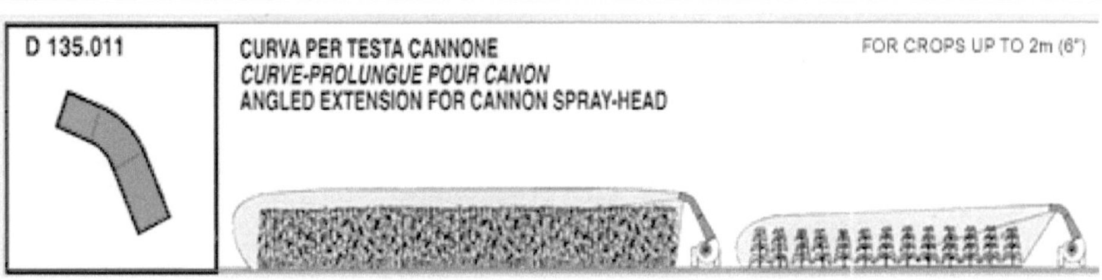


Spraying with two nozzles to spray shrubs

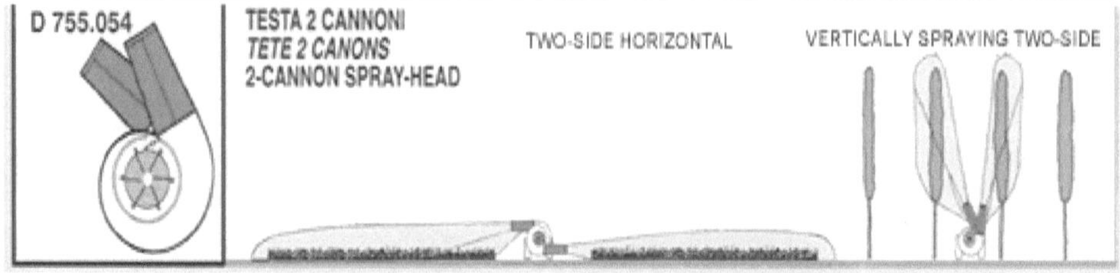


Spraying with two nozzles on both sides for annuals or tall trees

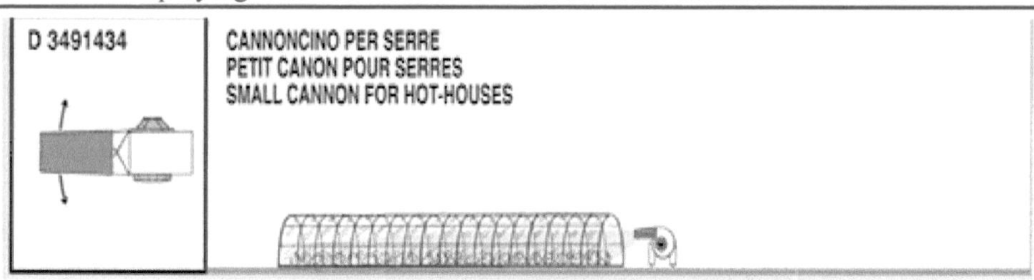


Spraying greenhouse with one nozzle

Fig. 10.31 Air assisted sprayer – Adjustment for spraying different crops

The canopy can be pruned for better penetration of spray (Fig. 10.24). The air assisted sprayer is suitable for spraying the crop on trellies and for spraying under bower (Fig. 10.25). Air assisted boom type sprayers are also available for spraying annuals especially vegetables (Fig. 10.30). Different type of adjustments for spraying the annuals, shrubs and perennials can be done in a air assisted sprayer as shown in Fig. 10.31.

Multi Head Air Assisted Sprayer has one hydraulic pump to supply insecticide solution to different set of nozzles. The insecticide sprayed by hydraulic nozzle gets further atomized by the air stream created by fan provided by the side of nozzle. Multi head air assisted orchard sprayer also known as tower sprayer is provided with 8 heads/ fans for spraying on both sides.

Fig. 10.32 Multi head air assisted orchard sprayer,8-heads ( tower sprayer)
http://oregonprogress.oregonstate.edu

Fig. 10.33 Multi head air assisted orchard sprayer for citrus– 6 heads
http://www.interlinksprayers.com.au/

Fig. 10.34 Multi head air assisted sprayer for spraying annuals / vegetable crops
http://www.interlinksprayers.com.au/

Fig. 10.35 Multi head air assisted sprayer for spraying crops on trellis
http://www.vinetechequipment.com

## 10.3 Ultra Low Volume Sprayers

These sprayers spray less than 5 l/ha. These are generally battery operated and are used to spray insecticide in concentrated form. The insecticide falls under gravity on battery operated spinning disc (high rpm) (Fig. 10.36) and gets atomized to finer spray particles (50-100 micron). The electric operated ULV sprayer (Fig. 10.37), has blow moulded tank of 4 liter capacity and a blower, which pushes the spray solution into the airjet and atomizes the same. It has an adjustable output of 0.1 to 13 litre per hour. The discharge metering valve allows for easy control of output and droplet size. It can be used for oil based chemicals, water based chemicals and wettable powders.

Fig. 10.36 Battery operated spin disc sprayer
aspee.com

Fig. 10.37 Electrical operated with blower
aspee.com

**Electrostatic sprayer**

The electrostatic sprayer is generally air assisted sprayer and it can be back pack mounted, portable, tractor mounted or air craft type. It achieves better coverage and lessen the environmental impact by allowing smaller amounts of chemicals to be used. The spray particles are charged during spraying and these gets deposited on the plant more efficiently than conventional sprayers. The lower side of the leaf get better coverage.

The heart of the air-assisted electrostatic sprayer is "air-atomizing induction-charging" nozzle.

Air and liquid enter the rear of the nozzle separately Figs. 10.49 to 10.51. The air moves through the nozzle under pressure and meets the liquid at the nozzle tip, causing the formation of spray droplets that are 30 to 60 microns in diameter. At the tip of the nozzle is a tiny electrode which applies an electrical charge to the spray. The electrical charging ( by 1 kV electrode) causes a natural force of attraction between the spray droplets and target surface, similar to the attraction between items of clothing created by the tumbling of a clothes dryer. The attraction to the target's surfaces relates to Coulomb's Law, which states that any two charged objects will create a force on each other. The charge on the droplets, though small, pulls the spray towards the target at 75 times the force of gravity. The spray droplets can reverse direction, moving against gravity, to coat all sides of an object.

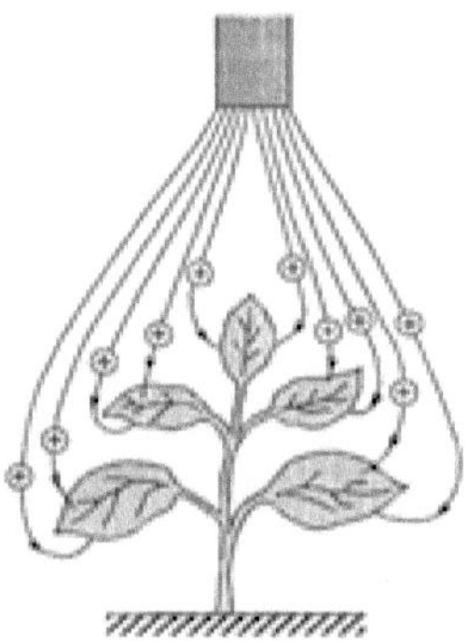

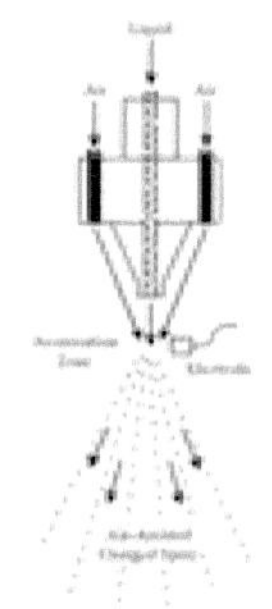

| | | |
|---|---|---|
| Fig. 10.38 Particle trajectory to the plant | Fig. 10.39 Back Pack electrostatic sprayer Electrostatic Spraying Systems | Fig. 10.40 air-atomizing induction-charging nozzle. |

Air is necessary in electrostatic spraying to create tiny droplets (30-60 microns, 900 times smaller than conventional sprayers) that can travel in the turbulent air stream into dense areas. The sprayer shown in Fig. 10.42 is air assisted mist blower with an is attached nozzle for charging the spray particles. Tractor operated electrostatic sprayers are also available.

**Aircraft Sprayer**

Fixed-wing aircraft with spray booms mounted below the lower wing(Fig. 10.53) have been used effectively to apply herbicidal sprays over extensive areas to control brush and weeds since the 1950s. Today's spray planes are designed specifically for aerial application of liquid and dry materials and meet strict safety requirements. They use radial-piston or turbine engines with a range in horsepower from 600 to 1300.

| | |
|---|---|
|  |  |
| Fig. 10.41 Back Pack electrostatic sprayer | Electrostatic Spraying Systems |
| 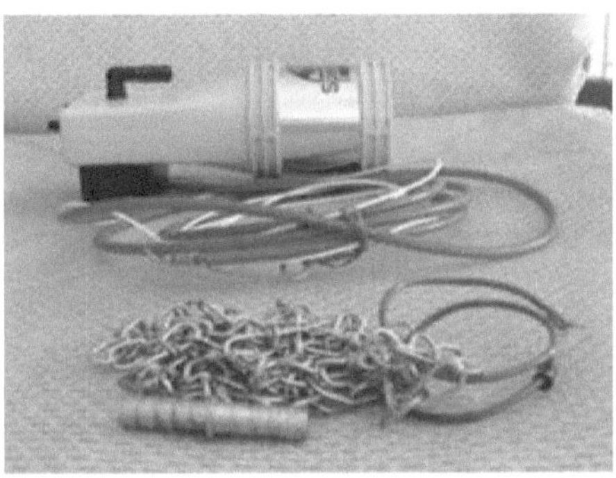 |  |
| Fig. 10.42 Air assisted electrostatic sprayer | http://spectrumsprayer.com |

Current spray tank capacities vary from 1500 to 3000 litres while older aircraft had spray tank capacities of 1200 litres or less. Aircraft speed can vary from 160 to 260 km/h but is about 300 km/h for brush and weed spraying. Spray boom is not more than three-fourths the length of the wingspan because the turbulent wingtip vortices cause drift. Spray pumps must have positive shut-off valves and nozzles must have check-valves to prevent continuous spraying or leakage over off-target areas. Hydraulic spray nozzles are the predominate type used on aircrafts and are available in many styles and sizes. GPS/GIS systems are helpful for precision application. The altitude for spraying is generally 60m and application rate is 2.5 liter/ha. The spraying capacity is upto 1300 ha/day.

Spraying equipment and techniques are designed to minimize drift while applying spray droplets of efficacious size to the plant canopy.

Aerial sprays are normally applied when wind speed is between 3 and 16 km/h and air temperatures are not above 32° C. Spraying should not be conducted in no-wind conditions because of inversions (warm air over cold air) or potential shifts in wind direction. Federal (EPA) and state regulatory agencies (USA) have strict regulations for aerial spraying, and applications must follow the specifications on the chemical's Product Label. Fixed-wing aircrafts are cost effective because they can spray large areas quickly and effectively. They have larger payload capacities and greater airspeeds than helicopters. Airstrips are required for landing, servicing, and takeoff. Excessive ferrying distances are wasteful and costly. GPS/GIS units negate the use of flagmen and can record flight patterns. Fixed-wing planes are not suited for spraying highly irregular shaped sites or mountainous areas.

| | | |
|---|---|---|
|  foxcouleeaviation.com Air Tractor, Inc |  foxcouleeaviation.com North Star Helicopters |  foxcouleeaviation.com North Star Helicopters |
| Fig.10.43 Air spraying by fixed wing aircraft | Fig.10.44 Air spraying by helicopter & use of truck as landing platform | |

Helicopters with boom sprayers are well suited for use in mountainous areas, steep terrain, highly irregular shaped sites, or remote areas They are highly maneuverable and apply sprays at much slower speeds than fixed-wing aircraft. Airspeed can vary from 50 to 150 km/h, but is about 70 km/h for brush and weed spraying. To maintain precision application rates, The aircraft should be equipped with variable-rate, flow-control units to compensate for changes in airspeed. GPS/GIS systems are helpful for precision applications. Spray tanks vary in size from 350 to 1200 litres, and the spray boom should not exceed 90% of the rotor diameter.

The conditions and regulations outlined in the fixed-wing aircraft section also apply to helicopters. Applications must follow the chemical's Product Label, and the applicator must be certified by the state. Batch trucks which carry clean water and chemical mixing tanks for helicopters usually have landing pads atop the truck for convenience and safety in servicing. Ground personnel must always be alert to the helicopter's moving rotors during servicing. Helicopters are more expensive to operate than fixed-wing aircraft, but they have the advantage of operating where fixed-wing aircrafts cannot. Helicopters do not need landing strips and are adapted to remote rugged terrain and irregular shaped sites. GPS/GIS units negate the need for flagmen in these remote sites and can record flight patterns.

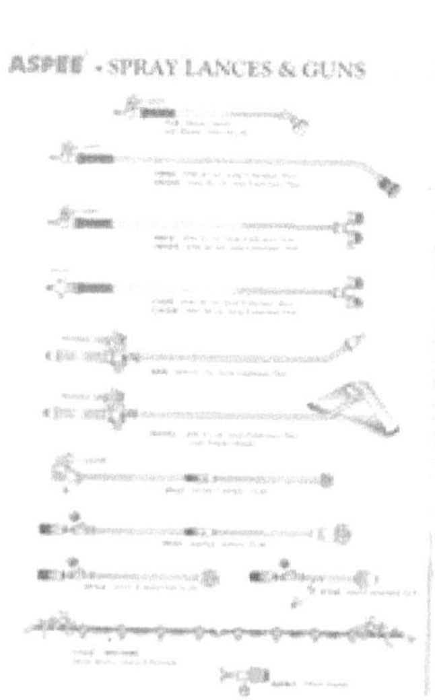


Fig. 10.45 Cut off valves and nozzles of different manual operated sprayers

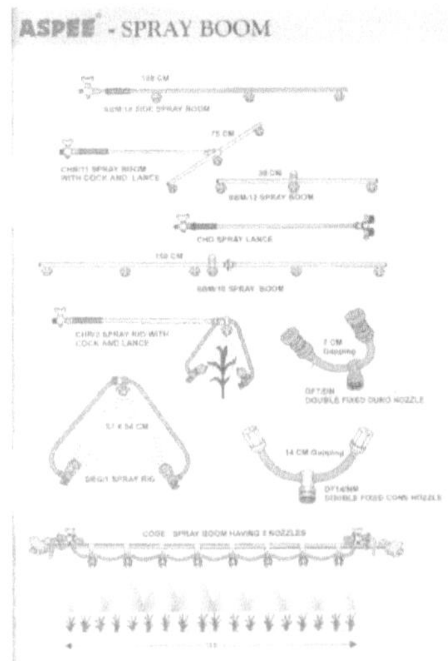


Fig. 10.46 Lances of different manual operated and spray guns of power operated sprayer

## Dusters

Manually operated dusters are shown in Fig. 10.45. The power operated knapsack duster and the tractor operated duster with spin disc are shown in

aspee.com

Fig. 10.47 Hand operated dusters

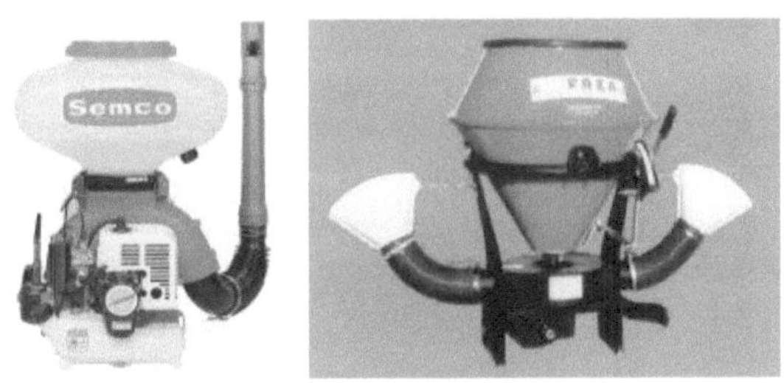


Fig. 10.48 Powered knapsack air assisted duster and tractor operated spin disc duster

## Kitchen Garden Sprayers

A number of kitchen garden sprayers are also available in the market. These may be pressure type or continuous pumping type (Fig. 10.49 &10.50).

| | |
|---|---|
|  aspee.com<br>Fig.10.49 Kitchen sprayer continuous pumping | 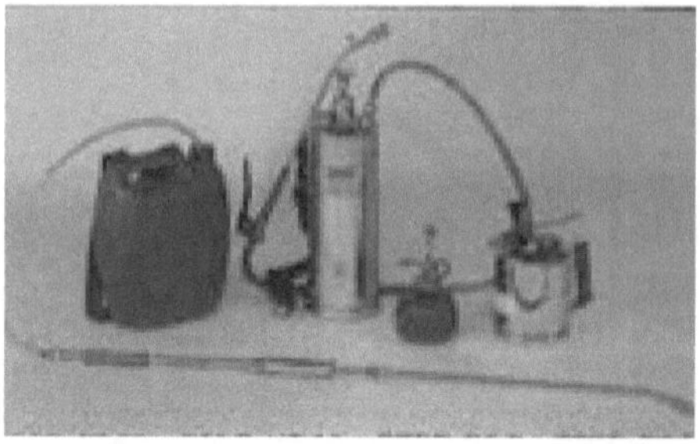 aspee.com<br>Fig. 10.50 Kitchen sprayer continuous type and pressure sprayer |

## 10.4 Economics

The approximate field capacity, labour requirement and cost of spraying of different kind of sprayers is given in Table. 10.1. This includes only cost of labour and machine and not the cost of chemicals.

## 10.5 Particle Size

The particle size in a spray may vary from 0.1 micron to 1000 microns The average particle size for a particular sprayer is obtained by collecting the spray in a thin layer of oil (Shell 33) or hydrocarbon solvent in a plate/dish. The particle sizes are measured by traveling microscope and the mean is worked out. The particles can also be photographed for measurement of diameters.

Another method of collection of spray particles is to put the blotting paper on the crop on tree leaves and spray the coloured water (Fig. 10.39). The diameter of dried impression is measured and the spray particle size is estimated by use of spread factor. The particles size distribution for some of the sprayers by using this technology is shown in Fig. 10.61. The particle impressions collected on the blotting paper can be magnified and particle diameter can also be calculated by projecting the impressions on screen. The scanners and computers are also used for measuring the particle size.

## Different kind of means are used to quantify the particle size.

Arithmetic mean diameter (AMD). This is arithmetic average of the diameter of all the particles (sum of diameter of 'n' particles divided by 'n')

Geometric mean diameter (GMD) – This is obtained by multiplying the diameter of 'n' particles and then taking the 'n'th root.

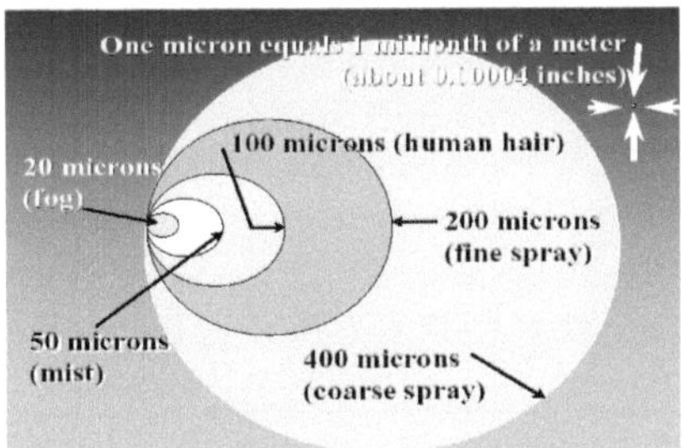


Fig. 10.51 Classification of particle size

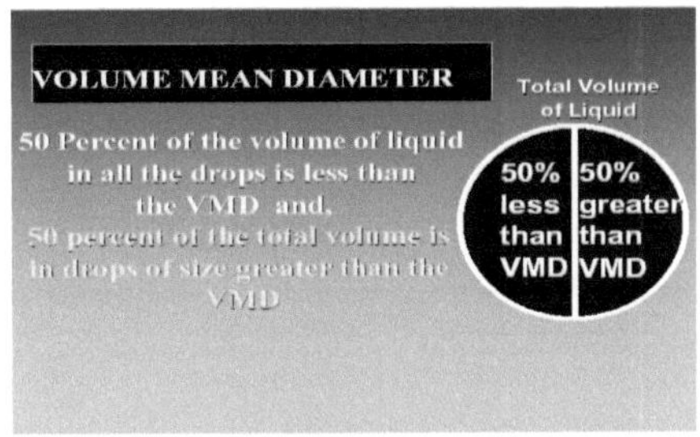


Fig. 10.52 Definition of volume mean diameter

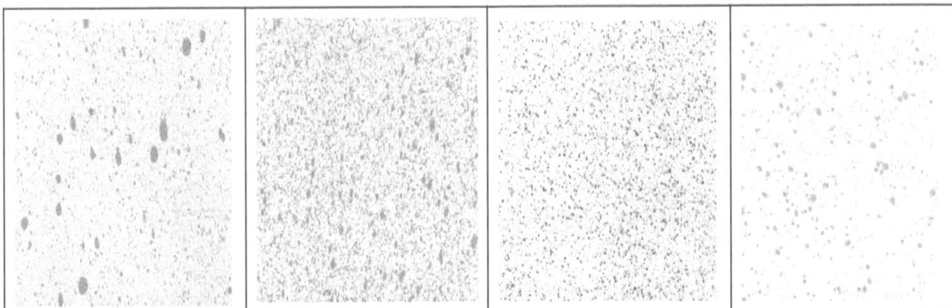

(a) (b) (c) (d)

Fig. 10.53 Spray particles of a. Manually operated knapsack, b. Rocker, c. Tractor operated air assisted (Tycoon), and d. Spin disc sprayers

Volume mean diameter (VMD) – This divides the droplet spectrum into two portions such that the total volume of all droplets smaller than the VMD is equal to the total volume of all droplets larger than the VMD. The volume mean diameter indicates that 50% of the liquid volume is in droplets smaller than this diameter and 50% liquid volume is in droplets larger than this diameter

Sauter Mean Diameter (SMD) – This represents ratio of total volume of droplets to the surface area of all the droplets.

$$SMD = \frac{\sum (Diameter)_i^3}{\sum (Diameter)_i^2}$$

The VMD is most common mean diameter for referring the particle sizes for sprayer or nozzle.

VMD>SMD>AMD>GMD

**Table. 10.1 Economics of Different sprayers**

| Sl. No. | Sprayer | Field Capacity (ha/day) | Labour required (nos.) | Cost (Rs./hectare) |
|---|---|---|---|---|
| 1 | Knapsack – Manual (Annuals) | 0.2 | 1 | 520 |
| 2 | Rocker/foot (Perennials) | 0.4 | 3 | 780 |
| 3 | Powered Knapsack/ Mist blower | 0.8 | 1 | 200 |
| 4 | Power tiller operated Power Sprayer (Perennials) | 2.0 | 4 - 6 | 420 |
| 5 | Tractor operated Boom sprayer (10m boom) (Annuals) | 6 - 10 | 1 | 280 - 420 |
| 6 | Tractor operated air assisted sprayer (Perennials) | 4.0 | 2 | 710 |
| 7 | Battery operated sprayer | 0.3 – 0.4 | 1 | 350 |

**Table. 10.2 Typical droplet size ranges for various pesticide applications[1]**

| Category | Average Droplet size in Microns | Examples of Uses |
|---|---|---|
| Fog | 0.1 – 50 | Thermal fog for pest control in greenhouses and other buildings |
| Aerosol | 1-50 | Ultra-low volume applications |
| Mist | 50-100 | High-pressure sprays |
| Fine Spray | 100-250 | Cone and fan nozzles for low-volume applications and air-blast sprayers |
| Medium Spray | 250-400 | Cone and fan nozzles for moderate volume low toxicity sprays requiring good coverage |
| Coarse Spray | 400-600 | Cone and fan nozzles for large volume and more toxic sprays |
| Minimum Drift Jet Stream Nozzle Spray | 600-900 | Aerial application near sensitive areas |
| Low Turbulence Nozzle Spray | 900-1,000 | Microfoil boom applicator for aerial spraying near sensitive areas |

[1]Ridgetown College, University of Guelph ,Canada
http://www.ridgetownc.uoguelph.ca/OPEP/On-Line%20Tool/pdfGPSC/22-ApllEquip.PDF

Different sizes of the particles are required for different pesticide applications

(Table. 10.2). Bigger size particles cover less area with more solution is to be sprayed. Small size particles cover more area but they have tendency to drift. Spray particles smaller than 200 micron are more prone to drift. Table.10.3 represents the drift of particles for spraying annuals and Table. 10.4 represents the drift of particles for spraying perennials. To control the drift the spraying operation shuld not be taken up when wind velocity is more than 8 kmph. The particle size has to be increased if the spray is to be done under windy conditions. Small size particles are also prone to evaporation before falling on leaves when ambient RH is very low. The particle size has also to be increased for spraying during sunny, windy and low RH conditions.

The chemical application rate and the particle size for different type of sprayers are given in Table. 10.5.

**Table 10.3 Distance Water Droplet Drifts While Falling 0.9m in 8 Kmph wind[2]**

| Category | Microns | Drift Meter |
|---|---|---|
| Very Coarse | 600 | 0.09 |
| Coarse | 500 | 0.12 |
| Medium-Coarse | 400 | 0.21 |
| Fine | 200 | 1.65 |
| Very Fine | 100 | 7.80 |
| Ultra Fine | 50 | 26.40 |

[2]David G. Wagner, Ph.D., P.E. Assistant Professor Extension Specialist in Precision Agriculture ,Agricultural and Biological Engineering, The Pennsylvania State University http://www.age.psu.edu/fac/Wagner/PrecisionAgriculture/Downloads/Presentations/ Spray DriftControl.PDF

**Table 10.4 Distance Water Droplet Drifts While Falling 3m in 5 Kmph wind[3]**

| Type of Droplet | Diameter (in microns) | Time Required for droplets to fall | Lateral distance traveled by droplets1 |
|---|---|---|---|
| Fog | 5 | 66 minutes | 5000 |
| Very Fine Spray | 20 | 4.2 minutes | 333 |
| Fine Spray | 100 | 10 Seconds | 13.2 |
| Medium Spray | 240 | 6 Seconds | 8.4 |
| Coarse Spray | 400 | 2 Seconds | 2.6 |
| Fine rain | 1000 | 1 Second | 1.4 |

[3]Wayne S Johnson, NEVADA Pesticide applicator's certification workbook, University of NEVADA Cooperative Extension University of NEVADA, RENO. www.ag.unr.edu/wsj/PAT/pat99mnl.pdf

**Table. 10.5 Chemical application rate and particle size (VMD) for different sprayers**

| Type of sprayer | Chemical application rate (l/ha) | Particle size (micron) |
|---|---|---|
| Manuallyi operated knapsack (HV) | 625 to 750 | 300 to 500 |
| Manually operated foot sprayer, rocker sprayer and power operated boom and spray gun type (HV) | 750 to 1000 | 300 to 500 |
| Manually operated knapsack mist blower (LV) | 100 to 200 | 150 to 200 |
| Tractor operated air assisted sprayers (LV) | 100 to 200 | 125 to 250 |
| Battery operated spin disc sprayer (ULV) | 2 to 3 | 5 to 50 |

# *Chapter 11*

# Fruit Harvesting

India is producing around 72 million tonnes of fruits per year.The harvesting of fruit trees is a cumbersome and time consuming process. Different methods are being followed based on the kind and value of the fruit. The fruits are generally plucked individually at natural break point by holding the fruit firmly and gently by hand. Due to shortage of labour some fruits like mango are being harvested by shaking the tree manually (Figs. 11.1 & 11.2). The fruits are allowed to fall on the ground and then picked up. This injures the fruit internally and the spoilage increases during ripening. The mango fruits are harvested by using conventional harvesters in different areas. In Karnataka this consists of a bamboo stick folded to form elliptical shape frame and tied on one end of a bamboo or any light wooden pole. The hand woven cotton net is tied below the frame to collect the fruits. The fruit is lowered in the net, held in frame and detached from the pedicel, while pulling the pole. The fruit gets harvested without pedicle and there is sap bleeding, which blemishes the fruit. The shine of the fruit is reduced and the fruit is susceptible to diseases like stem end rot. The conventional harvester shown in Fig.11.12 is fitted with a cutting notch to harvest the fruits with pedicel and a canvas collection sack to collect the fruits.

The mango harvester in North India generally consists of a bamboo pole fixed with a semicircular sickle blade and net at the distant end. The fruits are harvested by cutting the pedicel and the net is emptied after 4-5 fruits are collected in it. The labour also climbs the trees to harvest the fruits and throw these on a gunny bag ( Fig. 11.9). held by a person on the ground to reduce the injury to the fruit. This is a very slow process and dangerous for the person who climbs the tree to harvest the fruits. In some areas the fruits are harvested by shaking the trees and the fruits are collected on plastic sheet or canvas tarpaulin spread on the ground (Fig.11.10).

Timely harvesting of fruits is important for maintaining the transportability, quality and shelf life. The fruits are generally plucked / harvested individually at proper maturity from trees by hands. These are harvested at the natural break point formed at the junction of the stem and stalk when produce is mature. Labourer should gasp the product firmly but gently by hand and pull it upward as shown in Figures 11.1 & 11.2. Fruits are also harvested using different types of secature or shears (Figs.11.3 to 11.8). Wearing cotton gloves, trimming finger nails and removing jewellery such as ring and bracelets can help to reduce mechanical damage during harvest.

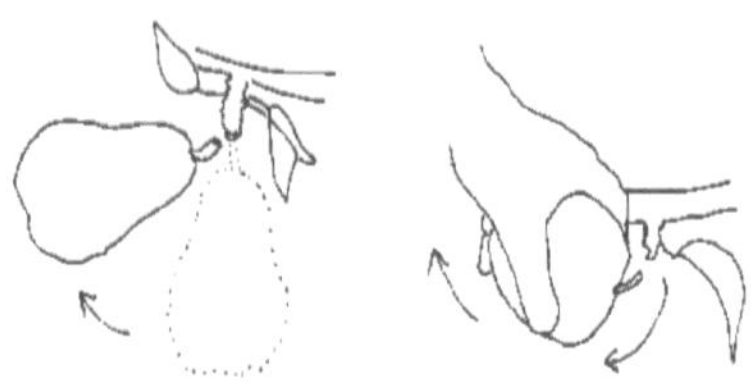

Fig.11.1 Harvesting by grasping in hand and pulling upward

FAO

Fig.11.2 Harvesting at the natural break point

FAO

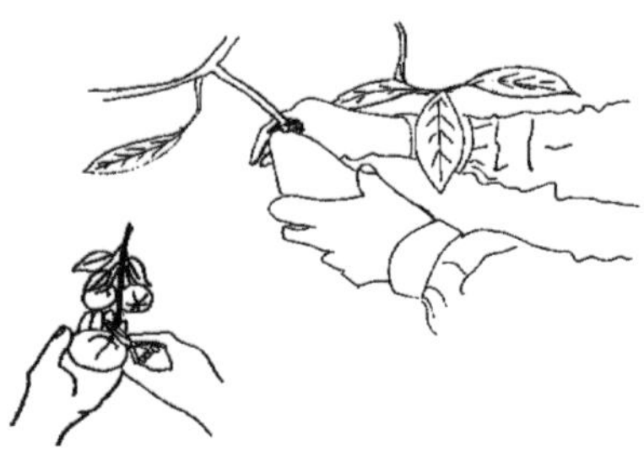

Fig.11.3 Harvesting using clippers

FAO

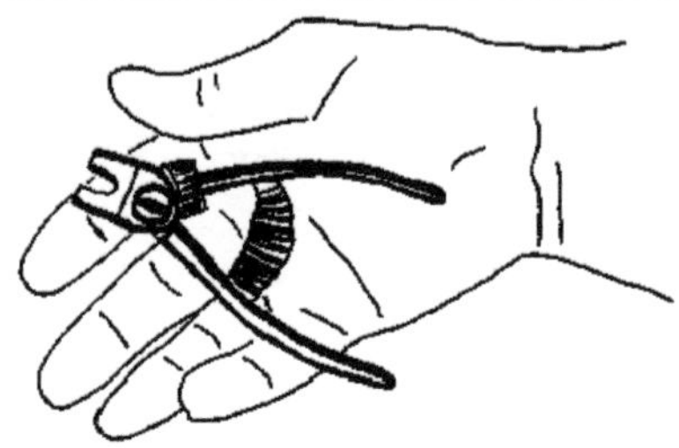

Fig.11.4 Clipper for citrus fruits

FAO

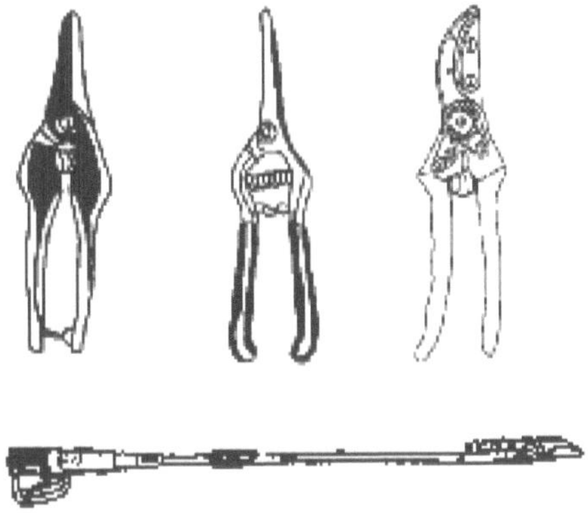

Fig.11.5 Straight blade, curved blade, cut and hold hand shears and Pole mounted cut and hold picking shears
FAO

Fig.11.6 Improved secateur to harvest orange fruits
(CIPHET, Ludhiana)

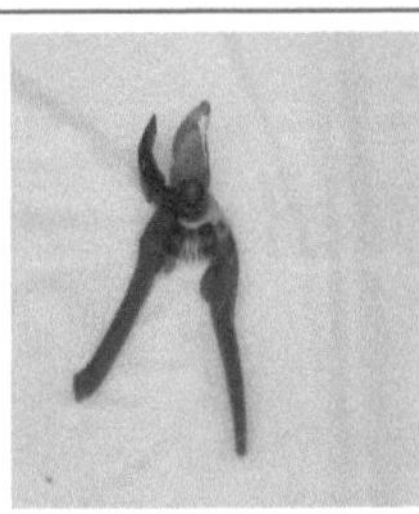

Fig.11.7. Cut and hold pruner/ harvester
IIHR Bangalore

Fig.11.8 Harvesting using cut and hold pruner/ harvester
IIHR Bangalore

The mango is harvested by shaking the branches of the trees and fruits are allowed to fall on the ground in the regions near by Kolar and other regions due to shortage of labour.Fruit harvesting aids for mango have been developed at Indian Institute of Horticultural Research (IIHR), Bangalore and Dr. Balasaheb Sawant Konkan Krishi Vidyapeeth (KKV), Dapoli (Maharashtra). These harvesters harvest the fruit with pedicel to increase the shelf life. The harvesting aids for sapota, guava and lime have also been developed at IIHR, Bangalore. However the harvesting capacity of these harvesters is very low as individual fruits are to be harvested and collected. In Thailand, vibrating harvester with 8.3/hp diesel engine for mango has been developed. The individual branches of mango are held in padded clamps and vibrated with amplitude of 76-102 mm at 11 – 13 cycles per second.. About 80-85% fruits are removed. The fruits are collected on a nylon catching net supported on portable stands. This harvester can be adopted in India for harvesting different fruits.

In Australia, a tractor operated mango harvester has been developed. The labourers stand on the platform on the trailer with harvesting aids for harvesting the individual fruits. The harvested fruits fall on the apron where these are washed with detergent and moved by conveyor to the rear end of the harvester to fill the bulk bin. This system of harvesting can also be easily adopted under Indian conditions.

Fig.11.9 Catching the fruits on gunny Bag
FAO

Fig.11.10 Spreading plastic sheet/ tarpaulin for fruit collection
FAO

Fig.11.11 Orange harvester with flexible pipe
CIPHET, Ludhiana

Fig.11.12 Conventional mango harvester
(Karnataka)

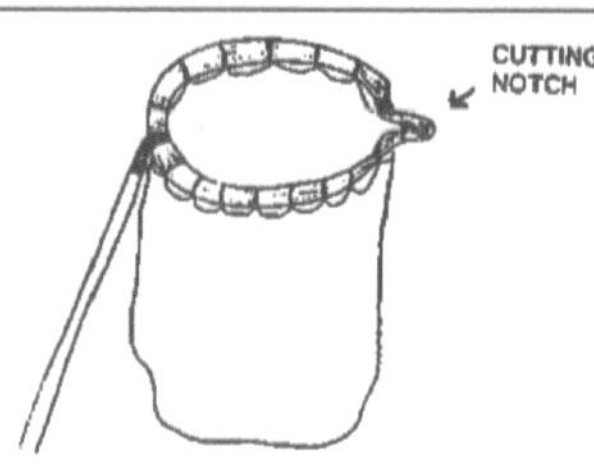


Fig.11.13 Harvester with canvas collection sack
FAO

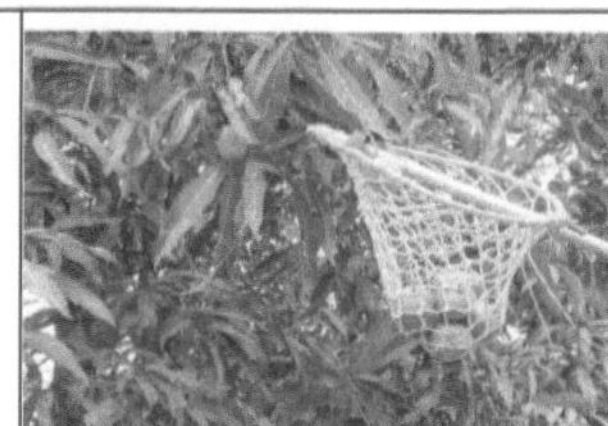

Fig.11.14 Mango harvester with blade
IIHR

Fig.11.15 Mango harvester with spring loaded blade
KKV Dapoli

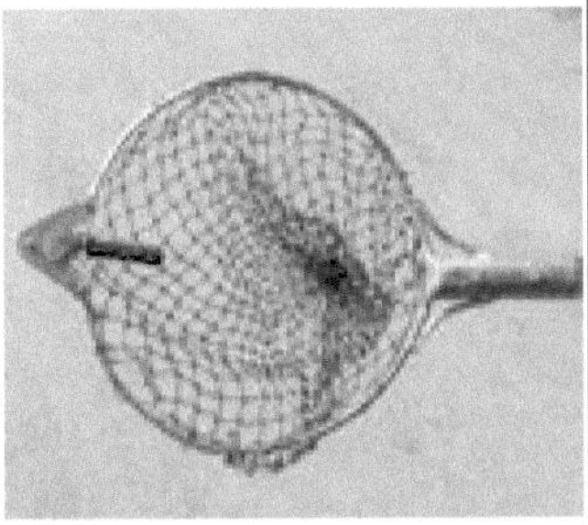

Fig.11.16 Mango harvester with serrated circular blade
KVK Dapoli

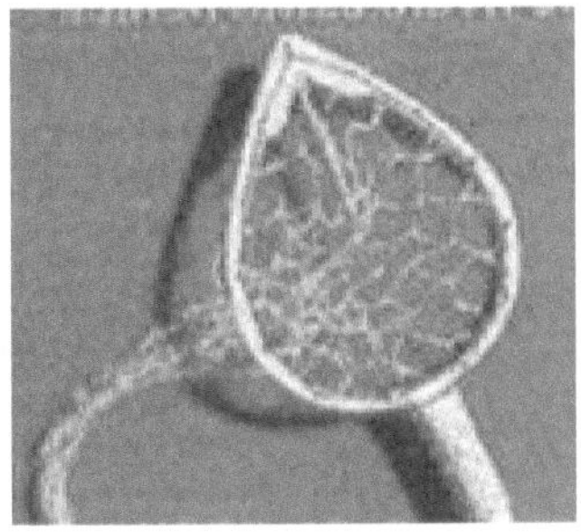

Fig.11.17 Mango harvester with collection chute
CISH, Lucknow

Fig.11.18 Mango harvesting Aid
(Australia)

Fig.11.19 Mango harvester in operation
IIHR

The canopy shaker harvesters with catching apron generally used for harvesting of orange in Florida, USA can also be modified for harvesting the fruits. The apple harvester with pneumatic shaking mechanism for branches can be also adopted. For adopting the mechanical harvesters the tree canopy has to be properly trained. The fruits harvested mechanically are generally utilized for processing. The mechanical banana harvesters can be adopted in Maharashtra.

The harvesters developed for different purposes in India and abroad are discussed here to increase the harvesting capacity and reduce the damage to the fruits.

## Manual Harvesting Aids

These have been developed to increase harvesting capacity of the person and to reduce the damage to the fruits. Generally one person is required to use these aids to harvest the fruits. The harvesting capacity depends upon the yield of fruit tree and type of fruit.

## Mango harvester

The conventional harvester used in Karnataka is shown in Fig.11.12. It harvests the fruits without pedicel. This consists of a bamboo stick folded to form elliptical shape frame and tied on one end of a bamboo or any light wooden pipe. The hand woven cotton net is tied below the frame to collect the fruits. The fruit is lowered in the net, held in frame and detached from the pedicel, while pulling the pole. The fruits get harvested without pedicel and there is sap bleeding, which blemishes the fruit. The conventional harvester shown in Fig.11.13 is fitted with a cutting notch to harvest the fruits with pedicel and a canvas collection sack to collect the fruits.

Mango harvester developed at Indian Institute of Horticultural Reaearch, Bangalore (Fig.11.14) is an improvement on conventional harvester and harvests the fruit with pedicel. The fruit is lowered in the net in the opening in between frame and aluminium pole . While pulling the harvester, the fruit is held between the cushioned frame and the pole and the blade cuts the pedicel of 1-2cm length to harvest the fruit. There is no sap bleeding, the fruits remain shining without blemishes and the shelf life of the fruit is increased by 3-4 days. The capacity of IIHR mango harvester is also more than the conventional mango harvester as it is light in weight. It needs less force for harvesting the fruit by cutting the pedicel by blade (1 kg for cutting pedicel in comparison to 4 kg by pulling). When 4-5 fruits are collected in net, the net is emptied. The frame is also made conical in shape for easy pushing of the harvester in the canopy of the tree. The harvesting capacity of the harvester is 94.0 and 98.2 kg/hr for Alphonso and Totapuri respectively in comparison to 52.7 and 44.2 kg/hr by tree shaking. The damage to the fruit by IIHR harvester was 6.4% in comparison to 9.2% by conventional harvester and 42.6% by tree shaking. Around 15000 IIHR mango harvesters are being used in Andhra Pradesh, Karnataka and Tamil Nadu.

Naveen/Dapoli harvester (Figs.11.15 & 11.16) have been developed at Dr Balasaheb Sawant Konkan Krishi Vidyapeeth ), Dapoli (Maharashtra). The KKV, Dapoli harvester (Fig.11.15 ) consists of 2 spring loaded blades fitted to the frame. The fruit is lowered in the net and the pedicel is sheered-off/cut when it is held between the blades while pulling the harvester. The fruits harvested with pedicel are suitable for export as no sap oozes out during harvesting. The damage to the fruit occasionally takes place when the blades touch the fruit. Dapoli harvester for mango (Fig.11.16) has a rotary serrated circular blade and cut the pedicel when it gets engaged in it while pulling the harvester. The harvesting of mangoes is shown in Fig 11.19. The net is emptied after harvesting 4 to 6 fruits. The harvester has to be conical in shape for easy pushing in canopy. The harvester has to be light in weight for easy handling.

The mango harvester shown in Fig.11.17 is developed at Central Institute of Sub-tropical Horticulture, Lucknow. It is provided with a long net chute to bring the harvested fruits to ground, without damage. However the chute increases the weight of harvester and gets entangled in the canopy of the tree.

The commercially available mango harvester in Maharashtra and Karnataka (Fig.11.24) is fitted with a "V" shaped blade for harvesting the fruits. In Australia, the mango fruits are harvested without pedicel with the harvesting aid (Fig.11.18). However the fruits are collected on catching frame and washed with detergent solution to remove the same.

<table>
<tr>
<td><br>Fig.11.20 sapota/ guava/ pomegranate harvester<br>IIHR</td>
<td><br>Fig. 11.21 Lime harvester<br>IIHR</td>
</tr>
<tr>
<td><br>Fig.11.22 Apple harvester<br>Bhardwaj, K.C. CIAE</td>
<td>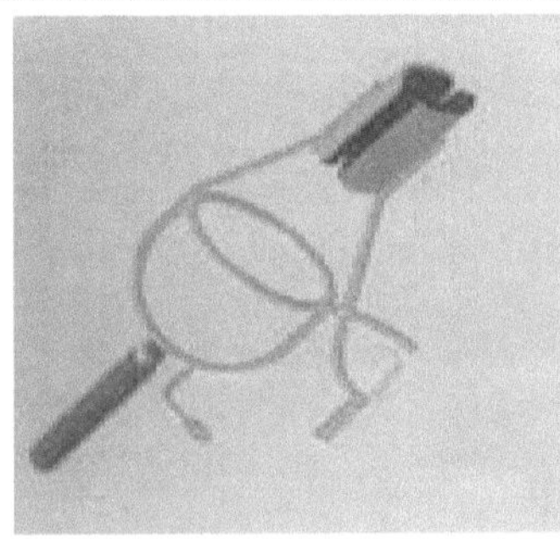<br>Fig.11.23 Papaya/pears harvester<br>Bhardwaj, K.C. CIAE</td>
</tr>
<tr>
<td>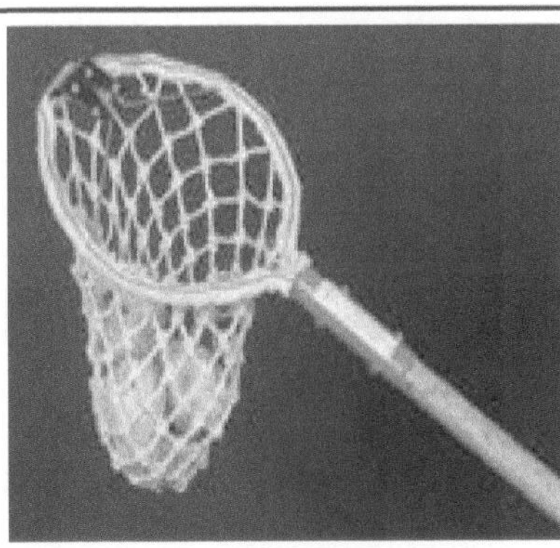<br>Fig.11.24 Mango fruit harvester<br>Jayshreedies</td>
<td><br>Fig.11.25 Orange / sapota harvester<br>Ningbo Beilun Daqi Hengxing Machinery Co., Ltd.</td>
</tr>
</table>

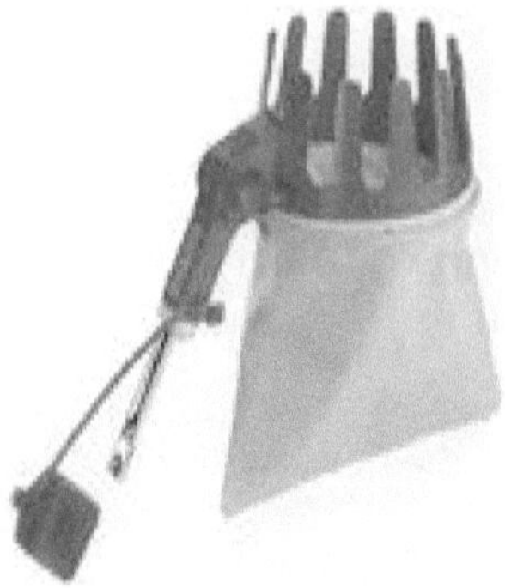

Fig.11.26 Sapota/guava harvester
WOLF-Garten Tools

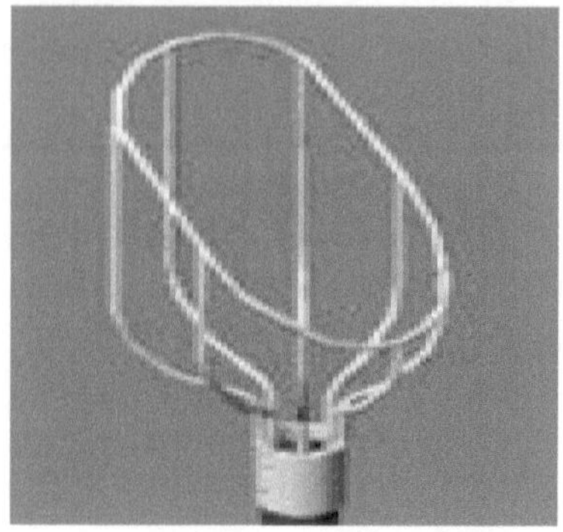

Fig.11.27Papaya harvester
FAO

**Lime harvester**

The lime fruits are conventionally harvested by a wire hook on a iron wire/ rod or wooden pole. The fruits are allowed to fall on the ground and collected manually. The lime plants have a very thick canopy and it is not possible to have the net for collection of the fruits. A lime harvester (Fig.11.21.) has been developed at IIHR, Bangalore. A small box of around 10cm dia and 20cm length is fitted with the hook. The harvester is pushed easily in the canopy, the fruit is held in the hook, harvested and collected in the box while pulling the harvester. The box is emptied after 8-10 fruits are collected in it. It has the harvesting capacity of 15-20 kg/hr.

**Sapota, Guava & Pomegranate harvester**

A sapota, guava and pomegranate harvester (Fig 11.20) has also been developed at IIHR, Bangalore. It consists of harvesting frame conical in shape and a pole. The harvesting frame is fitted with comb consisting of cushioned pegs to hold the fruit and harvest these while pulling the harvester. The fruits of required size only get harvested while the immature fruits are left on the tree. The capacity of sapota and guava harvesters is 50 – 100 kg/hr. In pomegranate harvester, the distance of pegs on the frame is more to suit the size of fruit to be harvested. The commercially available crown type harvester (Fig 11.26) has pegs in a circle and is suitable for harvesting sapota and guava fruits.

**Orange harvester**

Oranges are attached to the tree with pedicel and are to be harvested by cutting the pedicel without the rind break or damage to the skin. A improvised secateur (Fig.11.6) has been developed at the Central Institute of Post Harvest Engineering & Technology, Ludhiana. The fruits are harvested without pedicel and put into the flexible pipe (Fig. 11.11) to transport these to the ground. The flexible plastic pipe 10-12.5cm in diameter is fitted with a funnel type opening and tied at back of the labour/operator. The operator can also climb the tree/ladder to harvest the fruits and put the same into a plastic pipe to convey the fruits to the ground without any damage. The IIHR mango harvester can also be successfully used to harvest oranges. The commercially available harvester (Fig.11.25) is having a basket to hold the fruits and hooks for harvesting the fruits. It is suitable for harvesting oranges and sapota.

**Apple harvesters:**

It is a secature type harvester (Fig.11.22)and consists of one fixed blade and another moving blade. There is a net basket attached to collect the plucked fruits. The long handle facilitates reaching fruits from the ground. During harvesting, the fruit pedicel is adjusted to rest on fixed blade and pressing the lever at the grip end of the handle actuates the moving blade. The overall length of the tool is about 3 meters and weight of cutting head is about 1.3 kg. The process of harvesting is very slow, as the pedicel of the individual fruit has to be searched from a distance of about 3 meters. The harvester is also suitable for harvesting orange.

**Papaya / Pear harvester:**

It is a manual-harvesting tool with which individual fruit is just held between the two jaws and then twisted to shear off the stock (Fig.11.23). The jaws are held together by a tension spring on a pivot fitted on the frame. One of the jaws has lever bracket and rope arrangement for operating the jaw. The jaws are padded with rubber sheet to avoid any damage to the fruit while holding and twisting it. After its detachment, the fruit is released by pulling the chord into the ring. A cloth conveyer (chute) is provided below the jaws for collection of the harvested fruits to ground level. The handle of required length is provided to the harvester. It is suitable for harvesting papaya, pear and peach. The papaya harvester (Fig.11.27) is fitted with a basket fabricated out of metal rods and it can harvest individual papaya fruits.

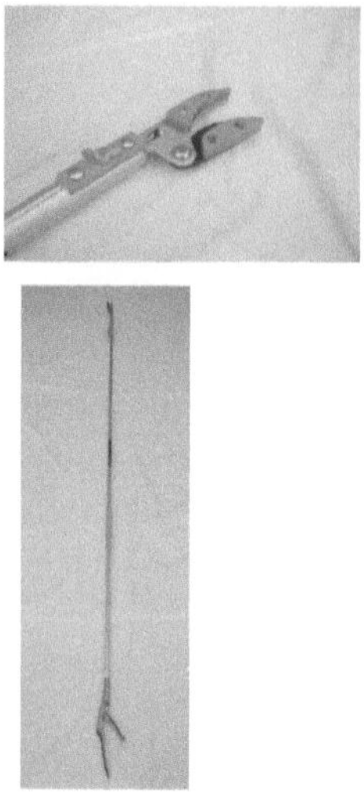

IIHR

b. padagro.com

c.E-tradersquare.com

d. Electric Fruit Picker
padagro.com

Fig. 11.28 Long reach cut and hold harvesters

Fig. 11.29 Long reach harvester with electric cutting

Fig.11.30 Harvesting litchi and mango with long reach cut and hold harvesters

**Long Reach Cut and Hold Harvesters**

The cut and hold harvesters have secateurs head fitted at distal end, which is able to hold the fruit after harvesting (Fig.11.28 a. and 11.30). These are suitable for harvesting most fruits including mango and litchi. The pedicel of the fruit is cut and it is lowered to the ground and put in the required container. The harvesters (Fig.11.28 b and c )are fitted with holding fingers /gripper at the distal end. The fruit is harvested by pulling the pedicel of the fruit after holding it. These are suitable for harvesting sapota. Harvesters with battery operated secateurs are also available commercially (Fig.11.29)

**Shaker harvester**

The telescopic pole harvester with battery operated revolving reel is available and is suitable for harvesting olives (Fig.11.31a) The beater harvester for olives is operated by 26 cc / 900 W petrol engine and produce 1150/beats per minute (Fig.11.31b) . The beater harvester has a harvesting capacity of 50-80 kg/hr. The shaker harvester for limb/branch is provided with 48.7 cc / 2.2 kW petrol engine, weighs 13.9 kg and the pole length is 2.83 metre. (Fig.11.31 c and d)

| | | | |
|---|---|---|---|
| 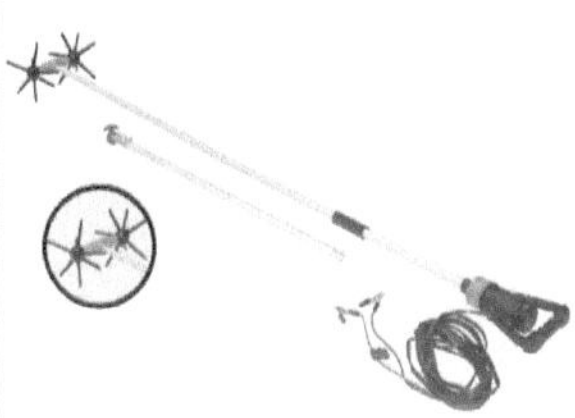 Electrical shaker harvester Zhejiang Zhongli Tools | 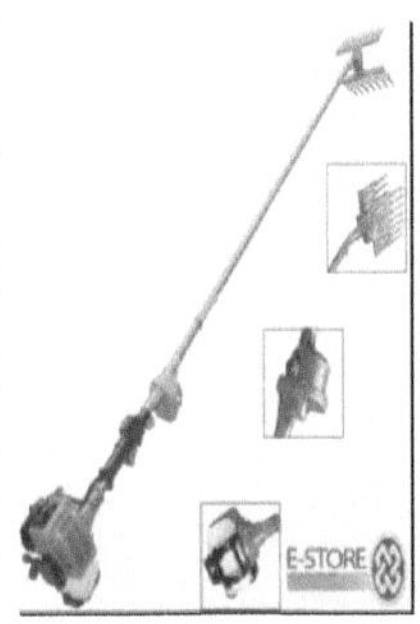 Olive shaker harvester China E-Store |  Shaker harvester for branch Stihl |  Shaking olive branch Stihl |
| Fig.11.31 Shaker harvesters | | | |

**Ultra Lightweight Coconut / Oil Palm Harvester**

The light pole aluminum harvesters with pole length of up to 12 meter are available for harvesting coconut and oil palm (Fig. 11.32). The aluminum pole is fitted with a sickle and the fruits are harvested by holding the bunch with the sickle and pulling the pole. The telescopic petrol engine (length 3.7 metre) operated vibrating sickle harvesters are available for harvesting the oil palm (Fig. 11.33). The weight of petrol engine operated harvester is about 7 kg. and the fruits can be harvested up to a height of 5 m. About 150 bunches can be harvested per day using this motorized harvester.

**Man Positioning System**

Generally the manual labour climbs the tree to harvest the fruits in India. However this has to be avoided as there is a risk for the person to fall down. Portable aluminium light weight tripod ladder (Fig.11.34 and 11.35) can be used as a man positioning system to avoid the climbing of labours on the trees to harvest the fruits. Ladders are very useful in pruning the tall trees also. The double pole telescoping ladder (Fig.11.36) and single pole ladder (Fig.11.37) are suitable aids for harvesting coconut, arecanut , black pepper and bêtel wine.

Fig. 11.32 Light pole harvester for oil palm and coconut

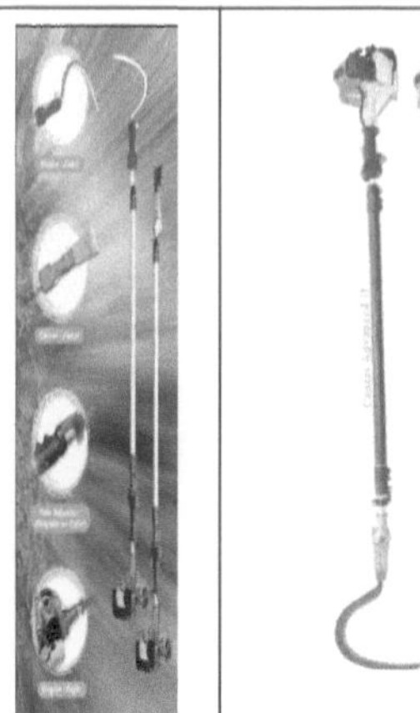

Fig. 11.33 Engine operated harvesters for oil palm
Cantas, Etani Sdn. Bhd.

Fig.11.34 Portable ladder for man positioning for citrus harvest
hort.purdue.edu/newcrop

Fig. 11.35 Tripod Apple Ladders
Oesco, Inc

Fig. 11.36 Single pole aluminum ladder
Dolphin Irrigation Systems

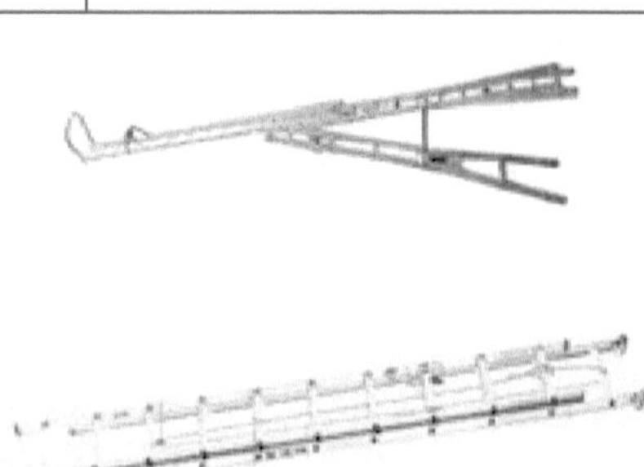

Fig. 11.37 Double pole telescopic Extension Ladders
http://www.kamsun.com

A self propelled hydraulically operated man positioning system (Fig.11.38) has been developed by Johnson Manufacturing Inc., USA. The labourer stands at different heights to pluck the fruits. Belt conveyers are provided on the side for the labour to place the harvested fruits. The fruits are conveyed to the field bin by the conveyor belt.

Self propelled elevating work platforms (Fig.11.39 and 11.40) are also available for pruning , spraying and harvesting tall trees. One operator stands in the cage / platform and controls the height and movement of the platform. The fruits are plucked manually and put in the bag provided on the machine. The bag can be emptied after lowering the platform to the ground level. The hydraulic operated chain saw and pruners ars also provided for pruning on these platforms.

Fig.11.38 Self propelled hydraulically operated man positioning system (Johnson Manufacturing Inc., USA)

Fig.11.39. Automatic mechanical man positioning system
hort.purdue.edu/newcrop

Fig.11.40 Self propelled platform
Afron

Fig.11.41 Self propelled platform
http://www.pillonsrl.it

Fig.11.41a Self propelled platform
http://www.goodfruit.com

Fig.11.41b Tractor trailor mounted platform
IIHR

Fig.11.42 Tactor mounted platform

Afron

Fig.11.43 Self propelled platform for palms

Afron

Fig.11.44 Tractor mounted platform for palms

KKV Dapoli

Tractor mounted and self propelled lift for harvesting of date palms developed in Israel. Tractor mounted platform developed at KKV Dapoli is shown in Fig. 11.44c and it can be lifted to a height of about 8.5 meter. It is suitable for harvesting and spraying of coconuts.

**Coconut climber**

The palm climber (Fig.11.45) designed and developed separately by M J Joseph Alias Appachan and P K Joseph consists of two metal loops of 10 mm MS rod having sub-loops, rubber belt, wire ropes, connecting clamp, MS plates etc. One loop is intended for the right leg called right leg loop and the other for the left leg. These are called the left leg loop. The left loop leg (main loop) is slightly bigger than the right loop. The top of the main loop is bent forward to form handle and just below this part, two metal plates are attached with holes with a long rubber belt. On the rubber belt a wire rope having rings on each end is fixed. The bottom most part is a plate and a clamp is provided above it. A long holed plate is fixed on the main loop, which is used as parking brake.

Fig. 11.45 Co-co nut climber
M J Joseph alias Appachan

Fig. 11.46 Co-co nut climber
TNAU Coimbtore

Fig. 11.47 Co-co nut climber
R- Tech

Fig. 11.48 Motorized co-co nut climber
Bhadra Industries

Ropes attached to the climber are used to fasten it to the palm. The ropes passes around the tree through hooks provided in the climber. A pedal is provided for resting the foot. The right side peddle is raised a little upward so that the grip of the right hand part of the climber will become loose facilitating upward movements by using the right hand and the right leg. Once the right part is moved up, the weight of the body rests on the right hand part of the climber. The whole process is repeated for the left hand side. In this way one can easily climb coconut or arecanut palms. The equipment costs about Rs 2,000. The labourers however do not feel safe 10 meter height.

The coconut climber designed by TNAU Coimbtore (Fig.11.46 and 11.47 ) has two rectangular iron frames with each frame fitted with two rubber bands, which hold the trunk of the tree when the weight is put on this . The rubber bands donot hold the trunk when the weight is not put on these bands. The labourer climbs the tree trunk by transferring his weight on lower frame ( by legs) and upper frame (by hip) and lifting the frames alternately.

The coconut climber (Fig.11.48) developed by Bhadra Industries is operated by 5 hp diesel engine. Two spring loaded rollers are provided across the trunk, the power to the rollers is provided by the engine. The operator has been provided the comfortable seat control for climbing or coming down. The machine operation is not strenuous to the operator and the spraying and harvesting of coconuts can be done safely and comfortably at much higher speed.

**Shaker Harvesters for Nuts and Apple**

Self propelled shaker harvesters are available for shaking the trunk of the tree (Figs. 11.49 , 11.50, 11.54, 11.56, 11.57, 11.58 & 11.60) or branches of the tree. Tractor oprated trunk shaker and limb/ branch shakers are shown in Fig. 11.51 and Fig. 11.52 respectively.

Their stroke and frequency of vibration can be adjusted. The fruits fall on the ground ( Figs. 11.50, 11.51) and are than collected by fruit sweeps (Fig. 11.61) The fruit pick up machines developed by Agricultural Machines Inc. Avon park, Florida pick up oranges at 15-20 tonnes/hr. at 97% efficiency 3 to 7% split peals from handling.The fruits are collected by horizontal or vertical rotors, conveyed to the bins through conveyor belt and leaves and dust is blown away by blowers. The fruits are also collected by the catching frames.

Fig. 11.49 Self propelled tree shaker
http://www.shakermaker.com

Fig. 11.50 Harvesting by shaking apple tree
http://www.shakermaker.com

Fig. 11.51 Tractor operated tree shaker
http://www.feucht-obsttechnik.de

Fig. 11.52 Tractor operated limb/ branch shaker

Fig..11.59 Mango harvester with manual aids
Granite Engg. Co., Australia.

Fig. 11.60 Self propelled shaker harvester with invertsd umbrella collection apron and belt conveyor

http://www.shakermaker.com

Fig. 11.61 Fruit sweeper

**Catching Nets /Surfaces**

To harvest mango generally one person climbs the tree with the harvesting aid and another person stands underneath to catch the fruit on the gunny bag to avoid damage and convey these to the ground safely. The portable frame and catching net can be used as shown in Fig.11.53. The catching net can be supported on 4 portable adjustable height stands. It is a retractable apron, which is pulled and spread under the tree. Fruits are harvested by trunk shaking and fruits fall on the apron. The apron is retracted by the tractor , fruits are gathered and transferred to bin. The fruits are collected at one place by adjusting the slope of the catching surface (Figs.11.55). The catching frames in Fig. 11.56 and 11.57 are inclined and self propelled. The tree trunk shaker is provided under one of the catching frame. The conveyor belt is also provided under the other catching net. Both the catching frames are brought to the sides of the tree. The trunk is shaken/ vibrated by the shaker. The fruits get harvested, fall on catching frames, roll and get collected on the conveyor belt. These are then transferred to the bin by the conveyor belt. Sufficient slope is given to canvas to collect the fruits at centre. The fruits are windrowed in the centre of catching frames. Tractor pulled catching frames along with vibration harvester is shown in Fig.11.52. The size of the catching frame/ canvas is large enough to cover the complete canopy of the tree.

The reverse umbrella type catching frames are shown in Fig.10.60. These can be tractor operated to open /close the umbrella and transport. The tree is covered completely underneath by opening the reversible umbrella. The fruits are harvested by shaking the trunk mechanically. The fruits get collected at centre and conveyed by the belt conveyor to the bin.

The mango harvester (Figs.11.59) has the catching frame, which covers the canopy of one side or half of the tree. It is pulled by tractor, the washing arrangement for fruits and conveyor belts for collection of fruits are also provided.

The stand for the laborer is also provided for standing and harvesting the fruits with hooks.

## HIGH CAPACITY HARVESTERS

Some of the harvesters used/developed for harvesting mango, apple, orange, blue berry, pineapple and banana are discussed below. These harvesters can be downsized and used for mechanization of harvesting of different fruits in India.

### Mango harvester with manual aid

A tractor operated harvester (Fig.11.59) has been developed in Australia (Bally, et.al, 1999). It consists of a frame, mango catching apron and conveyors for collecting the mangoes into the field bin. Seven to eight labourers stand on the frame of the harvester, which is pulled by a tractor. The labourers pluck the fruits with harvesting aids (Fig. 11.18) and these fall on the apron. The fruits are washed with detergent solution to remove the sap when they roll down the apron and are collected at the end by the conveyors (p.t.o. operated) and filled in the field bins.

### Shaker harvester for mango

A low cost shaker harvester has been developed at Asian Institute of Technology, Bangkok for harvesting mangoes. The limb of the tree is held in the padded clamps. It is shaken at a stroke of 76-102mm at a frequency of 11-13 Hz to achieve 80-85% fruit removal/harvesting. The fruits are collected on the nylon net supported on portable net stands below the tree. Two to four trees can be harvested in an hour depending upon the tree size.

### Shaker harvester for nuts

The tractor operated harvester (Figs.11.49, 11.50, 11.54 & 11.55). shakes the branches of the nut trees to harvest The stroke and frequency of the vibration can be adjusted and The fruits are collected by the catching frames.

Fig.11.62 Apple shaker (Pneumatic) harvester
(American Society of Agricultural Engineers, USA)

Fig.11.63 Orange harvester shaker/spiked drum
(Korvan Industries Inc., USA)

Fig.11.64 Mechanical harvesting of orange
(American Society of Agricultural Engineers,USA)

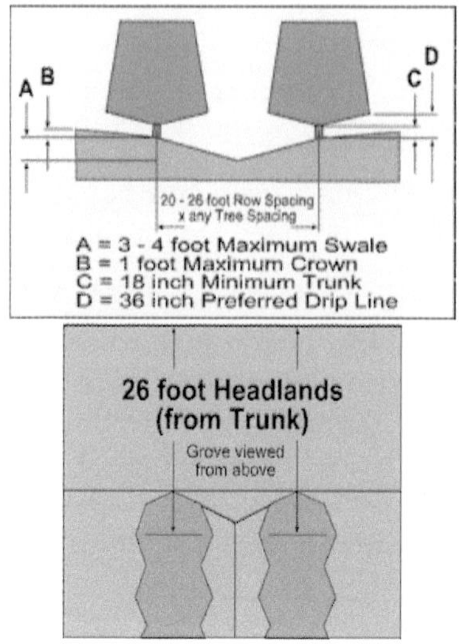


Fig.11.65 Planting arrangement for mechanized orange harvesting (
Korvan Ind.USA)

Fig.11.66 Spiked drum harvester for blue berries
(Littau Berry Harvester Co.Salem, USA)

Fig.11.67 Spiked drum harvester for blue berries
(Littau Berry Harvester Co.Salem, USA)

**Trunk shake catch harvester for orange**

The trunk shake – catch harvester (developed at Compton Enterprises Inc., Chico, California, U.S.A.) harvests 90-94% fruits at harvesting capacity of 10 tonnes/ hour. The orange tree is shaken for 5 to 10 seconds at a frequency of 8 to 10Hz. The shake – catch harvester has 2 sets of unbalanced masses totaling 125 kg. and an eccentricity of 22 cm. and an rotated in same direction at 8 to 10 Hz (480 to 600 rpm).

**Shaker harvester for apple**

The shaker type apple harvester has been developed in USA (Fig11.62). The canopy of the apple trees is trained in such a manner that shaker harvester can harvest fruits. The tree is generally trained into two branches (Y-shaped). The individual branch of the tree is shaken by pneumatically operated cylinder by giving one jerk. The fruits get detached and fall on the collecting apron and are collected and conveyed to the truck using belt conveyors.

**Canopy shaker harvester for orange**

A Spiked drum type harvester (Figs.11.63 & 11.64) has been developed in USA to harvest oranges. The planting of the trees is done at specified spacing and the canopy is shaped as shown in Fig 11.65. The tree is brought in between the spike drums of 6m dia. and the rotating drums are pushed inside the canopy, shake the branches of the tree and harvest the oranges (Fig11.64). The fruits fall on the catching frame/net and are conveyed to the truck by using belt conveyors (Fig.11.64). The harvested fruits are suitable for juice extraction. The canopy harvester developed by United States Department of Agriculture has 2 shaker drums with eight sets of horizontal whorls, 3.2 cm. diameter nylon spokes mounted on each of the two vertical shaker shafts. Each horizontal whorl set contains 16 spokes each1.5 meter long beyond the shaft/hub to which they are bolted. The vertical spacing between horizontal whorls is 38 cm and top set is about 3.7 meter above the ground. Oranges are removed by the 2 shaker drums revolving horizontally in opposite direction as the spokes free wheeled through the tree canopy. Upto 95% oranges were removed in small trees when the machines operated at 1.2 to 2.4 km/hour on one side of the canopy and the harvesting capacity is observed upto 25 tonnes/hour. The fruits fall on a conveyor belt provided below the whorls and trash is removed before filling these in tractor trailer (Peterson (1998) and Whitney (1999)).

**Canopy shaker harvester for blue berries**

A spiked drum shaker harvester has also been developed for harvesting blue berries

(Figs.11.66 & 11.67) (Richard, et.al, 1998). The fruits are collected on the catching nets below the harvester and conveyed to the storage bin through conveyor belts. The fruits are also graded by the machine and sorted manually by 2-3 labourers. The harvesting capacity is around 400 – 500 kg/hour. The fruits are suitable for fresh consumption (eating)/table purpose. The harvester machine can harvest the blue berris grown on trelliswith about 80% marketable yield. The harvester with spiked tooth type drums for blue berry can be easily modified for harvesting Ber and Aovla (goose berry).

## Pineapple harvester

Pineapples are generally harvested manually by a knife and are put into bags or baskets (Fig.11.68). As the fruit plant has spiny leaves and the fruits have crown with spiny leaves, the harvesting labour has to be very skilled and hare its wear protective gears like gloves, spectacles and clothing (Fig.11.69). The mechanical harvesting machine (Fig.11.70 to 11.72) is a collecting device for the fruits. Fruits are harvested manually the crown root portion is cut manually and placed on a conveyor belt. The damaged and small fruits are discarded in the field. The conveyor belt is pulled in the field by a lorry. The fruits are then put into field bins in the transport truck. The harvested pineapples are then taken to the packaging house for grading and packaging. Lights are also provided on the conveyor belt frame for working at night.

Fig.11.68 Harvesting pineapple Manually and putting in bag

Fig.11.69 Protective gears for Harvesting pineapple

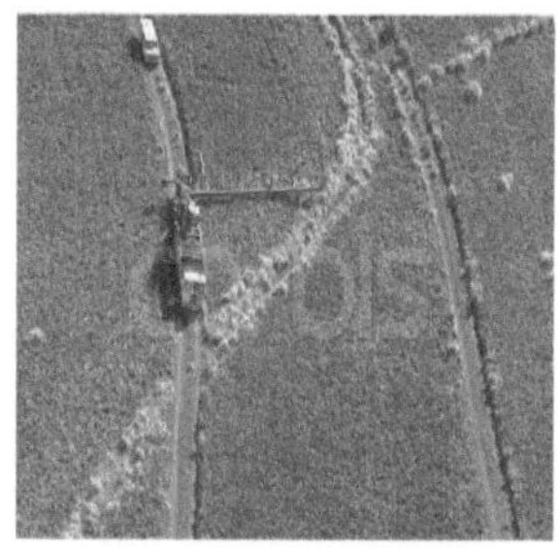

Fig.11.70 Pineapple harvester self Propelled
Dole

Fig.11.71 Harvesting pineapples manually and putting on conveyor Belt
*www.dole.com*

Fig.11.72 Pineapples on the conveyor beltand lighting arrangement for working at night
www.dole.com

## Banana harvester

The banana harvesting machine is a type of lifting crane (Fig.11.73). It lifts the bunch and harvests it from the plant by rotary circular blade. The bunch fruit is then hung on the rope conveyor for carrying to the packaging hall (Fig.11.74 to 11.76) . Generally a blue color polythene sleeve is put around the banana bunch for uniform ripening (Fig.11.74). The polythene sleeve is removed in the packaging hall (Fig.11.76) and bunches are cut into hands and packaged into CFB boxes.

Fig.11.73 Banana harvester

Fig.11.74 Banana harvester – the bunch is put in the plastic
Sleevebananachris.com

Fig.11.75(a) Conveying harvested banana bunches on Rope conveyor

Fig.11.75(b) Rope conveyor for banana bunches
FAO

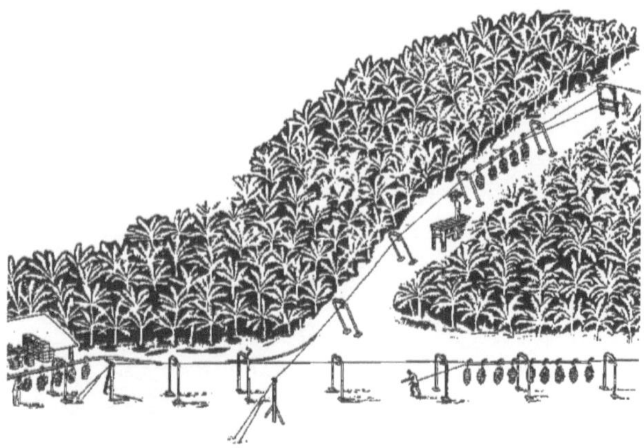

Fig.11.75(c) Rope conveyor for banana bunches

Fig.11.76 Rope conveyor for banana bunches in packaging hall

DOLE

## Grape harvester

Self propelled grape harvesters have become very popular for harvesting wine wine grapes in California (USA) and New Zealand raised on trellis (Figs. 11.77 to 11.84). The grape harvester moves over the row or trellis of grapes (Fig.11.82 to 11.84). The single trellis picking heads consist of two banks of nylon fingers which move from side to side shaking the grapes loose. The diameter and material of the fingers maximize the amount of fruit harvested. This design also reduces the head RPM required to remove the fruit.

The wines are shaken by the fingers in the harvesting head (Figs. 11.80 and 11.81). The fingers can be straight plastic or spring loaded metal rods or curved rods. The grapes fall on the collection conveyor and these are then conveyed to the tractor towed grape bin moving in the next row (Fig. 11.84). The trash and leaves are blown by blower. About one to two hectare of grapes are harvested in one hour.

Fig. 11.77 Grape harvester – self propelled
http://gregoireharvesters.com

Fig. 11.78 Grape harvester – self propelled
http://www.korvan.com/harvesters/3000.htm

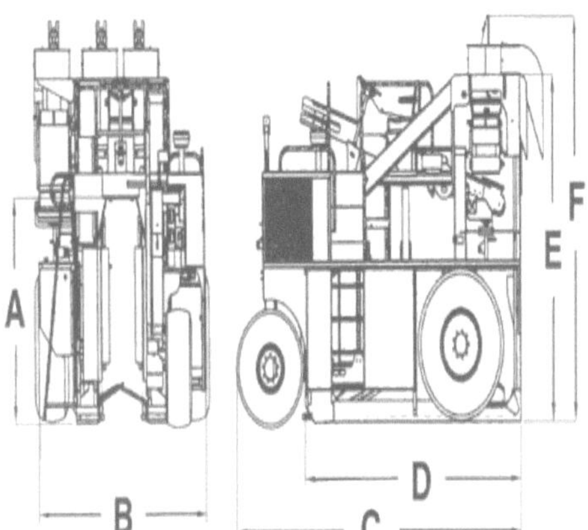


Fig. 11.79 Grape harvester – self propelled
http://www.korvan.com/harvesters/3000.htm

Fig. 11.80 Beater drum-with spring loaded fingers to shake the vines
http://www.wineoftheweek.com/stories/0104kerrharvest

Fig. 11.81 Beater drum-with spring loaded fingers
http://www.wineoftheweek.com/stories/0104kerrharvest
http://www.bei-inc.com/Lit/grape.pdf

Fig. 11.82 Harvesting the grapes-on trellies
kobrandwine.com/photo_gallery/cd_06/img0055.jpg

Fig. 11.83 Harvesting the grapes-on trellies
http://www.123rf.com

Fig. 11.84 Conveying harvested grapes to transport bucket
http://www.agha.org.au

# *Chapter 12*

# Vegetable Harvesting

Vegetables can be divided broadly in 2 categories (a) root crops and (b) surface crops for the purpose of harvesting. Root crops like potato, onion, carrot, radish, cassava, sweet potato need digging and removal of top leaves. Surface crops like tomato, brinjal, capsicum, cucumber, beans need selection of mature fruits before these are harvested by the labourers. Crops like cabbage and cauliflower need their head to be removed and collected. Most of the vegetables are harvested manually in India. The following operations are to be done sequentially while harvesting

- Selection of mature fruits/ stock
- Separation of fruits/stock from plant or soil.
- Collection of fruits/stock.
- Transportation.

All these operations need large amount of labourers as the volume of the fruits/roots handled is large. Generally manual harvesting takes 50% of the total labour requirement of cultivation.

Mechanization of harvesting is for reduction of labour in any of the above given operations. Crops like potato are easy to harvestand tractor operated potato diggers have become very popular in North India. Almost all the potato crop is being harvested by these diggers. The dug potatoes are left in the field and then are picked and collected manually. The harvesting of potato, onion and carrot (root crops) is generally done mechanically in European Countries and USA. This reducing the labour required for digging, removing the tops, removing the soil, blowing the leaves and collection of crop and its transportation is reduced.

In USA and Europe harvesting of surface crops like tomato, capsicum, brinjal and beans etc., is done manually for table / fresh purposes. However, it done mechanically for purposes of processing and canning. Also varieties have been developed in which fruiting takes place at one time in place of its spreading over long period so that the harvester move over the crop only once. By mechanized harvesting, the area under cultivation of vegetable crops has increased in USA and the cost of cultivation has reduced. The total processing of the vegetable crops has also increased . The excess labour relieved from manual harvesting and other operations has found job as a result of increase of area under cultivation and in also for vegetable processing industries.

In mechanized harvesting the crop is harvested in bulk and then selection of fruits is done by blowing the leaves and vines and sorting immature and over mature fruits. The manual sorting has been replaced by electronic colour sorters in some cases. The washing, grading and packaging is done in the packing house.

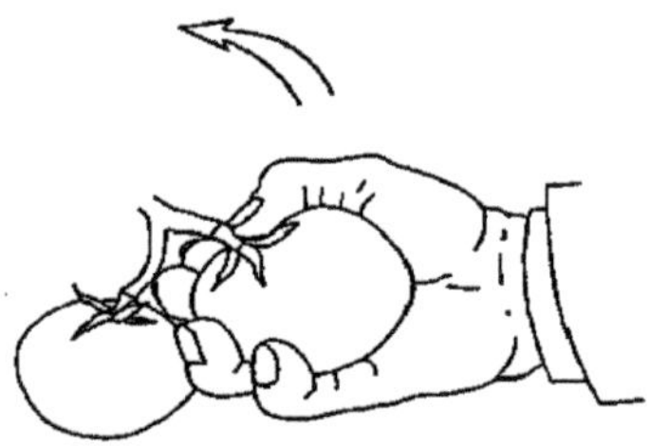

Fig.12.1 Harvesting at natural break point
FAO

Fig.12.2 Manual harvest of leaf vegetables
FAO

Fig.12.3 Bamboo basket

Fig.12.4 Gunny bag

Fig.12.5 Plastic crate

## 12.1 Manual Harvesting

Most of the vegetable crops are harvested manually or by using bullock drawn implements, in India. Potatoes are dug using Kodali or spade or animal drawn country plough and these are then collected manually. In North India digging of potatoes is done by tractor driven potato diggers (Figs. 12.16 and 12.17) and the dug tubers are then collected manually. Onion and other root crops like carrot, radish etc., are also dug manually or by using the animal drawn country plough. The surface crops like tomato, brinjal, cucumber etc., are harvested by hand (Fig. 12.1) at the break point of the pedicel and collected in heaps in the field. The leafy vegetable like (Palak) spinach and methi are either plucked manually (Fig. 12.2) or harvested with sickle. These are then packaged in bamboo baskets (Fig 12.3), gunny bags (Fig.12.4) or cloth. Some crops like onion, cabbage and cauliflower are also transported in bulk without any package. The use of plastic crate (Fig. 12.5) for packaging vegetables and their transport is being popularized by the National Horticultural Board and Safal Market of National Dairy Development Board. The use of Plastic crates reduces injury/damage to the produce by compression and transport.

A large number of harvesters have been developed for different crops and these are in use in developed countries. These reduce labour in at least one of the functions (a) selection, (b) separation, (c) collection or (d) transport. By use of these the labour efficiency increases and the labour requirement is reduced. Efforts have been made in different agricultural universities, institutes and by private firms to develop suitable and tools implements for harvesting vegetable crops which are discussed in this chapter.

Drängen Lay-Down Work Carts

DRANGEN Manned Planter, Harvester and Weeder
Drängen Lay-Down Work Carts

Fig.12.6a. Self propelled Harvesting Aid for pumpkin , cabbage and cauliflower ( labourers in lying position)

Fig.12.6b. Self propelled Harvesting Aid ( labourers in sitting position)

## 12.2 Harvesting Aids

The fruits of crops like pumpkin, muskmelon, cucumber are spread over the field and are plucked manually and transported. These fruits are generally big in size and large volume is harvested. The volume of produce for cabbage and cauliflower is also very large. The self propelled harvesting /fruit collection aid (Fig. 12.6a) is operated by an engine. Seats are provided for one or more labourers to lie down, so that they can easily harvest crops while the machine is in motion. The labourers keep the fruits on conveyors fitted on the machine, which in turn fill the produce in the field bins. These machines reduce the labour required for collection and transportation of the produce.

Self propelled harvesting aids (Fig.12.6b) help labourers to work in sitting position or in a more comfortable posture. These self propelled platforms can also be used for transplanting seedlings of vegetable or ornamental crops. The height of the platform / seat is 30 to 40 cm above the ground and these move at a speed of about 1 km/hr.. Tractor drawn conveyor belt (Fig. 12.7 and 12.8) used to transport the manually plucked fruits to tractor trailers. The length of the belt is upto 15 metres. While labourers work in lying down position ( Fig. 12.7) while harvesting cucumber,. they work in standing postion for harvesting cabbage, cauliflower and pumpkin (Fig.12.8.). Cabbage fruits are harvested and cleaned manually and kept on the belt conveyor for transport to packaging station provided on the machine (Fig.12.8 a). Self propelled harvesting cum packaging platforms (Fig.12.9) are however suitable for very big farms.

Fig.12.7 Tractor pulled ( labourers in lying position) harvesting aid for pumpkin , cabbage and cauliflower

http://www.kox.sk

a. http://www.freshpik.com

b. http://www.weeningbrothers.com

Fig.12.9 Self propelled harvesting cum packaging platform
http://www.valleyfabrication.com

## 12.3 Potato Harvesters

Potato is cultivated in about nineteen lakh hectares in India and it is being cultivated almost in all the states. The digging of potatoes in some of the states is still done by using hand tools like kodali and spade. In some of the states the bullock drawn country plough or desi plough is used to break open the ridge. The potatoes come to the surface of the field, from where these are picked by the labourers manually. In Punjab almost all the potato crop is harvested with tractor operated potato digger (with elevator). These are picked manually by women labourers from the field. The potato harvesting combines are used in Europe and USA are provided with belt conveyors for manually sorting out the clods and stones etc. and filling the potatoes in field bins. The haulms of the potato plants have to be cut and left in field for drying before digging. The haulm cutter may have rotary blade or rotating chains (Fig 12.18 to 12.20). The design of the different harvesters is detailed below.

### Potato Digger with Fixed Blade and Extension Rods

A Bullock drawn potato digger (Fig. 12.10) has been developed at Punjab Agricultural University, Ludhiana. This is provided with V-shaped blade with extension rods at the rear wings. The blade opens up the ridge and pulverizes the soil around potatoes. The extension rods help in bringing the tubers to the surface and separation of soil and dirt from potato tubers. This implement is suitable for digging potato tubers after removal of vines in one row. The vines are generally removed manually by sickles or scythe. It works at a speed of about 2 to 3 km/hr and its field capacity is about 0.08 to 0.12 ha/hr. It takes about 300 to 400 man hours to collect the tubers from the field. The tractor operated potato digger is generally fitted with 2 blades similar to these shown in (Fig. 12.10) and the operating speed is 5 to 6 km/hr. The field capacity is about 0.3 ha/hr. The power tiller drawn harvesters (Fig. 12.11) have the field capacity similar to bullock drawn harvesters.

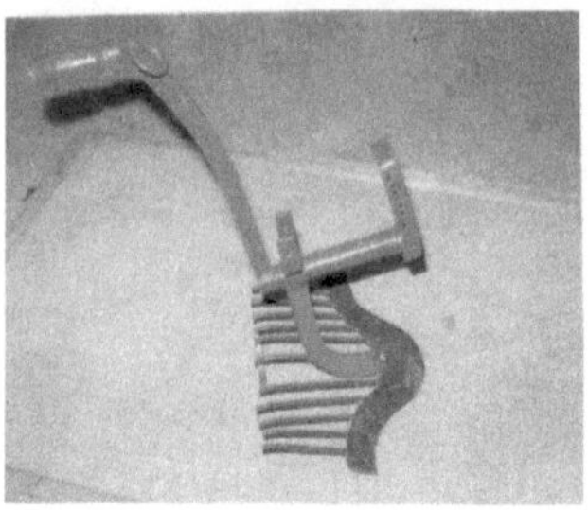

M.M.Pandey, CIAE

Fig.12.10 Animal operated potato digger

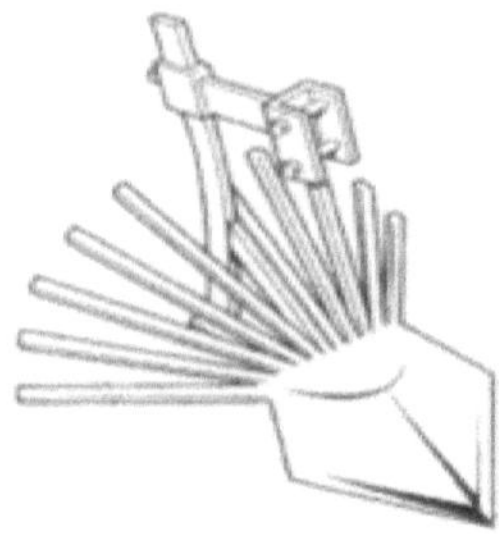

http://www.honda-uk.com/tillers.htm

CTM UK

Fig.12.11 Power tiller drawn fixed blade potato potato digger

Weifang Langpak Machinery Co

Fig.12.12 Power tiller operated potato digger with elevator conveyor

Green Land Engineers

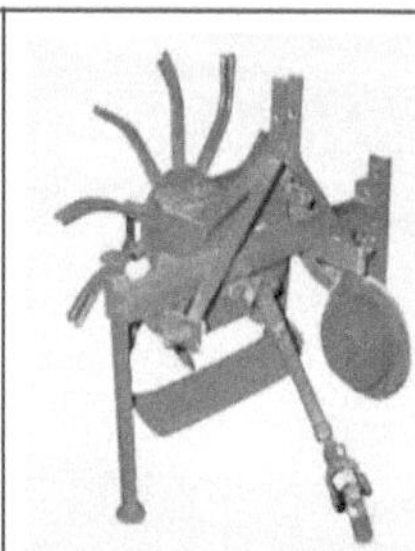

Speed Link

Fig.12.13 Tractor drawn spin type potato digger

Fig.12.14 Tractor operated oscillating blade type potato digger
http://www.spedo.it

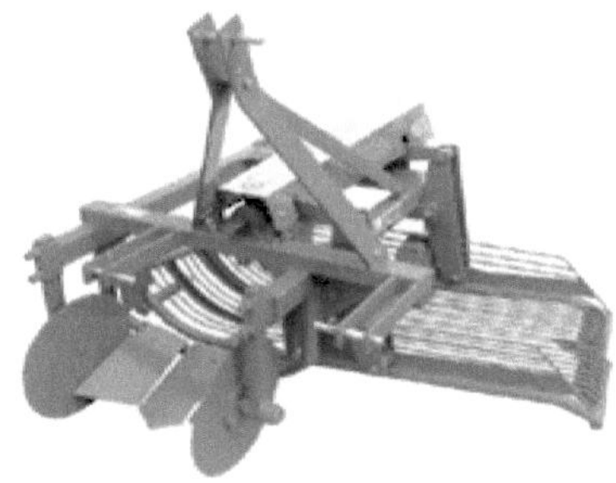

Fig.12.15 Tractor operated folding blade type potato digger
http://www.spedo.it

Tractor operated potato digger for one row/ridge is having one fixed blade to move in the same direction has the tractor. The spin type potato digger (Fig. 12.13) is also provided with pto operated rotary rods. These rods separate the potatoes from soil and window these on one side.

## Potato Diggers with Oscillating Blade

This tractor operated harvester (Fig. 12.14) has the blade with an extension rods. The blade oscillates horizontally ( in the direction of motion of tractor) by the tractor pto. The oscillation of the blade reduces the draft requirement of tractor. The oscillation of extension rods also helps in better separation of potatoes from the soil. The oscillating potato digger is generally mounted to front of power tiller. The blade of potato digger (Fig 12.15) consists of two parts. Both the parts oscillate to compress and loosen the dug soil and help in pulverization of soil and better separation of tubers.

Droli industries

Fig.12.16 Tractor operated two row potato digger with elevator conveyor

CPCRI Jallandhar

Fig.12.17 Tractor operated two row potato harvester in operation

## Potato Digger cum Elevator

This tractor rear mounted single row potato digger consists of a digging blade, endless rod chain conveyor and gauge wheels. The elevator conveyor is driven by HTO. The blade dug opens the ridge and pulverizes the soil. The elevator conveyor transports and shakes the soil and potatoes to separate these from soil.

The single row potato digger is mounted offset to the tractor. The operating width is 60 cm. and working speed is 1.5 to 2.5 km/hr. It is suitable for digging and exposing of potatoes. It saves about 75% labour and operating time and 50% on cost of operation as compared to conventional method of manual digging with spade or desi plough or tractor operated cultivator. The single row digger elevator needs about 25 hp tractor. Tractor operated 2 row digger elevator (Fig. 12.16) is provided with 2 digging blades and 4 rotary coulters for better collection of dug soil and potatoes and transfer to elevator conveyor. It needs about 35 hp tractor, and operating speed is 1.5 to their 2.5 km/hr and field capacity is 0.30 to 0.35 ha/hr. The operation of potato digger and exposed potatoes is shown in Fig 12.17. The potato digger can be provided with windrow attachment to collect these in one windrow. The dug out potatoes are picked and collected manually (Fig. 12.17).

Fig.12.18 Potato haulm cutter

http://www.grimme.de

Fig.12.19 Potato haulm cutter

http://www.grimme.de

Fig.12.20 Blades of Potato haulm cutter

Asa-lift

Kongskilde Juko Ltd

Fig. 12.21 Potato harvester with haulm cutting, conveyor-cum-separator and sorting platform for clods and trash etc.

Fig. 12.22 Potato harvester with circular endless rod conveyor-cum-separator and sorting platform for clods and trash etc.

http://www.sanei-ind.co.jp

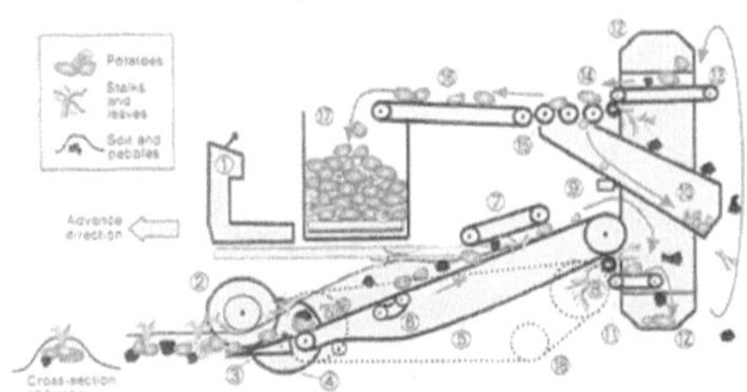


Fig. 12.23 Potao harvester with cylindrical rod conveyor-cum-separator and sorting platform

http://www.sanei-ind.co.jp

Fig. 12.24 Potato harvester with octagonal endless rod conveyor-cum-separator

www.grimme.de

Fig. 12.25 Potato harvester with circular endless rod conveyor-cum-separator clods and trash etc.

Hirabhai, Khedut, Gujarat and TV9 Gujarat

## Potato Combine Harvester:

In Potato diggers, the potatoes are dug by the digger blade, separated from the soil by elevator conveyor and left in the field in exposed position for manual pickup. In tractor pulled combine harvester the length of endless rod chain conveyor is increased to improve the separation of potato from soil (Fig. 12.21), This potatoes are then transferred to belt conveyor for sorting out these manually, platforms are provided on sides of the belt conveyor for labourers to stand and separate any clods or trash from the potatoes. The potatoes are then filled in field bins or gunny bags at the outlet of the belt conveyor. Shade is provided over the conveyor belt for the labourers to work in the sun. In some of the harvesting combines circular rotary blade is provided for cutting the haulms in front of the digger blade. A compession roller is also provided for pressing the ridge to pulverize the soil in front of the digger blade (Fig. 12.21), this helps in easy separation of the potatoes from soil and reduces the clods in soil. The potato harvester (Figs. 12.22 and 12.23) is provided with cylindrical rotary rod conveyor for separation of soil from potatoes. The blower is also provided in some of the combine harvesters for blowing away loose soil, vines and leaves for separation and cleaning of potatoes. The potato harvester (Fig. 12.24) is provided with additional octagonal endless rod conveyor to separate the soil and transfer the potatoes to another elevator conveyor to fill the clean potatoes in field bins. The potato harvester under development (Fig. 12.25) in Gujarat is provided with 2 cylindrical conveyors cum separators for soil.

The vines are cut from the potato plants before the operation of potato digger and vine/haulm cutter (Figs. 12.18 to 12.20) is used for this purpose. This may have horizontal rotary blades or vertical rotary chains assembly.

Fig.12.26. Single bed Onion Harvester cum Windrower (tractor operated)
http://www.asa-lift.com

Fig.12.27. Single bed Onion Harvester cum Windrower(tractor operated)
Univerco

Fig.12.28. Single bed Onion Harvester cum Windrower
http://www.asa-lift.com

Fig.12.29. Three bed Onion Harvester cum Windrower
univerco.net

Fig.12.30. Onion topper , tractor front mounted
http://www.asa-lift.com

Fig.12.31. Onion topper , tractor rear mounted
http://www.standen.co.uk

Fig.12.32. Tractor front mounted onion topper and rear mounted digger cum windrower
http://www.onions-potatoes.com/harvest/harvest-mechanized.php

## 12.4 Onion Harvesters

Onion is generally grown on ridges or rows by transplanting seedlings in most areas in India. It is also sown by broadcasting in Northern Karnataka. Onions are to be cured after digging. The broadcasted onion crop is generally dug by using the *kodalis* (hand hoe) and is left in the field (for 2 to 3 days) with leaves for drying and curing. These are then picked and detopped manually and filled in bags. The onions raised in rows on ridges are generally dug using the animal drawn country plough or ridger plough. The picking is done manually after curing. In U.S.A and European countries, the onions are grown on beds of about 1.2 meter width. The onion diggers are used to dig and windrow the onions. In some onion harvesters/diggers the onions are left in the field on soil surface. After curing for 4-5 days, the tractor pulled onion lifter collects onions from the field, detopping is done and then filled in field bins. Which have sufficient side openings for aeration. These filled bins are then left in the field for onion curing for 4 to 5 days. The cured onion bins are then picked and taken to the packaging houses for grading and packaging. In some cases the bins of freshly harvested and detopped onions are brought to the packaging house and cured by hot air. Different kinds of onion harvesters available commercially are discussed below:

### Onion Digger-cum-Windrower

Single bed harvester for 100 to 110 cm. bed width is shown in Figs. 12.26 to 12.28. It consists of a straight digging blade, 2 circular coulters, pto operated chain link elevator cum agitator and windrower attachment to collect the harvested onions in a row of about 60 cm width. The onions are dug by the blade, passed on to elevator-cum-agitator and collected in about 60 cm wide windrow. The onions are dug along with the tops. The Windrowed onions are left for curing and drying in the field. Three row onion diggers cum windrower is shown in Fig 12.29. In some cases the onion leaves / tops are cut using tractor front or rear mounted detoppers (Fig 12.30 &12.31). The detopper and digger cum windrower are fitted at front and rear of the tractor (Fig. 12.32), The detopped and windrowed onions are then left in the field for curing.

Fig.12.33 . Single bed Onion Harvesting Combine

Top Air

Fig.12.34 . Single bed Onion Harvesting Combine

sgaequip.com/onion_harvesters.php

Fig.12.35. Single bed Onion Harvesting Combine(tractor operated)

Lee Shuknecht & Sons

Fig.12.36. Single bed Onion Harvesting Combine(tractor operated)

Lee Shuknecht & Sons

These machines also require loader-cum-detopper/finisher. The loader consists of a (rotary cylindrical) header to collect the cured onions. These are then detopped and filled in the field bins.

The main parts of a onion combine harvesters are : Digging blade, Chain link elevator conveyor, Detopping unit, Blower, Onion outlet, Header, Power source

The single bed harvesting combine is provided with one digging blade (Figs. 12.33 to 12.36) and is generally operated by tractor pto. The self propelled combine harvester is provided with 2 blades for 2 beds and rotary header for better collection of onions (with leaves) for feeding to digging blade and elevator conveyor. As the onions are conveyed above the elevator conveyor, these are agitated and the soil gets loosened and separated. The loosened soil falls through the gaps in conveyor.

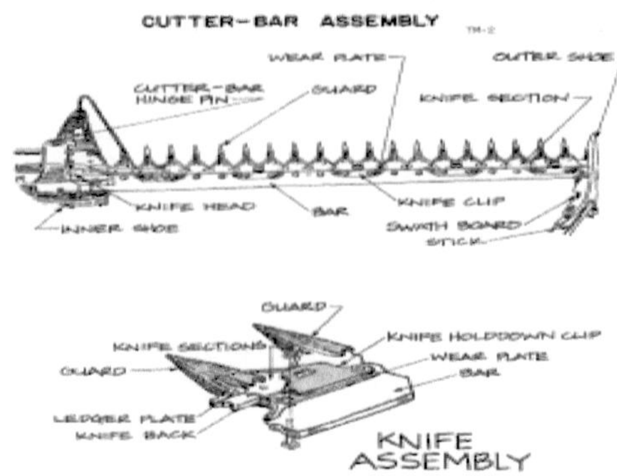


Fig.12.37. Cutter bar for dettoping / leaf cutting
http://cast.csufresno.edu/agedweb/agmech/graphics/Safeoperation2.gif

Fig.12.38. Onion detopper/ Finish Topper
Lee Shuknecht & Sons

The detopping unit receives the onion with leaves and cuts the leaves. The detopping unit may be of cutter bar type, in which, the leaves of onion are made to stand vertical by the air from blower and the reciprocating cutter bar (similar to the cutter bar of harvester for wheat and paddy (Fig. 12.37) cuts the leaves. The leaves are then blown by the air from the blower. The detopping unit may have rods or pipes rotating in the opposite direction (Fig.12.38 ) The leaves are held and pulled by the rods and detopped by the circular blades fitted below the rods. The leaves are then blown away by the air from the blower. The leaves/ tops get rubbed and partially removed by coming in contact with rotary rods/pipes. The loose soil, dirt and leaves also get removed by the air blast. The detopped and cleaned onions then fall on the belt conveyor and are their filled in field bins (Figs. 12.33) or tractor trailer or truck (Figs. 12.35 and 12.36). The single bed tractor operated combine needs a tractor of about 50 hp and can harvest 30 to 50 tonnes/hr. Self propelled, 2 to 4 bed harvesters need an engine of 85 to 110 hp and have harvesting capacity of 50 to 120 tonnes/hr. These harvesters are suitable for 100 to 110 cm beds or the onion grown on ridges. Mostly the onions are collected in the field bins by these harvesters. The field bins are then left in the field for drying and curing of onions and further transport. However, in some cases the onions are lifted from the field without detopping, and are then detopped in packaging houses using onion detopper for cured onions (Figs. 12.39).

## 12.5 Garlic Digger

The garlic diggers are similar to onion diggers (Fig. 12.39). The garlic is dug by the blades from the ridges rows, detopped (by rotary rods/pipes) and filled in field bins.

Fig.12.39 Two-Row Garlic Digger

http://www.newhouse-mfg.com

## 12.6 Tapioca/ Cassava/ Turmeric Harvester

The clumps of cassava/ tapioca or turmeric roots are dug by tractor operated digger with one blade (Figs. 12.40) Single row cassava digger cum elevator is provided with endless rod conveyor to remove the soil from cassava tubers

Weifang Xiaer Imp.&exp.co.,lt

Fig.12.41. Single Row Cassava Digger

Shanghai Sheng Qian Industry Co., Ltd

Fig.12.42. Single Row Cassava Digger cum Elevator

## 12.7 Carrot harvesters

Carrots need digging of roots and cutting of tops. Self propelled and tractor operated carrot harvesters are shown in Fig. (12.42) and Fig. (12.43). The digging blade opens the ridge and pulverizes the soil around the carrot. The leaves are held in feed rollers (Fig. 12.43 and 12.44) or V belts or flat belts and the carrots this get pulled from the pulverized soil. In some harvesters the roots are cleaned by rotary brush and the leaves are cut by 2 rotary blades revolving in opposite direction. The cleaned and detopped carrots then fall on a belt conveyor. These are then collected in plastic crates (Fig.12.42 ) or field boxes or tractor operated tipping bucket or tractor (Fig12.43). The tractor operated machines are operated by pto and depth of digging is controlled by the hydraulic system.

Fig.12.42 Self propelled single row carrot harvester

http://www.iseki.co.jp/english

UNIVERCO

http://www.asa-lift.com

Fig.12.43 Tractor operated single row carrot harvester

## Harvesters for Surface Crops

Mechanical harvesters are commercially available for cabbage, cauliflower, tomato, capsicum, chillies, cucumber and leaf vegetable like spinach and celery. Mechanical harvesters are also available for harvesting dry peas. These harvesters are suitable for one pass harvesting of the crop. One time harvest varieties of the crops have been developed so that most of the fruits mature at same time. The fruits are hardy enough to handle the mechanical operations, without much damage. Mechanically harvested fruits are generally used for processing. The harvesting of vegetable crops for table purpose or fresh cooking in kitchen is however done manually ever in the developed countries. The harvesters for green beans, peas have also been developed but those are not popular commercially.

## 12.8 Cabbage Harvester

Cabbage is generally grown on ridges or beds. Tractor operated (Fig. 12.45) and self propelled harvesters (Fig. 12.44) are available commercially. These can be single or multi row harvesters. The single low harvester consists of rotary header to align the heads in the row, feed rollers (having helical spiral elements) to hold the heads and uproot it and elevate/convey these upwards. When the heads are being pulled from the soil, these sit firmly on the rotary feed roller set and the root is cut by the disc cutters. The heads are then guided over the feed rollers to belt conveyor or field bin. The heads are inspected by a labour for sorting out the unwanted, over mature or under mature heads. The speed of the feed rollers is synchronized with the speed of the harvester in the field.

National Agri Res Center for Hokkaido Region, JAPAN
cryo.naro.affrc.go.jp/.../ cabbageharvester.jpg

Tainan County Agri. and Forestry Station, Taiwan

Fig.12.44 Self propelled cabbage harvester

http://www.univerco.net/en/frame.htmlUniverco http://www.asa-lift.com

Fig.12.45 Tractor operated cabbage harvester

## 12.9 Tomato Harvester

The work on development of a mechanical harvester for tomato was started in the university of California, Devis, during 1950 , due to the requirement of large labour force and high cost of harvesting tomato. Coby Lorenzen of Agricultural Engineering, Department and Jack Hanna of the Vegetable Crops Department, started work for developing a system for mechanical harvesting of tomatoes for processing. Hanna began breeding tomato varietis that could withstand the stress of mechanical handling, ripen uniformly and detach easily from the plant during machine harvest. Commercial use of the new variety and harvester began in 1962. About 262 harvesters were being used during 1964 and 25% crop harvested by harvesters. Four year later 95% of the crop was harvested by the machine. The production of tomato in California was 20 lakh (2 million) tonnes during 1960, which increased to about 1 crore (10million) tonnes during 2000. The harvester had a labour requirement of 2.9 man hours/tonne during 1964 and 0.4 man hour/ tonne during 1990-2000, compared with 5.3 man hours / tonne for hand harvest. One machine can harvest about 300-350 hectares in one season. The machine can also work at night as sufficient light is provided, where the labour does the manual sorting of tomatoes on the machine..

The machine is tractor operated (Figs. 12.46 to 12.48) or self propelled (Fig. 12.50). It has a header for collection of vines and a cutting blade to cut the roots. The plants are passed on to the elevator conveyor. Necessary guides are provided for smooth movement of the vines. The vines are then fed to the stripper or shaker where tomatoes are separated from the vines/ plants. The vines are then moved behind and discharged in the field. The blower (Fig. 12.49) is used to blow away dirt, leaves and vine pieces from the stripped tomatoes. Tomatoes are then passed on to sorting (belt) conveyor for manual sorting of dirt, broken and moldy fruits. About 20 labourers are required as sorter on the machine and necessary platforms are provided on both sides of sorting conveyor for standing and working of the labours. Shade / roof is provided above the machine or sorting conveyor for ease in working in the sun. Necessary lights are also provided over the sorter conveyor for working at night. The sorted tomatoes are then passed to the elevator conveyor for filling in field bins or tipping type bucket moving along the harvester. In some of the recently developed machines electronic sorters are used for colour sorting and the manual sorting labourers are reduced to six. The tomato harvester with manual sorting system reduces the cost of harvesting to 33% of the total cost and with electronic sorter reduces to 16% of the total cost. The machine can harvest about 0.4 to 1.0 hectare/hour.

Fig.12.46 Tomato harvester tractor operated
PIK RITE, INC.60 Pik Rite Lane Lewisburg, PA 17837, USA

Fig.12.47 Tractor operated tomato harvester working in field
PIK RITE, INC.60 Pik Rite Lane Lewisburg, PA 17837, USA

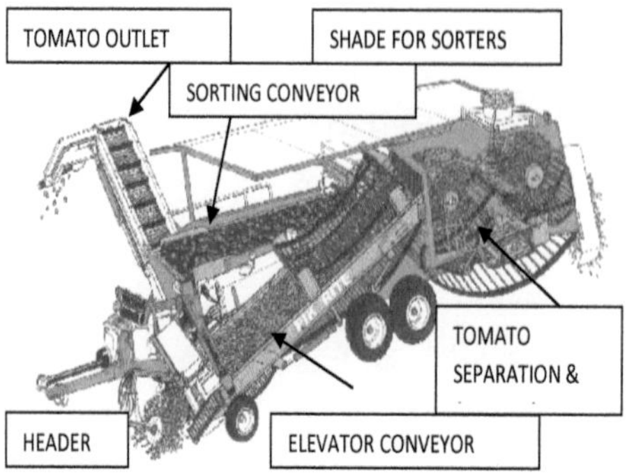


Fig.12.48 Tractor operated tomato harvester (components)
PIK RITE, INC.60 Pik Rite Lane Lewisburg, PA 17837, USA

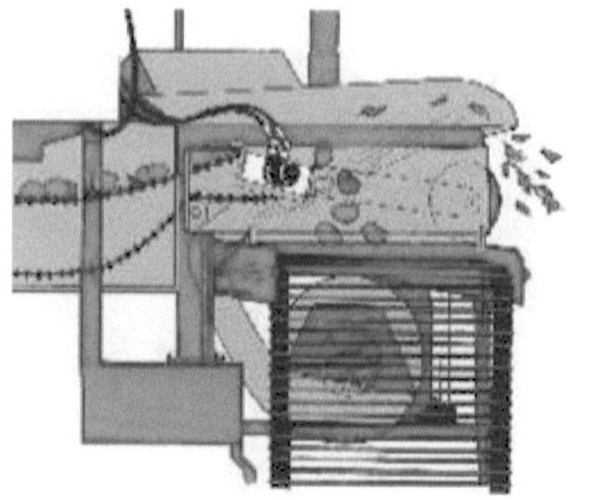

Fig.12.49 Blower for leaf & vine separation
PIK RITE, INC

Fig.12.50 Self propelled tomato harvester with tractor operated tipping bucket for collection
California Agriculture, May-June 2000

## 12.10 Pepper and Chilli Harvesters

About 195 prototypes of chilli harvesters (Figs.12.51 to 12.56) have been developed by 75 groups. The plants are collected by the header (Fig.12.51 to 12.53). The plants in rows are uprooted by cutting the roots and conveyed by elevator conveyor. The pepper or chillies are separated from the plants by stripper or shaker. The harvested pods (fruits) then fall on a conveyor and the leaves are blown away by blower. The fruits are then inpcted by electronic sorter while further sorting is done manually by about 6 labourers (Figs. 12.53 to 12.55) standing on the side of sorting conveyor platform. The sorted pods/fruits are then filled in field bins on the tractor trailer moving parallel to the harvester.

Fig.12.51 Chilli harvester
http://www.pikrite.com/chili.html

Fig.12.52 Chilli harvester header
http://www.pikrite.com/chili.html

Fig.12.53 Pepper harvester
http://www.pikrite.com/pepper.html

Fig.12.54 Capsicum harvester sorting conveyor
http://www.pikrite.com/chili.html

Fig.12.55. Chilli harvester sorting conveyor
http://www.pikrite.com

Fig.12.56.Filling the capsicum in field bins
p://www.pikrite.com

In some cases the cotton harvester have also bee modified. These harvesters strip the fruits from the plants by stipper rolls, fitted in the header . The pick up or gathering mechanism is generally provided before the stripping rolls and it can be powered by brushes or rotary rubber disks gathering unit. The loss reduction shields may also be provided by the sides of gathering brushes or discs to avoid the falling of fruits in the field. The stripper rolls 10 to 15 cm diameters revolve in opposite direction (200 to 1000 rpm) and a gap of about 1.3 cm is provided in between the rolls for paprika/capsicum plants to pass through these. The stripping rolls may be provided with closed helix for better stripping and conveying of the fruits (pods). The guide cover is provided over the stripper rolls for guiding these to the separation unit. In some of the harvesters blower is provided to convey the stripped capsicum (pepper) to the separation unit. In cleaning unit the leaves, plant tissue and dirt are blown away and the fruit/pods are conveyed to the sorting unit. When electronic colour sorter is used about 6 manual labours are required for the final sorting otherwise 24 labourers are required for sorting the culled and diseased fruits. The clean fruits are then filled in field bins or tractor trailer through elevator conveyor (Fig. 12.56). The spacing in the headers can be adjusted to harvest the paprika in rows sown at 20 to 50 cm apart. The harvesters operate at a speed of about 1.5 to 8 km/hour and the harvesting capacity is upto 10 tonne/hour or 0.4 ha/hr. The damage to fruits and field loss may be 3 to 5%. These harvesters are suitable for harvesting paprika, , capsicum and chilli for processing and colour extraction.

## 12.11 Cucumber Harvester

Cucumber harvesters have been developed by various firms, which work on the principal similar to tomato or paprika harvester. These may uproot the vines and strip the cucumbers or collect the vines on strip rollers and harvest the cucumbers. Further sorting may be done electronically and manually.

## 12.12 Green Leaf Harvesters

These are generally tractor operated and are suitable for harvesting the leafy vegetables grown on beds (Fig. 12.58). The leaves are cut above the ground by the continuous straight blade. The elevator conveyor transports the harvested leaves to the plastic crates provided at the rear of the harvester. The trays when filled are removed manually. The rack for storing the empty trays is also provided on the machine. These harvesters are suitable for harvesting spinach, fenugreek / *methi* and coriander etc. More than one harvesting is possible by the machine for crops like spinach. The self propelled harvester (Fig. 11.55) has a regular cutter bar and the pneumatic system to suck the insects and make the leaves in standing position before harvesting. The leaves are harvested by cutter bar and conveyed to the bin by reel and belt conveyors after harvesting. The length of cutter bar is 2.25 or 3 meter in different models. The harvesters are fitted with 140 hp engine and the harvesting capacity is 3 to 4 tonne per hour.

Fig.12.57 Tractor operated spinach /green leaf harvester
http://www.seedspider.com/

Fig.12.58 Tractor operated spinach /green leaf harvester with pneumatic arrangement for erecting the leaves and sucking the insects
John deere

# *Chapter 13*
# Packaging of Horticultural Produce

The packaging line for fruits and vegetables generally consists of the following units.

- Washers
- Drying unit by rotary brushes
- Waxing and polishing unit
- Grader
- Sorting table/conveyors
- CFB box assembling unit from knocked down box
- Weighing unit for CFB boxes
- Palletizing unit

This chapter discusses some of the equipments used in packaging line which are in use or can be used in India.

## 13.1 Washing Equipments

The field produce has to be thoroughly washed and graded before packaging into the containers.

### Soak Tanks

Generally the produce has to be soaked in water for 3 minutes to loosen the dust, microbes and pesticide marks etc. This is generally done by soaking the produce in soak tank where water is circulated and the produce is agitated

(Fig. 13.1 and Fig.13.4). The filtering of water is also done during circulation to remove the dust particles. Sodium hypochlorite solution may be added into the soak water to control possible thermophilic microbes buildup. A residual chlorine content of 6 to 8 parts per million may be maintained at all times. The soak water is generally chilled to remove the field heat. The produce is generally dumped into the soak tank and removed by the belt conveyors having perforations to take out the produce without water for next stage of cleaning (Fig.13.3). The detergent solution can also be added in the soaking water if required. In some of the soak tanks, rotary brushes are fitted to clean the produce and these are known as brush washers. When the paddle wheels are fitted for agitation and cleaning of the produce these are known as immersion washers. Fig.13.9 and Fig.13.10 show the two stage immersion washers. The combination of a water stream with a paddle system makes the washer suitable for floating as well as sinking products. An air agitation system intensifies the washing result. At the bottom of the tank is a lift-out filter that filters out most small particles from the wash. An over-flow side screen filter also helps in the removal of small unwanted material. If the produce is for processing, the second stage washing is done in rotary barrel type washers and then washed by spray washers. If the produce is for packaging, the produce is cleaned by spray washers only after it is brought from the soak tank.

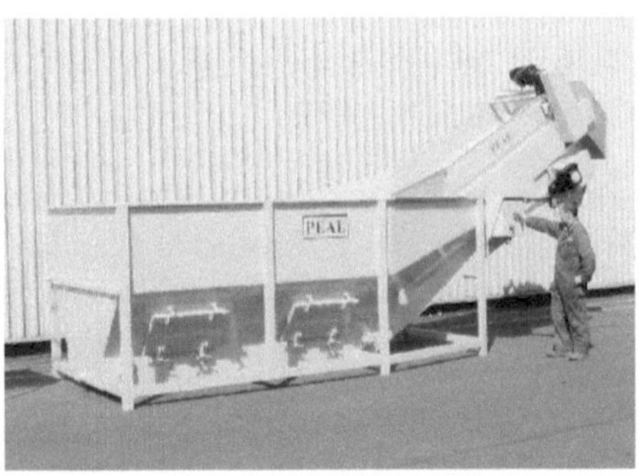


Fig. 13.1. Soak TankPeal Engineering, UK, Tong Engineering Ltd,

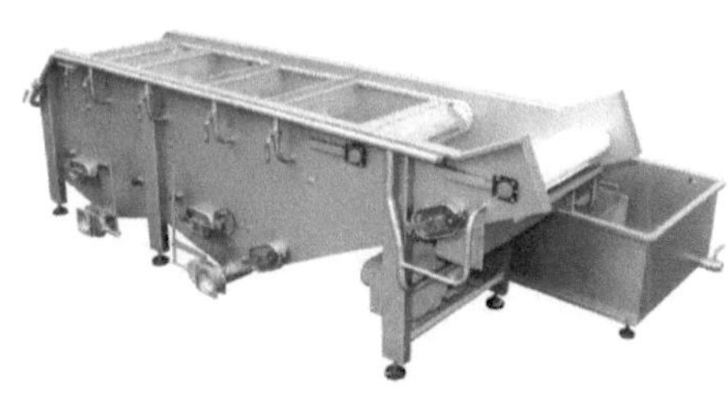

Fig.13. 2. Soak tank with paddle system and conveyor belt with perforations http://www.sormac.co.uk

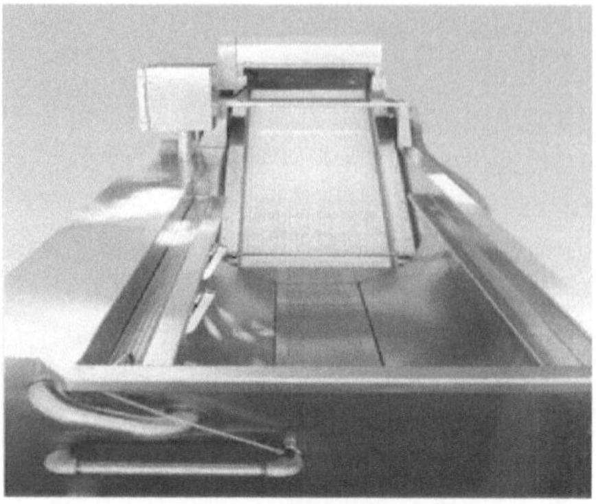

Fig.13. 3. Soak tank with conveyor belt with perforations
http://www.charliesmachineandsupply.com

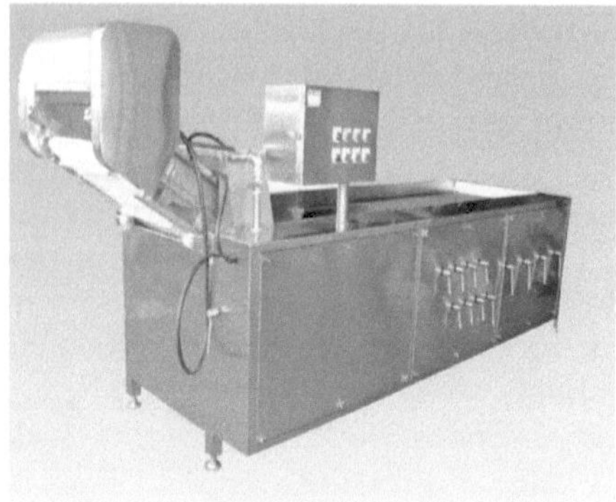

Fig.13. 4. Soak tank with conveyor belt with perforations
http://www.charliesmachineandsupply.com

## Rotary barrel washer

The barrel/drum washer is a multi product washer typically for carrots, cucumbers (gherkins), shallots etc. The washer drum has a rotational action which produces a very thorough cleaning action. The product flow through the drum is controlled by an adjustable outlet door and in combination with the variable rotational speed enables to optimise the cleaning and washing of all products. It consists of a cylindrical barrel generally lined with rubber material for cushioning. with holes on periphery (Fig. 13.5 and Fig. 13.6). The agitators are also fitted for proper agitation of the produce (Fig. 13.7). The barrel is partially submerged in water which is circulated and sprayed over the rotary barrel to assist in cleaning. The produce from the soak tank is fed to the rotary barrel washer, where it rotates and tumbles the produce in water. The external impurities get removed from the produce by rubbing of the produce with each other and the lining of the barrel. The water is circulated and filtered. and changed when required. In some of the models, the cylindrical rotary brush is provided in the center which also rotates and helps in cleaning (Fig. 13.8). The rotary barrel washer can have a capacity from 1 tonne to 60 tonne per hour.

Fig.13.5 Rotary barrel washer
http://www.haith.co.uk

Fig.13.6. Rotary barrel with holes
(Alard Equipments, USA)

Fig.13.7. Rotary barell with agitators
(Alard Equipments, USA)

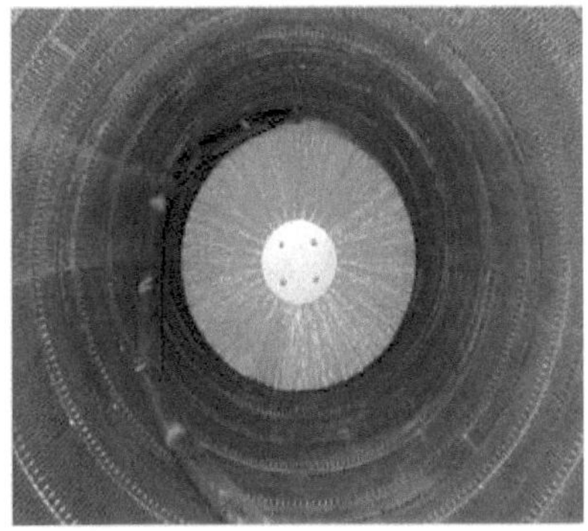

Fig.13.8. Rotary barell with cushion and central brush
http://www.haith.co.uk

Fig.13.9. Two stage immersion washer (soak tank with paddle wheels
(Alard Equipments, USA)

Fig.13.10. Paddle wheel of a immersion washer
(Alard Equipments, USA)

## Spray washers

The produce from soak tank or barrel washer is conveyed to spray washer (Fig. 13.11 to 13.13). It consists of a roller conveyor and spraying nozzles to spray from all sides. This is also equipped with crop conveyor with feed adjustment, and water tank with water recirculation pump. The individual fruits are made to rotate and water is sprayed under pressure over this by nozzles. The washed fruits are then sent to sorting table, which consists of roller conveyors where the produce rotates during conveying. The cull and uncleaned fruits, if any, are picked manually and discarded (Fig. 13.14). The cleaned fruits are also sent for grading in the packaging line.

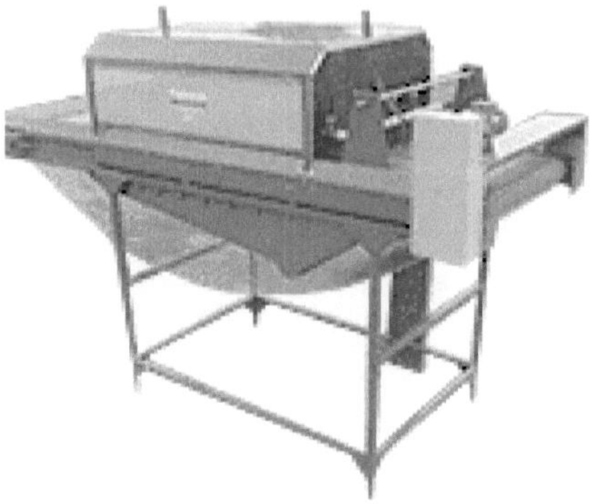

Fig.13.11. Spray washer
(Haith Tickhill, UK)

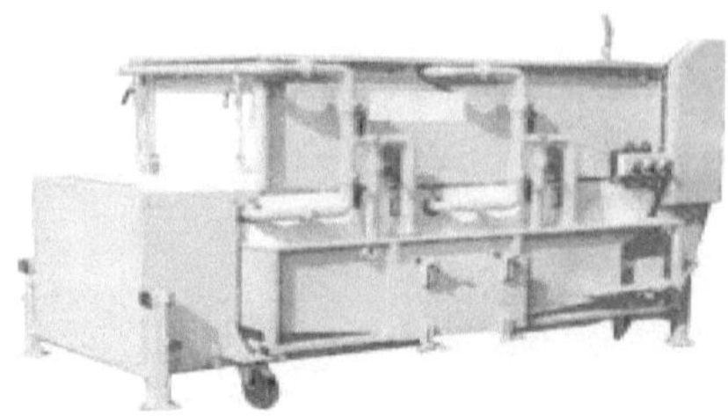

Fig.13.12. Vegetable Spray Washer
Global Agri Tech Engineers

Fig.13.13. Spray washer ( spraying nozzles)
http://www.sunkistresearch.com/english/

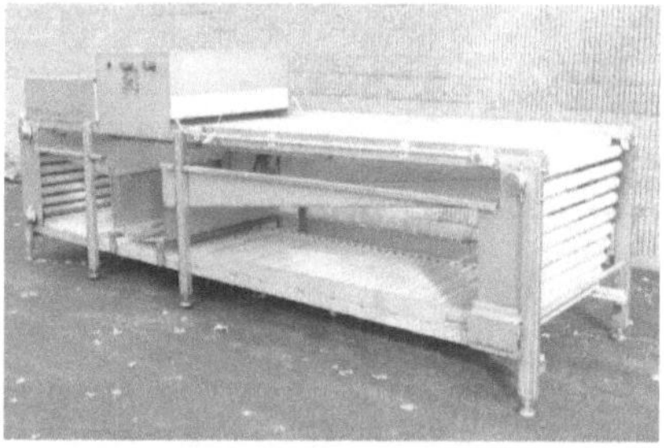

Fig.13.14. Spray washer cum inspection table
(Alard Equipmetns, USA)

Fig.13.15. Brush washer
(Haith Tickhill, UK)

Fig.13.16. Brush washer with brushes for cleaning at bottom and top
http://www.kwautomation.com.au

| | |
|---|---|
| <br>Fig.13.17. Multi-roller brush bed for dry brushing or wet washing of vegetables<br>http://www.haith.co.uk | <br>Fig.13.18. Citrus brush washer with spray nozzles<br>JBT FoodTech |

## Brush Washer

Brush washer cleans the exterior of fruits using a foaming detergent combined with brushes to remove dirt and debris. As the fruit contacts the rotating brushes, it begins to spin, The tumbling effect maximizes the fruit surface exposed to the cleaning spray and the brush's scrubbing action. The brush washer is constructed of stainless steel equipped with a set of brushes, intake rollers and discharge roller.

## 13.2 Grading equipments

The horticultural produce is to required be graded properly based on the size, weight, colour or maturity. Different kinds of graders developed by research institutes and commercial firms are available commercially.

### Size Graders

| | |
|---|---|
| 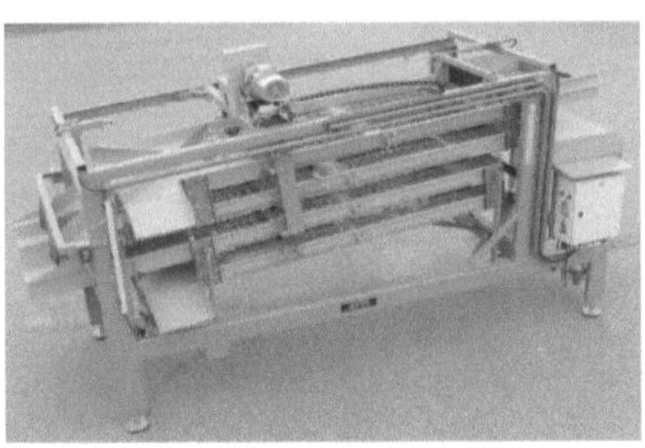<br>Fig. 13.19 Oscillating screen grader<br>wwwhttp://www.zeta-zr.cz | <br>Fig. 13.20 Potato grader-oscillating screen<br>CPRI , Jallandhar |

**Oscillating screen graders**

Oscillating screen graders (Figs.13.19 and 13.20) have been in used to screen under sized produce. However, these have now been modified for grading into 3 to 4 size grade. The top most sieve has the biggest opening while the lower most sieve has the smallest opening. The sieves are oscillated by a crank shaft. And can be changed to grade spherical fruits and vegetables like potato and onion. The material is generally fed to a set of sieves by a belt conveyor. The sieving capacity is upto 10 tonnes/hr.

| | |
|---|---|
|  Fig.13.21. Three stage screen type grader http://www.tongpeal.co.uk |  Fig.13.22. Three stage screen type grader ( grading potatoes) http://www.tongpeal.co.uk/products |
|  Fig.13.23. Four stage screen type grader ( grading potatoes)http://www.haith.co.uk |  Fig. 13.24. Different types of screens for different produce http://www.haith.co.uk |

## Screen graders

The screen graders consist of more number of screens moving at a constant speed. The smallest size of the screen is fitted at the feeding end while the biggest size near the outlet end as shown in Figs. 13.21 to 13.23. The produce is continuously fed by elevator conveyor to the grader feeding end and the grades are collected in the respective outlets as shown in Fig. 13.21. Different types of screens for different produce are shown in Fig. 13.24. These are suitable for guava, onion and other fruits and vegetables.

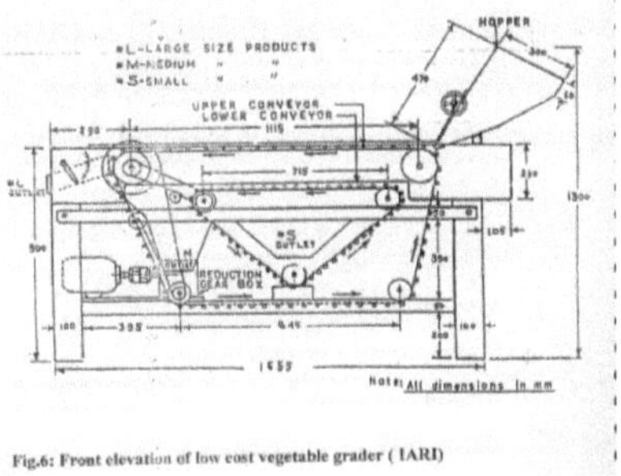


Fig. 13.25a. Fixed opening IARI Potato grader

Fig. 13.25b. Fixed opening IARI Potato grader

IARI New Delhi

## Fixed opening – Roller type graders

These graders consist of the endless belts, a feed hopper and a frame (Fig. 13.25). The two endless fixed roller belts having regular opening of size 40 cm x 5 cm and 40 x 2.5 cm move one over the other with the same speed and the same direction with gentle shaking. The shaking helps in changing the orientation of the products thus increasing the grading efficiency. The ungraded material from the feed hopper being fed to one end of the upper belt , the product of the size larger than 50 mm is carried over by the belts to the other end of the belt where it is discharged into the bags attached with the holder or for packing into other packages. The product smaller than 50 mm size falls on the lower belt which has 25 mm opening. This belt retains the product of size ranging between 25 mm and 50 mm and allows the product of less than 25 mm to fall through the gap between the PVC rollers on lower belt. This graded material collected separately. The medium size product is carried away by the lower belt, discharged and collected at the end of the lower belt. This machine separates the product into three grades of desired size. Belts of different spacings of PVC rollers can be used for sizing the product. This grader is suitable for grading potatoes, and needs ¼ hp electric motor to operate. The optimum belt speed is 0.16 m /s and its capacity around 1.8 tonne per hour.

## 13.28 Rotary barrel grader

In this machine, a cylindrical barrel is provided with holes or slots for grading round or cylindrical fruits (Fig. 13.26 to 13.29) fed from one end , the barrel is rotated around the central axis, the fruits smaller than the opening passe through this. Fruits get graded in 2 grades small size and ii. bigger size. when one barrel is used with one size of holes throughout periphery (Fig. 13.28 and 13.29). The barrel grader is used to grade sapota in the Navsari district of Gujarat state. The machine needs around 1 hp motor and the grading capacity is 2 tonne per hour. Sometimes , the fruits clog the openings and are to be dislodged. This again falls in the barrel. The damage to the fruit is more than the other types of graders. The barrel in Fig 13.27 has small size opening near the feeding end and bigger size opening near the outlet end and the produce can be graded to the desired numbers of grades.

The produce can be fed to barrel using conveyor belt (Fig 13.27) to increase the grading capacity. The grader in Fig 13.30 has two sizes of circular holes along the barrel length and is used to grade cashew in 3 grades. The grader in Fig 13.31 has three sizes of rectangular holes along the barrel length and is used to grade the produce in 4 grades. There are three sieves made of wire mesh of different openings mounted one above the other on a rotating shaft (Fig. 13.32). The fruit to be graded is fed on the innermost sieve. While rotating, smaller fruit will fall on to the second sieve & in further rotation, still smaller fruit will fall on to the third sieve. This way the fruits are graded in four sizes. The grader with 1 hp motor can grade up to 1 tonne produce per hour.

Fig.13.26. Rotary barrel/drum type grader
http://www.globalagritech.in

Fig. 13.27. Barrel with longitudinal expanding slots
CPRI , Jallandhar

Fig.13.28. Barrel of a rotary grader with circular holes

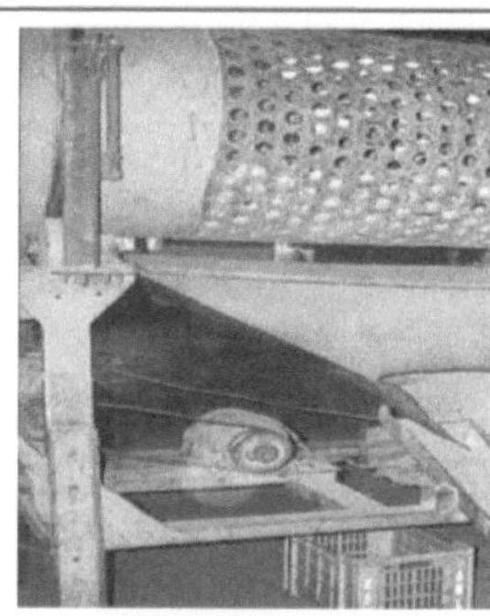

Fig. 13.29. Barrel type grader - Grading sapota

Fig.13.30. Barrel of a rotary grader with 2 size circular holes
http://www.tatcnyumbu.org

Fig.13.31. Barrel of a rotary grader with 3 size rectangular holes
http://dodmanlimited.co.uk

Fig.13.32. Three concentric barrels of a rotary grader for 4 size grading
http://www.digvijayindia.com

## Divergent aperture roller type grader

In this grader, the fruit is fed into two rollers rotating in opposite direction and having an expanding opening (Figs. 13.33 to 13.35). The fruit moves in the divergent opening due to rotation of the rollers. It passes through the opening when the size of the opening is equal to size of the fruit. The grader an shown in Fig.13.33 is for grading the fruits into three sizes. This kind of grader is most suited for round shape fruits like apple, oranges and conical vegetables like carrots.

Fig. 13.33. Divergent opening (Roller) three stage grader

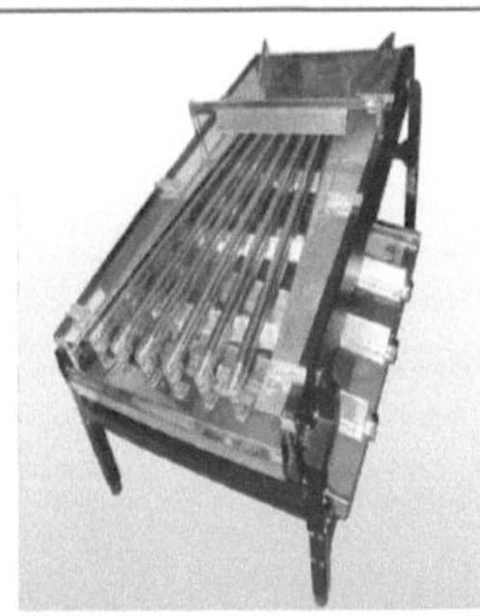

Fig.13.34. Divergent aperture rollers
http://www.intellimac.com/

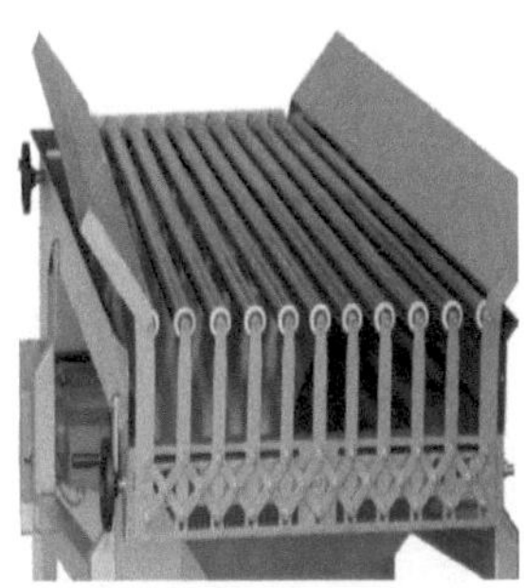

Fig.13.35. Divergent aperture rollers ( with adjustable aperture)
http://www.sormac.co.uk

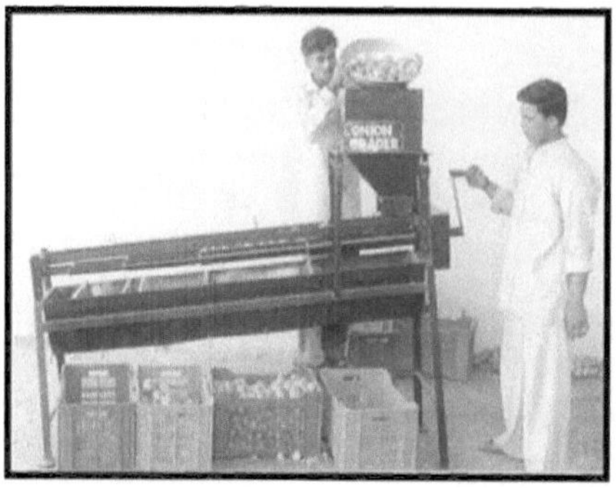


Fig.13.36 Manually operated onion grader (divergent aperture 2 rollers)
NRC onion and garlic, Pune

Fig.13.37 Motorized onion grader with divergent aperture rollers ( 4 rollers )
NRC onion and garlic, Pune

The size of aperture can easily be changed based on crop requirement (Fig. 13.35). The distance between the rollers can be adjusted continuously with a hand wheel between 10 to 35 mm. The adjustment for size variation alters along the full length of the rollers. The rollers are preset at an angle creating a natural inclination which transports the product without fall towards the roller discharge end. The product which is undersized falls through rollers and can be transported out of the machine or delivered directly into a hopper. With ½ hp motor upto 5 tonnes onion can be graded.

The size graders developed by NRC for Onion and Garlic, Rajgurunagar are becoming popular among farmers and traders. The manually operated grader has a capacity of 0.5 tons/hr/person while the motorized version of the grader has a capacity of 1.5 to 2.0 t/hr.

## Expanding opening double belt grader

This consists of two endless belt sets moving in oppsite direction. The distance between the belts increases along the length and the belts are inclined to vertical portion by 35° to 45° to hold the fruits. The ungraded produce is fed from the end having smallest opening. Fruits get singulated and get held in the space between the belts and move along the direction of motion of the belt. Small size fruits get graded by passing through the small opening near the feeding end. The medium sized fruits get graded in the middle of the belts as per the opening size. The larger sized fruits get graded by passing through the largest opening near the far end of the belt. The graded fruits can be collected in the outlets. The number of grades can be adjusted as per the number of openings. Fruits are made to fall on padded collection units to avoid the damage. The grader is most suitable for round fruits and can be adjusted for different sizes by adjusting the opening between the belts.

## Expanding opening single belt grader

In their grader one endless belt is moving and a channel is created above for continuous movement of the fruits by fitting to planks on the side (Fig.13.38). One side plank has a divergent opening, small opening near the feeding end and large size opening near the far end. The fruits are fed near the small opening and as these move along the belt, small sized fruits pass through the opening near the feeding, end medium size, fruits get graded through medium size near the center of the belt and large sized fruits get graded at large size opening near the far end of the belt. This grader is most suitable for round fruits like orange, etc.

Fig.13.38. Expanding opening single belt grader

CIPHET, Ludiana

## Diverging chain / belt grader

Diverging chain grader (Fig.13.39) is used for diameter grading of elongated products with particular emphasis on the grading of cucumbers, and carrots. The operating principle of this grader is based on the number of chains which move at the same speed. The internal distance between the chains increases as the product is conveyed along the grader. The product which is positioned between the chains will pass through at the corresponding point relative to the product diameter. Rotating brushes and discs assist in aligning the product between the diverging chains. The product collection area post grading is directly under the grading chains and is divided into sections which correspond to the grading sizes required. These sizes are adjustable within certain limits, so that the required grading sizes can be adjusted. The product is discharged from the machine either by conveyor belts or directly into pallets/containers The diverging belt graders can be manufactured with capacities from 1 ton up to 150 tons per hour.

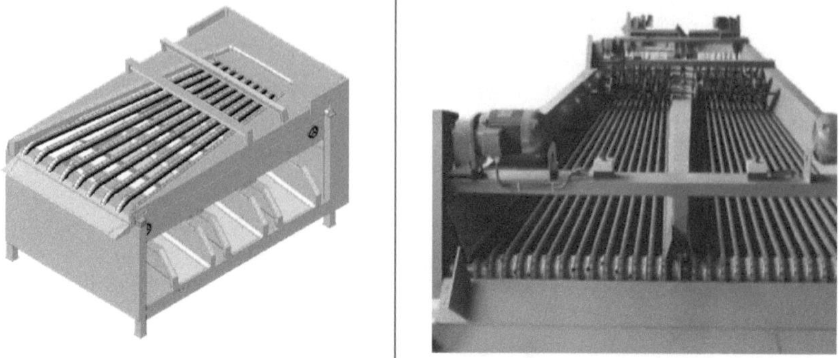

http://www.intellimac.com http://www.sormac.co.uk

Fig.13.39 Diverging Chain Grader

## Expanding opening rotary disc type grader

This grader has a rotary disc and the fruits to be graded are put at the cetnre of the disc by a conveyor belt. All along the circumference of the disc a divergent opening is created in vertical plane. Fruit collection platforms for different size grades are provided around the revolving disc (Fig. 13.40). The fruits fed at the centre of the rotating disc come in contact with the peripheral divergent opening, pass through the opening as per size and get graded based on the diameter of the fruit.

## Expanding aperture gravity fed grader

In this divergent aperture grader, two divergent apertures are provided (Figs. 13.41 and 13.42). The length of the aperture is 2.4 metre and the opening can be adjusted as per the type of fruit and sizes grade. The fruits are fed from the feeding end, roll under gravity, pass through the opening as per the size of fruit and get collected in the respective grades. The capacity of 2 channel manually fed grader is about one tonne/hr. The grader is suited for round fruits and vegetables like sapota, tomato, lime and mango.

Fig. 13.40 Expanding opewwning rotary disc type grader
NRC Citrus

Fig. 13.41. Expanding aperture gravity fed fruit grader
IIHR Bangalore

Fig. 13.42. Expanding aperture gravity fed IIHR Fruit Grader
IIHR Bangalore

## Roller grader / Expanding spool grader

The grader consists of a frame, an elevator, feed conveyor, an intermediate receiving conveyor, a sizing conveyor with rubber spools/ rollers and two identical driving rollers with helical grooves of gradually increasing pitch (Figs. 13.43 to 13.45). The helical groove rollers with increasing pitch are fitted along the length of the grader. The rods with the spools/ rollers are carried forward as the driving helices rollers rotate. The gap between the individual spools/rollers goes on increasing gradually according to the pitch of the helices of the driving rollers.

This leads to sizing of fruits as these are carried forward over the sizing conveyors. The produce can be graded into different grades by providing necessary collecting outlets/conveyors below the spools/rollers. This type of graders are extensively used commercially for grading apple, orange, onion, potato etc.

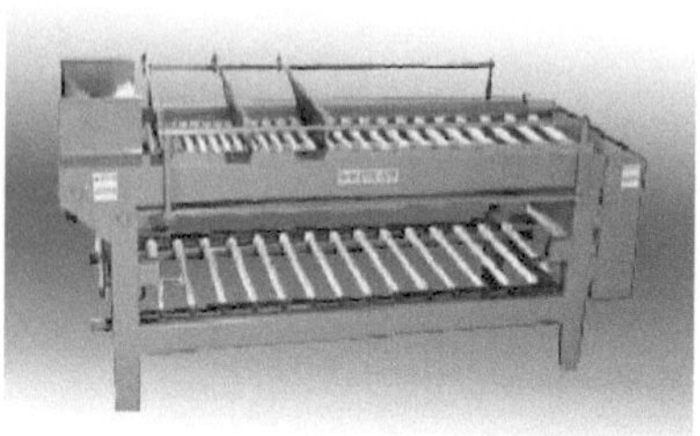

Kerian Machines Inc.

Kerian Machines Inc.

Fig. 13.43. Roller grader / Expanding spool grader

Fig. 13.44. Expanding opening roller grader with feeding conveyor

Fig. 13.45. Expanding opening / roller grader in operation

## Radial Grader /Expanding opening slat type grader

Radial grading machines (Fig. 13.46) are equipped with lateral placed grading rolls out of stainless steel. The distance between these rolls changes in steps over different grading stages. Along the length of the conveyor the gap between flats increases as the roller spacing increases to drop through the smaller fruits first and bigger fruits later. One end of the stainless flat is hinged to a rod connecting two tracks of chain. The lower end of the flat is attached with resting wheel / roller and these resting wheels move on a track. The height of track can be adjusted by using appropriate mechanism to make the grader suitable for different kind of fruits. A number of conveyor belts are placed under the machine, one for each grading size. This type of grader is being extensively used for grading sapota in three grade sizes in Navasari, Gujarat State.

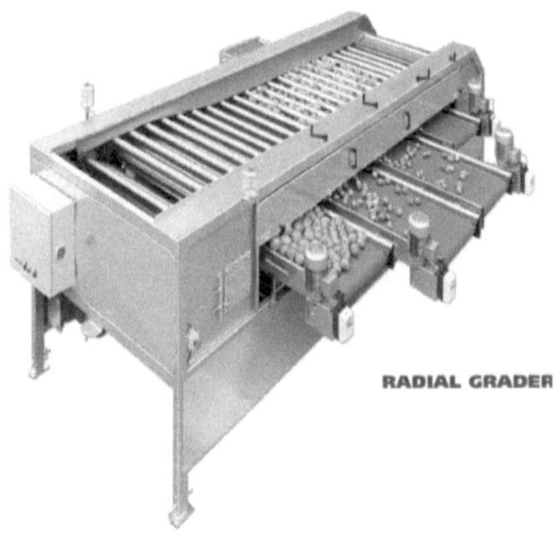


Fig. 13.46, Radial Grading Machine
Vishvkarma Machine Tools, http://fruitgrading.net
http://www.allroundvp.nl

CIAE Bhopal

## Mechanical weight graders

In this grader, there is a endless chain of cups as shown in Fig. 13.47 and Fig. 13.48. The grader may have one to ten channels to convey the fruits. Different kinds of cups are used to hold and grade fruits as shown in Figs. 13.44. 13.48. 13.54. 13.55 and. 13.56. The mechanical weight sensing mechanism is fitted all along the chain below cups carrying the fruits. The fruits are fed into the cups by the conveyors and singulation unit as shown in Figs. 13.53 and 13.54. The fruits move in the cups all along the chain. The fruits with larger weight get graded near the feeding end by titling of the cups when the weight of the fruit is more than the tipping/holding capacity as per the adjusted capacity of the sensing unit. As the remaining fruits move along the length of the conveyor lesser and lesser weight fruits get graded and collected in different grades as per the adjustment of the sensing units. The fruits of different grades are then passed on to the sorting tables as shown in Figs. 13.47 and Fig. 13.48 These are then packed in CFB boxes manually by labourers and any cull or damaged fruits are sorted out. The weight of the filled CFB box is maintained as per the requirement and the box is then closed. In packing apple and other high value produce, the cushioning material like pulp tray or styro foam trays are also used and the required number of fruits per CFB box are filled. In some of the packing houses fruits of the different colours are sorted out manually and packed in different boxes for the same weight grade.

http://www.aweta.nl

Fig.13.47 . Tomato weight grading line with sorting tables

http://www.aweta.nl

Fig.13.48 . Tomato weight grading line with sorting tables

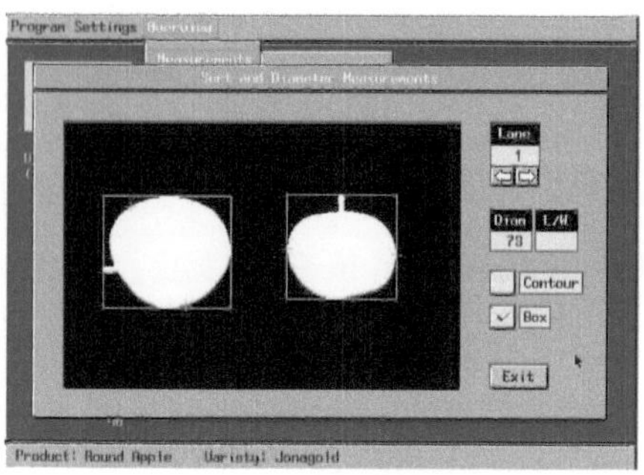


http://www.aweta.nl

Fig. 13.49. Sensing the dimensions

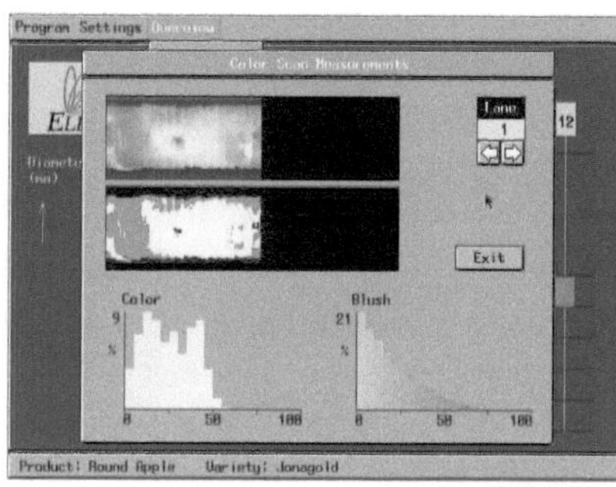


http://www.aweta.nl

Fig. 13.50. Sensing the colour

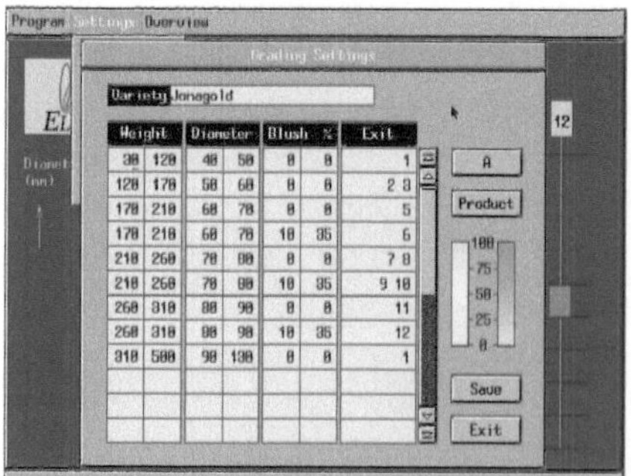


http://www.aweta.nl

Fig. 13.51. Feeding the Grading Data

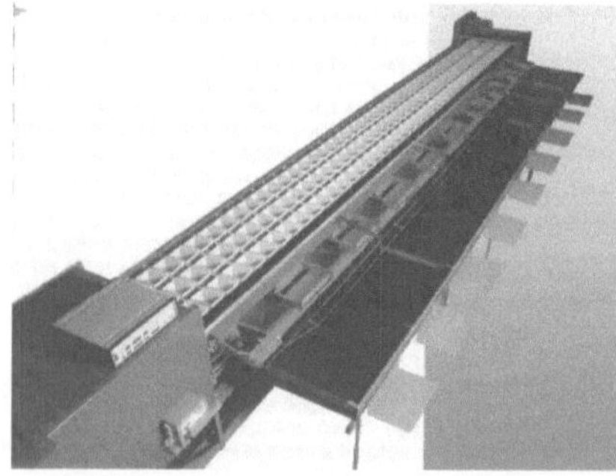

http://www.aweta.nl

Fig. 13.52. Overall view of a computerized grader with cups

http://www.aweta.nl
Fig. 13.53. Feeding the produce to grading conveyor

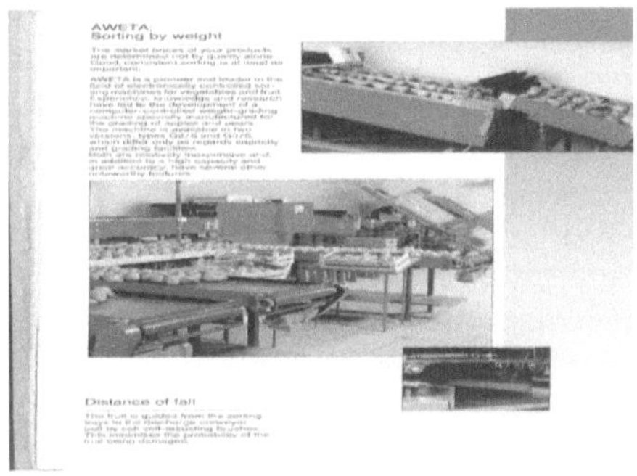


http://www.aweta.nl
Fig.13.54. Singulating the fruits for cups

http://www.aweta.nl
Fig.13.55. Cups for conveying for pine apple grading

http://www.aweta.nl
Fig.13.56 Cups for grading garlic

http://www.aweta.nl
Fig.13.57. Cups for muskmelon grading

http://www.aweta.nl
Fig. 13.58. Packaging of weight graded muskmelons

## Computerized graders for size, weight and colour

The fruits can be graded as per size, weight and colour or a combination of these. The size of the fruit is sensed by the camera as shown in Fig. 13.49. The colour of the fruit is sensed by the computer as shown in Fig. 13.50. For determining different colour grades about 10 fruits of required grade are fed on the roller conveyor chain (Fig. 13.59 and Fig. 13.60), underneath the camera of the computer for calibration. The weight of the fruits is also sensed near the feeding end. Different required of grades for size, weight and colour are fed to the computer as shown in Fig. 13.48. The fruits move on the chain with cups as explained in mechanical weight grader (Figs. 13.47., 13.48., 13.52 to 13.57 and 13.61 to 13.62). These get discharged into different outlets for different grades as per the data fed to the computer. For colour grading, roller conveyors are used as shown in Fig. 13.59 and Fig. 13.60. The computer can also discharge equal number of fruits for the same grades in different opening if required. Different grades get collected into different outlet as shown in Fig. 13.61 and Fig. 13.62. These are then filled manually into CFB boxes. The computer can also signal when the required number of fruits for required weight or number for a CFB box get discharged in the outlet. The colour grader can also discard fruits with unwanted colours or those have disease shots or mechanical damage.

Fig. 13.59. Roller Conveyor for Colour and Weight grading
http://www.aweta.nl

Fig. 13.60. Rolls for Colour and Weight grading
http://www.aweta.nl

Fig. 13.61. Computerized grading and packaging line for red capsicum
http://www.aweta.nl

Fig. 13.62. Computerized grading and packaging line for capsicum
http://www.aweta.nl

## Waxing

During washing of fruits and vegetables the natural coating of wax gets removed. Some of the fruits and vegetables like apple, banana, peach, avocado, orange, cucumber, capsicum, tomato and melons etc. are waxed with edible wax to replace the natural waxes. Waxing generally reduces the physiological weight loss, improves the appearance, decreases transpiration, reduces the shrinkage and increases freshness. The wax is dissolved in a suitable solvent and melted wax is sprayed on the fruits. This is subsequently brushed mechanically until a film of desired thickness is obtained. Very small quantity ie., a drop or two of wax are applied on each fruit. Washing, waxing and drying systems give the fruit a superior appearance in the market. The wax is dried in temperature controlled hot air tunnel ( Fig. 13.66). The commercial waxing machines are shown in Figs 13.63 to 13.65.

Fig. 13.63 Waxing machine for apple
http://article.wn.com/view/2012/07/11/Whats_for_Dinner_Lemony_Harvest_Grains_with_summer_veggies_c/

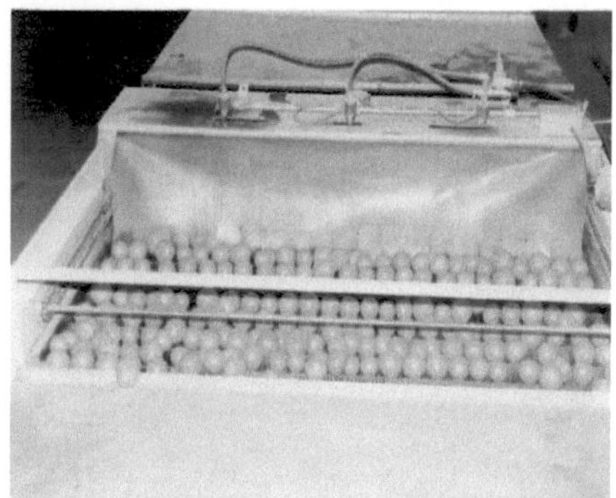

Fig. 13.64 Waxing machine for orange
http://olimpias.gr

Fig. 13.65 Waxing machine for orange

Fig.13.66. Drier for waxing

## Packaging line

A packaging line generally consists of washer, waxer, grader, sorting tables, weighing equipments, carton assembling, filling cartons, closing the cartons and palletization along with necessary conveyor belts. The simplest packaging line may consist of washing of fruits and the sorting and grading platform (Fig 13.67). The orange packaging line at National Research Centre for Citrus at Nagpur consists of grader of expand opening (rotary disc), waxing and drying unit and sorting conveyor (Fig 13.68) for packaging of oranges.

Fig. 13.67 Washing and grading platform for small growers
www.marketfarm.com

Fig. 13.68 Sorting and packaging oranges

Fig.13.69 Hand packing fruit in cartons

Fig.13.70 Automatic packingmachine with the suction cups holding a layer of orange
http://www.sunkistresearch.com

## Filling the cartons

Cartons are generally available in knocked down conditions and these are opened and assembled. The fruits are filled manually (Fig 13.61, 13.62 and 13.69) and the isolation and cushioning partitions are kept in between the fruit layers. Automatic carton filling machines are also available (Fig 13.70). One layer of fruits (as per carton size) is picked by the suction cups and placed in the carton. The partition or cushioning tray is kept manually in between the 2 layers. The filled cartons are weighed and may be kept on the conveyor belt (Fig 13.71). The filled cartons are labeled and closed by a tape.

## Palletization

A pallet is a standard wood structure of 90x90 or 100x100 or 100x120cm for mechanically lifting of cartons by a fork lift. The corrugated fibre boxes or cartons are kept on the pallet in interlocking position and sufficient space is left in between the cartons for air circulation (Fig.13.72). The palletized cartons may also be stabilized by putting plastic film around these.

Fig.13.71 Filled carton on conveyor for closing
http://www.sunkistresearch.com

Fig.13.72 Palletization
http://www.sunkistresearch.com

## Packaging mangoes for export

Mangoes for export are harvested with 1 to 2 cm long pedicel. These are then desapped in the desapping hall and then placed on bamboo sticks with pedicel facing down ward. The pedicel is then cut by secatuer and the sap is allowed to fall on the old newspapers. After desapping the fruits are washed with water. These are then graded manually on weight basis. Hot water treatment is given hot water treatment plant to control the anthracnose and fruitfly in Fig. 13.73 and Fig. 13.74. The water may be heated with electrical heating elements or with a diesel or waste fired boiler. The temperature control and recording system has to be fully automatic or computerized as per US Standards. The water /mango ratio has to be 5:1 for better stabilization of temperature. The temperature and duration of treatment required for control of anthracnose is 52°C for 10 minutes. The treatment of 46°C for 65 to 95 minutes is required for control of fruitfly for fruit weight of 250gm to 700gm. The treatment for fruitfly also controls anthracnose in the fruits. The hot water treated fruits are to immediately given cold water treatment to bring these to room temperature. These can be then packed in cartons, palletized and stored in cold storages at 12 °C.

| | |
|---|---|
|  Fig. 13.73 Hot water treatment plant (capacity 250kg/hour) to control fruitfly and 1 tonne/hour to control anthracnose IIHR Bangalore |  Fig 13.74 Hot water treatment plant (capacity 3 tonne/hour to control fruitfly) |

# *Chapter 14*

# Post Harvest Management and Storage

In India most of the horticultural products like fruit and vegetables are sold fresh for eating or cooking. However, the tendency of consuming processed fruits as tetra packed juice, jam, jelly and fruit bar is increasing. The dehydrated vegetables are becoming popular for use in noodles and readymade soups. Some of the horticultural produce need primary processing at farm. These include plantation crops like coconut, areca nut, black pepper, turmeric etc. The processing system can be continuous for high capacity industry or batch type for small industry. The on-farm processing and processing for preservation of fruits, vegetables and spices have been discussed here. The literature for processing of fruit and vegetables is available in plenty and there are large number of manufacturers for manufacturing the machines for processing of fruits and vegetables. The pickles and chutney are the speciality of Indian sub-continent and are given major emphasis in this chapter. Major unit operations for processing are (a) Washing/ cleaning, (b) Grading, sorting, (c) Peeling, (d) Slicing/ cutting (size reduction), (e) Mixing, (f) Heating, pasteurization, evaporating, concentrating etc., (g) Canning, (h) Packaging ( bottling / pouch filling, sealing etc.) and (i) Drying.

The post harvest equipments developed for economically important horticultural crops by the Institutes of Indian Council of Agricultural Research and State Agricultural Universities and their commercial variants available in the market have been discussed here.

## 14.1 Mango

### Raw mango peeler

Raw / unripe mangoes are peeled for making chutney and amchoor ( dried powder of unripe mango). The batch type and continuous type peelers have been developed at Indian Institute of Horticultural Research, Bangalore. In batch type peeler (Fig. 14.1) 8 mangoes are fed and the rotation of mangoes in the elliptical chamber formed between concave wheel (with sharp projections), guide and casing peels the mangoes. The peeled mangoes are taken out through outlet by manual rotation of guide wheel . The machine is fitted with 0.5 hp electrical motor. The peeling capacity is 120 kg mangoes per hour.

In continous raw mango peeler (Fig. 14.2) unripe mangoes are fed through the inlet. Concave wheel has sharp projections and revolves at 250 rpm, with 2 hp motor. The guide drum revolves at 2 rpm and maintains the continuous feeding and output of the mangoes. The peeled mangoes come out continuously through outlet. The peeling capacity is 200 kg mangoes per hour. The peelers can be modified as grater by changing the shape of the concave wheel.

IIHR Bangalore

Fig. 14.1 Batch type type raw mango peeler

IIHR Bangalore

Fig. 14.2 Continuous type raw mango peeler

## Raw Mango Slicer

The mango slicer (Fig. 14.3) slices the mango longitudinally. The mangoes are cut into 4, 6 or 8 slices based on the size. The width of the slice is 20 mm, the thickness of the slice is as per the thickness of pulp and the length of the slices is as per the length of the mango. The mangoes are cut by the blades made out of AISI 420 grade hardened stainless steel (cutlery grade). The feeding of mangoes to the cutting blades is under gravity and a particular orientation to facilitate cutting of mangoes into desired sized slices. The mango is cut into 2 slices (cheeks) in first cut by upper set of blades and further cut into 4,6or 8 slices by the lower set of blades rotating at 750 rpm. The cut slices fall on nylon mesh and the cut pieces of kernel are separated from pulp. The cut kernels/stones can also be sieved off. The power requirement of the slicer is 6 hp ( four motors of 1.5 hp and 750 rpm). The slicing capacity is 0.8 tonne/hr.

## Raw Mango Cube Cutter

The raw mango slices are fed to cube cutter (Fig. 14.4) by 4 labours . The cube cutter cuts the raw mango slices by a set of rotating blades set, spaced at 2.0 cm into cubes/ pieces of approximately 2.5 cm X 2.0 cm x 1.6/1.8 cm. The blades are fabricated out of AISI 420 grade hardened stainless steel (cutlery grade). The remaining parts of the cube cutter are fabricated out of AISI 304 grade stainless steel. The outlet consists of nylon net and the cut cubes fall in 50kg containers. The machine is powered by 5 hp motor. The cutting capacity is 0.5 tonne/hr.

IIHR Bangalore

Fig. 14.3 Raw mango slicer

IIHR Bangalore

Fig. 14.4 Raw Mango Cube Cutter

## 14.2 Coconut

### Coconut dehusker

The hand operated coconut dehusker weighs 7 kg and has a capacity of 70-110 nuts/h. The pedal operated coconut dehusker (Fig. 14.5) weighs 12 kg and has a capacity of about 70-170 nuts/h. The capacity of power operated (electrical motor) (Fig. 14.7) coconut dehusker is 500 nuts/hr. with a efficiency of 91%.The dehusking machine(Fig. 14.8 ) peels off the coconut husk from coconut fruit to obtain dehusked coconut fruit via hydraulically controlled dehusking devices. An operator is required to handle the machine during the dehusking process.

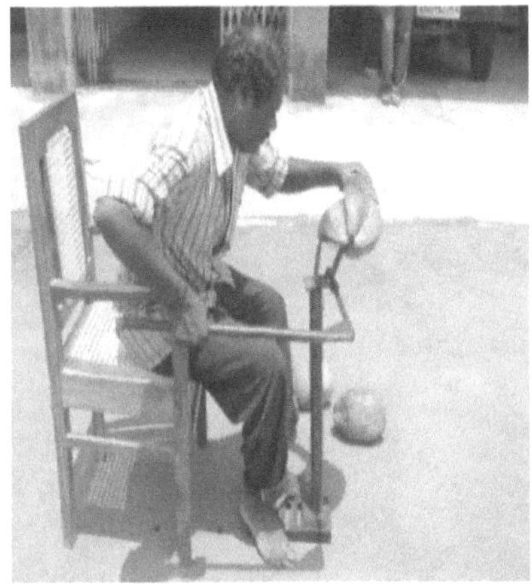

CARI Portblair

Fig. 14.5 Pedal operated coconut dehusker

Fig. 14.6 Pedal operated coconut dehusker
CARI Portblair

Fig. 14.7 Power operated coconut dehusker
Kovai Classic Industries

| | |
|---|---|
| 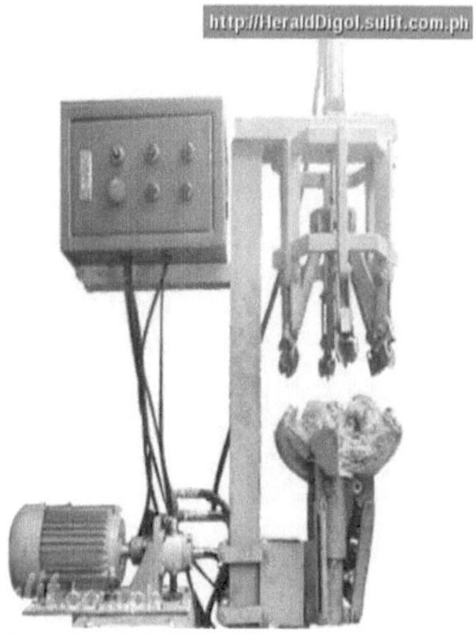  |  |
| http://sulit.com.ph | http://www.flickr.com/photos/me-mawo/3173772178/in/photostream |
| Fig. 14.8Coconut d ehusking machine with hydraulic cylinder and husk spliter | |

## Coconut Splitting Device

A manually operated splitting device (Fig. 14.9) developed at Central Institute of Plantation Crop Research split the coconut by pressing the blade on the shell . It can split 100-120 coconuts per hr.

### Coconut De-Shelling Machine

The commercially available coconut deshelling machine (Fig. 14.10) is operated by 1 hp motor. Two revolving wheels have serrations which act as blades and cut / split the shell. It has a capacity of 300-500 nuts/hr.

CPCRI Kasargod

Fig. 14.9 Coconut splitting device

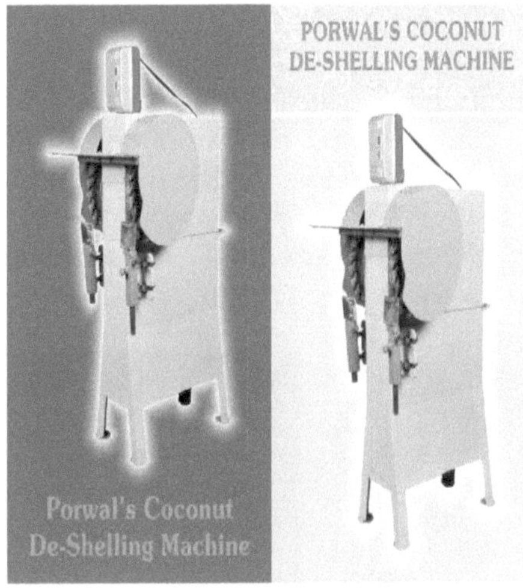


Porwal Coconut Machines

Fig. 14.10 Coconut deshelling machine

## Tender Coconut Punch

A simple tender nut (Fig. 14.11) punch has been developed. This is for making a hole in coconut for drinking or removing the coconut water. The hole is made after removing the husk.

## Tender nut Cutter

A cutter (Fig. 14.12) is developed for tender nuts. It consists of a semicircular blade fixed on a frame and operated by a handle.

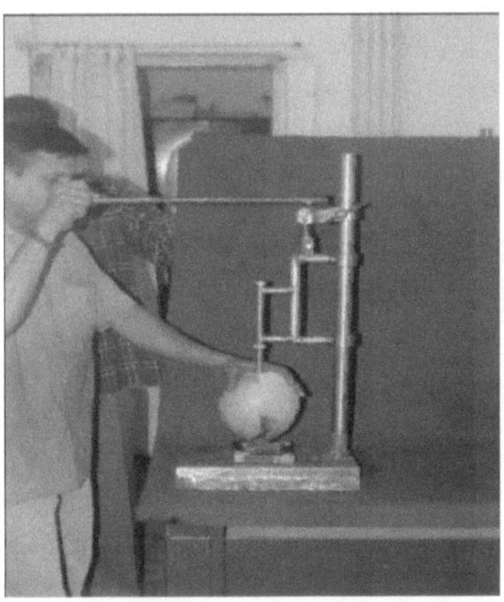

Fig. 14.11 Tender Coconut Punch

CPCRI Kasargod

CPCRI Kasargod

Fig. 14.12 Tender nut Cutter

CPCRI Kasargod

## Snowball Tendernut Making Machine

Snow ball tender coconut (Fig. 14.13) is globular tender coconut kernel containing tender coconut water inside and it is scooped out with the help of specially devised tool after cutting the shell of tender coconut of 7-8 months maturity by using the newly designed snow ball tender coconut machine and thereafter the SBTN is made free of the adhering testa, packed hygienically to use either fresh or after refrigerated storage. Fresh kernel of matured coconut containing reasonable amount of water is to be used for this process.

CPCRI Kasargod

Fig. 14.13 Snow Ball Tender Nut

## 14.3 Potato

### Potato Peeler

CIAE Bhopal has developed two models of potato peeler, a hand operated (Fig. 14.14) and a pedal operated one. It is a batch type equipment for peeling outer skin of potatoes of any size for further processing to produce value added products like chips, wafers etc. It consists of main frame, handle, rotating drum with notches, water inlet, top cover etc. A batch of 8 kg is fed in the drum and peeling is completed in 8 minutes. The capacity of the unit varies from 30-32 kg/h. The peeling efficiency is 77% and peeling losses 4%. The pedal operated potato peeler is developed for small entrepreneurs for use where electricity is not available.

The perforated stainless steel drums during rotation remove skin of potatoes as in manual peeler. The capacity of the peeler is 188kg/h.

### Batch type peeler (abrasive peeler)

Abrasive peeler (Fig. 14.15) removes the skin from potatoes and onion etc. washed potatoes are poured into the drum of the potato peeler from the top, the inner of the drum is coated with abrasive carborundum. In some cases bottom plate is coated with abrasive carborundum. As the base of the drum rotates the potatoes strike against the abrasive inner wall which removes the skin, this processing requires flow of water through a pipe to remove the skin from the walls, the peeled potatoes are removed from the side of the drum through a gate. The diameter of the base plate for various capacities is 50 to 90 cm, batch capacity 7 to 25 kg and motor horse power requirement is 0.5 to 2 hp. In plc / electronic controlled peelers, filling time, peeling time and discharge time are adjustable with timers.

## Potato Slicer

CIAE Bhopal has also developed a pedal operated potato slicer (Fig. 14.16) which is suitable for small entrepreneurs where electricity is not available. The weight of the unit is 46 kg and capacity 1881 kg/h. Electrical operated potato slicers of different capacities are also available commercially.

Fig. 14.14 Potato peeler
CIAE Bhopa

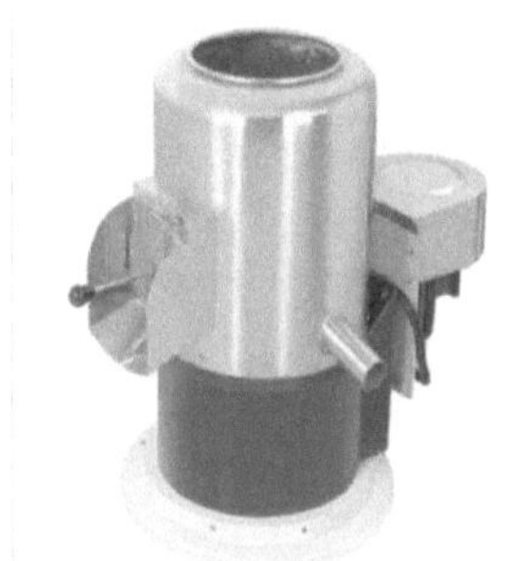

Fig. 14.15 Abrasive potato peeler
**Master Enterprises**

Fig. 14.16 Potato slicer
CIAE Bhopal

## 14.4 Onion

## Multiplier Onion Peeler ( Batch Type)

CIAE Bhopal has developed a batch type peeler (Fig. 14.17) for multiplier onion. It consists of aluminum drum of 300 mm diameter and 290 mm height seated over a 300 mm disc. The drum is rotated at 300 rpm. The opening near bottom in drum serves as outlet. Multiplier onions are fed in batches of 2 kg after end cutting and soaking in water. Corrugated rubber sheet acts as abrasive substance for peeling. The capacity of the unit is 30 kg/h. The cost of peeling is Rs. 1.2/kg.

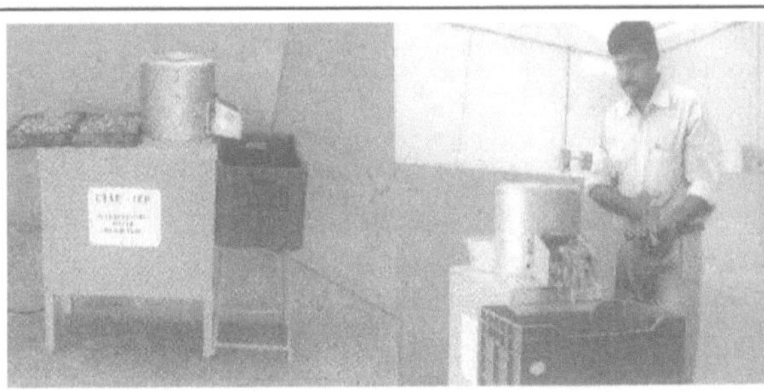

Fig. 14.17 Batch type peeler for multiplier onion
CIAE Bhopal

## 14.5 Garlic

### Garlic Bulb Breaker

The garlic bulb breaker (Fig. 14.18) consists of two rubber rollers rotating in opposite direction. The garlic bulbs are fed into the rubber rollers. The rollers compress the bulbs and the cloves, centre stem, root portion and wrappers get separated due to shearing action during compression. The rpm of the drums is about 20 for optimum performance. The Capacity is 50 kg per hour and efficiency is 95%.

### Garlic Clove Peeler

The clove peeler (Fig. 14.19) consists of the rubber roller and the concave is fabricated out of GI mesh. The size of opening of GI wire mesh is 1.7 mm square and the thickness of wire is 0.45 mm. The concave has an angle of 120°g and the minimum clearance is about 2 mm. The peeler can peel the cloves dried at 65° C for 30 minutes. About 83.41% of the cloves were fully peeled. The capacity of the hand operated peeler is about 11.66 kg per hour. The peel has to be sucked through aspirator. Capacity is 12 kg per hour and efficiency is 85%.

Fig. 14.18 Garlic bulb breaker
IIHR, Bangalore

Fig. 14.19 Garlic clove peeler
IIHR, Bangalore

## 14.6 Tapioca/ Casava

### Cassava Peeling Knife

Peeling is the most labour intensive unit operation in the process of starch extraction. A cassava peeling knife (Fig. 14.20) of a novel design has been developed. Peeling capacity is about 110 kg/h. The knife costs can be sharpened and blade can be replaced.

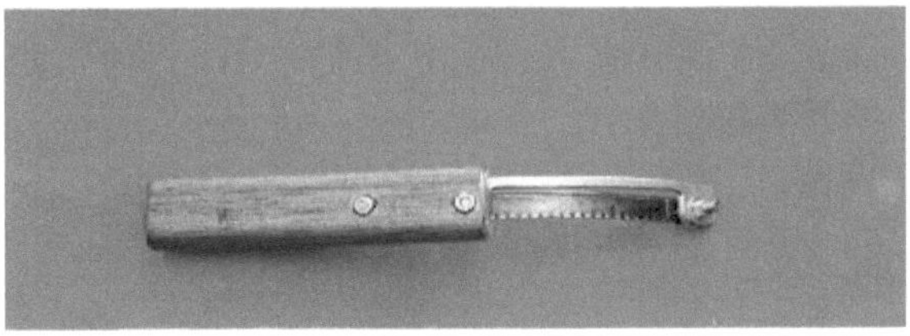

Fig. 14.20 Cassava peeling knife

CTCRI Trivendrum

## Cassava Chipping Machine

The machine consists of two concentric mild steel drums, the annular space between which is divided into compartments for feeding the tubers, supported on four legs. A rotating disc at the bottom of the drum carries the knives assembly. A pair of H.S.S (High speed Stainless Steel) bevel gears is provided to operate the machine manually with a crank arm (Fig. 14.21). The pedal operated chipping machine (Fig. 14.22) has provision of a pivoted pedal for transmitting the power to the cutting disc through a suitable belt and pulley drive mechanism. The motorised chipper (Fig. 14.23) is operated with a 0.5 hp single-phase motor through a suitable belt drive. Tubers are fed into the compartments from the top and the chips are collected at the bottom. Chips of 1.0-7.0 mm thickness can be produced. The capacity of hand-operated cassava chipping machine is 40-120 kg/h, pedal operated chipping machine is 80 - 770 kg/h and motorised chipping machine is 280-1100 kg/h.

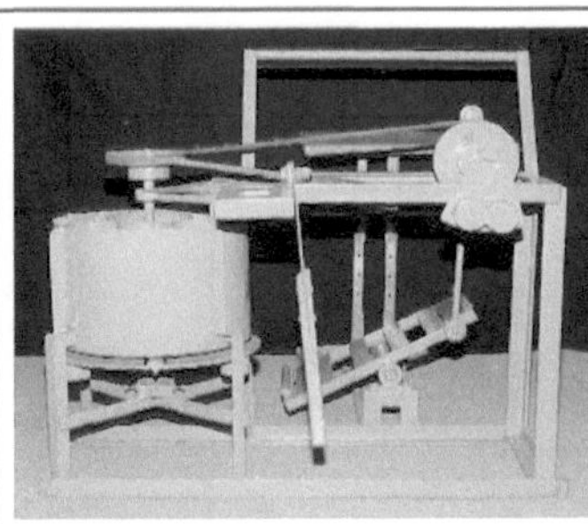

CTCRI Trivendrum

Fig. 14.21 Hand-operated chipping machine

Fig. 14.22 Pedal -operated chipping machine

Fig. 14.23 Electric motor -operated chipping machine

## Starch Extraction Machinery

For Extraction of starch from the tubers at the production site the starch extraction machinery has been developed (Fig. 14.24) . The major components of the machine are a hopper to feed the tubers, crushing disc or cylinder with nail-punched protrusions rotating inside the crushing chamber to crush the tubers, sieving tray to remove the fibrous and other cellulosic materials, stainless steel or plastic tanks to collect the sieved starch suspension, tuber storage chamber, handle and wheels for easy transportation from place to place and a frame to support these components. An electric motor (0.75 hp) attached to the frame is used as the energy source to operate the machine.Capacity of the machine is 200 kg/h for cassava, 135 kg/h for sweet potato and 120 kg/h amorphophallus .

## Cassava Rasper

A power-operated rasper (Fig. 14.25) for the small scale processing of cassava roots into starch has been developed. The main components of the machine are hopper, top and bottom covers, rotating drum, chute and pulp collecting box, fabricated in mild steel. The capacity of the rasper is 350 - 400 kg/h.

## Granulator for Cattle/Poultry Feed

A drum type centrifugal granulator (Fig. 14.26) for tapioca starch has been developed. The machine can be operated both manually and by 0.75 hp electric motor. Capacity of the machine is about 20kg/h.

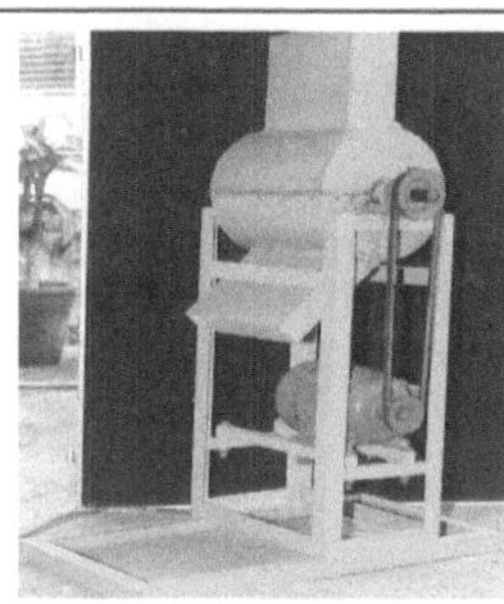

CTCRI Trivendrum

| | | |
|---|---|---|
| Fig. 14.24 Starch extraction machinery | Fig. 14.25 Cassava rasper | Fig. 14.26 Granulator for cattle/poultry Feed |

## 14.7 Turmeric

## Turmeric boiling:

The turmeric boiler in (Fig. 14.27) has a capacity of about 100 kg turmeric. The water is first made to boil and the crates filled with turmeric are kept above the boiling water and covered with a lid. It takes 10-15 minutes to steam / boil the turmeric. The portable turmeric boiler (Fig. 14.28) has four tanks and the capacity is 600 kg/hr.

| | | |
|---|---|---|
|  |  |  |
| | TNAU Coimbatore | Raviraj Engineering & Fabri-mech, Sangli |
| Fig. 14.27 Turmeric boiler | | Fig. 14.28 Portable turmeric boiler |

## Turmeric polishing:

The turmeric is polished in rotary hexagonal drum to give a surface finish (Fig 14.28). The hexagonal drum has spikes inside it to polish the turmeric. A small quantity of powder turmeric is also coated on the polished turmeric to give a better look. It has a capacity of 100 kg/batch with 4 % polish is obtained in about 45 min .

| | |
|---|---|
| IISR, Calicut<br>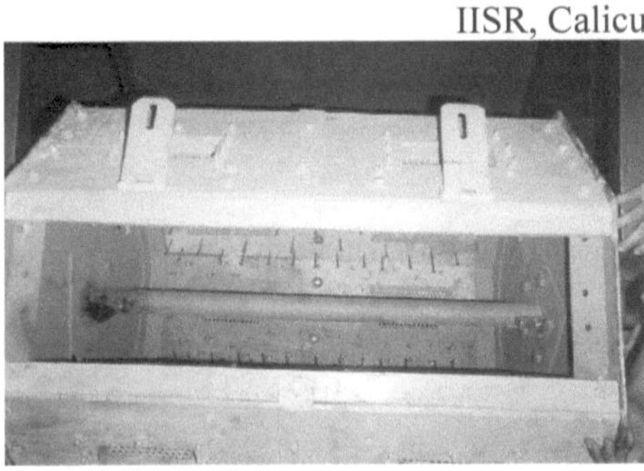 |  |
| Fig. 14.28 Turmeric polisher<br>Jaidi Ramakrishna, Nikalpur | |

## 14.8 Pomegranate

## Hand operated Pomegranate Aril Remover

| | |
|---|---|
|  |  |
| Fig. 14.29 Hand tool for separation of arils<br>CIPHET Ludhiana | Fig. 14.30 Pomegranate aril remover<br>IIHR, Bangalore |

| | |
|---|---|
|  |  |
| Fig. 14.31 Electric motor operated aril remover<br>Digvijay Industries | Fig. 14.32 Electric motor operated aril remover<br>CIPHET, Ludhiana |

To facilitate in easy separation of arils (juice enclosed seed) from Pomegranate Tool (Fig. 14.29) consists of fruit holders having knife arrangement that penetrate into the peel. Due to rotating action of the holders, fruit breaks into two irregular halves as a simultaneous effect of tension with twist over peel surface. During this action, whole fruit experiences a shearing effect; and inside arils become loosen which provides an opportunity of easy separation. About 30-45% arils are separated in the process of irregular breaking. Tool accommodates medium size (5-8 cm) of fruit. Salient features of the hand tool for easy separation of arils are: (i) Simple in design and fabrication. The material of fabrication is primarily wood and steel knives (ii) Drudgery is negligible and it is safe in use (iii) Tool is suitable for all varieties of pomegranate (iv) No damage to arils because the knives only penetrate into the peel.

The pomegranate fruit is cut into two halves and one half is kept upside down on the fruit holding platform so as the fruit skin is faced towards the hitting head (Fig. 14.30). The handle is operated and the hitting head is moved in reciprocative motion to hit the fruit. With in two or three hitting , the arils are removed from the fruit and a collected in container through the chute. The capacity of the machine is 50 kg per hour in comparison to 18-20 kg per hour in manual operation. The efficiency of the machine is 96 % . The capacity of the machine is 50 kg per hour.

The pomegranates are cut into halves. These are put into rotating drum having holes on the surface (Fig 14.31). The cut pieces are made to tumble with help of baffles fitted inside the drum. The arils get separated and come out from the holes on drum surface. The rind comes out from other side of the drum. The aril removal is a continuous process-It is fitted with 1hp motor and capacity of the machine is 500kg/hr.

CIPHET-Pomegranate Aril Extractor (Fig 14.32), is a mechanical and continuous system for processing of whole pomegranate of any size, shape and variety; and for recovery of clean, whole and undamaged arils.

The machine is capable to process the whole pomegranate at a rate of approximately 500 kg/hr with extraction capacity of 90-94% and with little damage of arils that is about 1-2%. The Pomegranate breaking and arils separation mechanism is highly efficient with little damage and waste. Peels and other extraneous matter are separated out and it is yielding clean arils that can be used for further processing or for fresh eating.The working principle of CIPHET-Pomegranate Aril Extractor" is exclusively novel. The machine comprises the main components such as Pomegranate Fruit Breaking unit, Drive unit, Collection Trays where separated arils and extraneous matters are received, Vibrating Sieve unit; and the clean arils and extraneous matter Collection Troughs. The fruits are continuously broken by fins provided on rotary drum, in the process aril and rind/peel get separated. The fruit breaking unit breaks the fruit in such a manner that the major portion (85-90%) of arils is safely separated out. The rest of the arils that is about 10-15% remains attached with the broken peels are further getting separated over the vibrating screen.

## 14.9 Black Pepper

## Black Pepper Thresher

Dried pepper spikes are to be threshed to separate berries. Two types of pepper threshers (Fig. 14.33) are available. One of the design is using rotary wooden hammers as cylinder to thresh the spikes (capacity 400 kg/hr.) with damage to 5 % berries. The other design has a rubberized / rubber coated cylinder to thresh the spikes and has a capacity of 700 kg/hr with reduced damage to the berries.

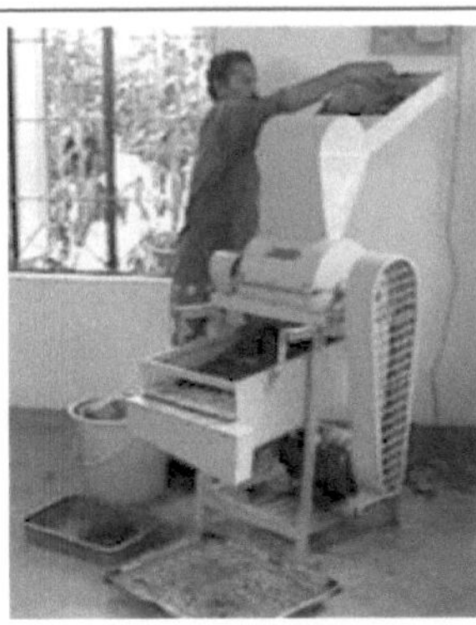

IISR, Calicut

Sagar Engineering

Fig. 14.33 Pepper thresher

## Pepper drying

Electrical Black Pepper Drier: It has a capacity of 100 kg/batch with 3 nos 500W electrical heater (Fig. 14.34). It takes about 19 h to dry black pepper from initial moisture content of 70 % to 10 % .

Fig 14.34 Electrical Black Pepper Drier IISR, Calicut

## Black Pepper Grader

It has a capacity of 15 kg/batch and grades in to 4 grades of the size <3mm, 3-3.5 mm, 3.5-4 mm, > 4mm with Garding efficiency more than 98 %.

## 14.10 Arecanut

### Semi-mechanized Arecanut Dehusker

The dehusking assembly (Fig. 14.35) consists of two sharp edged flaps, one being stationary and the other movable, operated by a pedal through linkage mechanism. The unit has a hopper to hold 20 kg nuts. The outer shell of the freshly harvested nuts is pierced on to the upper edge of flaps and is split open by the pedal mechanism. The unit is made of Mild Steel body mounted on angle iron stand and the dehusking mechanism is made of spring steel and is suitable for dehusking freshly harvested mature green areca nuts. The dehusking capacity is 20 kg /hr, when 4 persons work on the machine.

### Mechanized arecanut dehusking

The arecanut dehusker (Fig. 14.36) operated with an electric motor can dehusk dry and wet arecanut and had a capacity of about 300 – 400 kg/hour. The electric motor operated dehusker (Fig. 14.37) is for dry arecunt with a capacity of 300-400 kg/hour. The arecanut dehusker for green arecanut (Fig. 14.38) picks up single nuts and dehusk these by two jaws. The capacity is about 100 – 150 kg/hr.

UAS Bangalore

Fig. 14.35 Semi-mechanized arecanut Dehusker

Dharma Technolgies Tumkur

Fig. 14.36 Mechanized arecanut dehusker for dry and wet arecanut

Fig. 14.37 Arecanut dehusker for dry arecanut

Sri Durga Engineering Works, Koppa

Fig. 14.38 Arecanut dehusker for green arecanut

Agrobot, Bangalore

## 14.11 Cardamom

### Cardamom Garbling Unit

It is a hand operated unit has a capacity of 5 kg/batch with the Garbling efficiency more than 98 %. It reduce drudgery to the labourers percentage broken is less than 5%. Power operated garbling unit is also available in the market

Hand operated rotary type cleaner cum Grader for cardamom

It is manually operated machine suitable for cleaning and grading into two or three grades. It can be used for other crops also by changing the sieves.

## 14.12 Oil Palm

### Fibre Decorticator for Extracting of Oil Palm EFB Fibre

On an average 250 kg of Empty Fruit Bunch (EFB) is produced per MT of FFB processed which is mostly being wasted creating problem in disposal. A fibre extracting decorticator (Fig 14.40) is developed which can extract and grade fibre continuously. Separation of fibre from EFB is achieved through the rotating action of the beaters welded on to the shaft provided inside the decorticator. The fibre is graded into three categories viz. short with less than 50mm length, medium with 50-100mm length and long with greater than 100mm length The fibre thus extracted is suitable for preparing rubberized mattresses, cooling pad etc. The mattresses prepared from oil palm EFB fibre are graded as extra firm as per Indian Standards.

Fig 14.39 Cardamom garbling unit
IISR Calicut

Fig 14.40 Oil palmEFB fibre decorticator
DOPR, Padavegi, West Godavari

### Mobile Oil Palm Waste Shredder

Power tiller mounted mobile waste shredding unit is developed for *in situ* disposal of fronds. About 10.5 MT of fronds (leaves) are produced per ha per year in a plantation. Effective disposal is a must which otherwise leads to hindering the movement of harvesters and collectors in the plantation and acts a breeding ground for insects, rats, reptiles etc. A fond (leaf) shredding unit is developed which consisted of a chaff cutter with specially designed blades. The shredded material is a good mulching material; to smother weeds in addition to improving the nutrients and physical properties like water holding capacity of the soil. The capacity of the shredder is 1 MT per hr, fuel consumption is 3.0 litre per hr, maximum power requirement is 12 hp and operational cost is Rs 160 per hr.

### Shredding:

The coconut / oil palm fronds and other organic waste can be shredded in the high capacity shredder (Fig. 14.41). The capacity is about 1 tonne/hr.

Fig. 14.41 The coconut / oil palm fronds shredder

Kovai Classic IndustrieS

## 14.13 Dryers

### Power Operated Curry Leaf Dryer

CIAE Bhopal has developed a curry leaf stripper in collaboration with a manufacturer M/s Ponmani Industries, Coimbatore. The unit is powered by 1hp electric motor and the capacity is 40-50 kg/h with stripping efficiency of 95%. CIAE Bhopal has developed a forced flow type dryer (Fig. 14.42) for bulk drying of curry leaf. The capacity of the unit is 50 kg/h. It consists of drying chamber, heating chamber and a blower driven by a 2 hp electric motor. Moisture content decreased from 67 to 5% in 6h whereas in conventional system it took 3 days of sun drying and 5 days of shade drying for the same reduction in moisture content.

### Solar Dryer (Bamboo Frame)

The bamboo frame solar dryer (Low cost dryer) has been designed and fabricated (Fig. 14.43). The dryer is 3.0 m long and 1.6 m wide with drying capacity of 110 coconuts per batch. The temperature in solar dryer varied from 36 to 52°C and relative humidity ranged from 50% to 74% inside the dryer. The drying time required was 4.0 days.

### Solar Dryer (Greenhouse Type)

Greenhouse type solar dryer with MS frame having floor area of 2.4 x 1.2 $m^2$ was designed and fabricate for drying of different commodities (Fig. 14.44). It is suitable for drying coconut, black pepper, mushroom, green chillies, jack fruit bulb and fish. The temperature inside the dryer varied from 30°C to 44°C. The drying time varied from 16-32 hours for different crops.

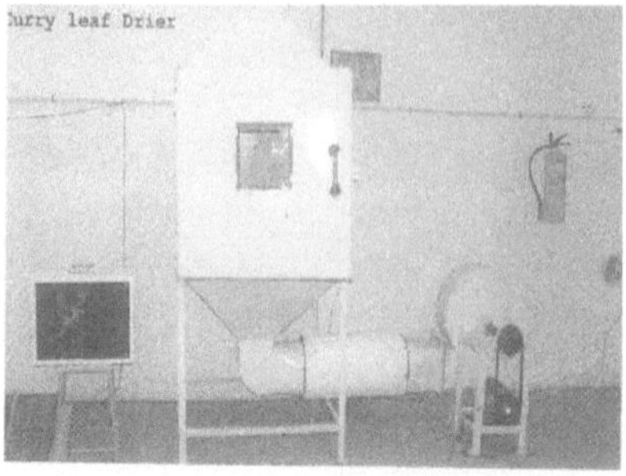


Fig. 14.42Power Operated Curry Leaf Dryer

CIAE Bhopal

Fig. 14.43Low Cost solar dryer(Bamboo Frame)

CARI Portblair

Fig. 14.44 Solar Dryer (Green house type with MS frame)

CARI Portblair

## Shell Fired Copra Dryer

A copra dryer has been developed(Fig. 14.45). This works by indirect heating and natural convection principles using coconut shell as fuel. This dryer requires less amount of fuel, makes copra in short time and less expensive too. The capacity of the dryer is 1000 nuts per batch. The dryer developed has two heating chambers (Fig. 14.46) , which were arranged in parallel. Specially developed rolling in type fuel trays were used for burning the fuel.

## Biomass Fired Dryer

Biomass fired dryer having capacity for 1000 nuts of coconut is developed (Fig 14.47) The size of drier is 2.4m x 1.2m x 1.70 m. Coconut shell is used as fuel .The drying bed consists of a MS square wire mesh, kept at a height of 1.20m above the ground level. 1000 coconuts are dried with 50kg shell. The drying time for separating kernel cups from the shells after 8-10 hours of drying. Coconut shells obtained from one batch of 1000 nuts are sufficient for drying them completely. No extra fuel is required for drying.

Fig. 14.45 Shell fired copra dryer
CARI Portblair

Fig. 14.46 Heating chamber with fuel tray
CARI Portblair

Fig. 14.47 Biomass fired dryer
CARI Portblair

## 14.14 Evaporative Coolers

### Evaporative Cooled Room for Storage of Fruits and Vegetables

This evaporative cooled room (Fig. 14.48) is for on-farm storage of fruits and vegetables after their harvesting. It is a masonry storage room. The walls of this structure are different from that of an ordinary room. Each wall has three layers. The outer and innermost layers are made of bricks and the middle layer is comprised of river sand. The roof is extended on all sides to provide shading from direct sunlight. At the top of the middle sand layer, just below the roof, a drip system is laid all around and connected to an overhead PVC water tank of 300 litres. A drop in temperature of 7 to 10 degrees C and increase in relative humidity to 85-95% has been achieved in EC room. The EC room has a capacity of 2 tons of fruits and vegetable. It increases shelf life of fruits and vegetables like kinnow, cauliflower, spinach, tomato, potato, peas, okra and cabbage.

## Evaporative Cooled Storage Structure

For short-term storage of fruits & vegetables in hot and dry region CIPHET has developed ECS of 5-7 tonne capacity (Fig. 14.49) and it is a bigger unit than EC room. It has special design of roof, orientation and uses wetted pad as cooling medium which can achieve 20°C lower temperature than outside temperature and relative humidity as high as 70 – 99 % depending upon the outside temperature. The perishable harvest /commodities can be stored with significant increase in shelf life.

a

Fig. 14.48 Evaporative cooled room for storage of fruits and vegetables

CIPHET, Ludhiana

Fig. 14.49 Evaporative Cooled Storage Structure

CIPHET, Ludhiana

## Two stage Evaporative Cooler

Two stage Evaporative Cooler is to provide the cooling for storage of fruits and vegetable, animal shed and greenhouse etc. The cooler was portable and the dimensions were 1.5m x 1.0 x 2.0m as length, breadth and height respectively. It can reduce the temperature up to 22 degree C at low relative humidity ambient conditions. The developed two stage evaporative cooler (Fig. 14.50) could drop the temperature up to about the wet bulb depression and to 90 % relative humidity. The effectiveness of the two stage evaporative cooler ranged from 1.1 to 1.2 over the single stage evaporative cooler.

**Hydro Cooler cum Washer for Fruits**

The washer (Fig. 14.51) has been developed by using the principle of water slide. As fruits roll down through the pipes, high pressure jets fitted at strategic places spray water on the sliding fruits. As a result the fruits get thoroughly washed till they reach the outlet at the bottom. The machine has water storage capacity of 200 litres which is circulated by a 0.5 hp pump and needs water change every 90 minutes depending upon the quantum of dust & dirt sticking on the produce. The capacity for the potato was found to be 1.8 tonne/h with two water slides. The machine is being attached with an elevator for automatic loading and grader. Field heat removal and washing of the fruits and vegetables is important to increase their shelf life. Washing with bio-anti fungal solution reduce the microbial load and increase the shelf life of fruits.

| | |
|---|---|
|  | 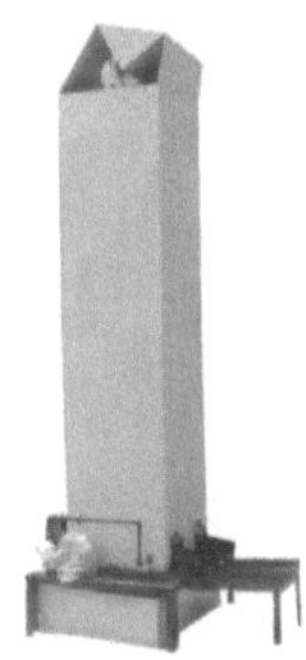 |
| Fig. 14.50 Two stage Evaporative Cooler<br>CIPHET Ludhiana | Fig. 14.51 Hydro cooler cum washer for fruits<br>CIPHET Ludhiana |

## 14.15 Miscellaneous Machinery

## Low cost basket centrifuge for minimal processing of vegetables

A Basket centrifuge (Fig. 14.52) consisting of a detachable perforated cylinder, rotating at 500 RPM is developed to remove surface water from the minimally processed vegetables like spinach and mustard.

### Machinery for mushroom spawn prodution

1. Grain Cleaner – (Manual / Motorized): The manual operated grain cleaner (Fig. 14.53) has a capacity of 12 kg/ hour and motorized grain cleaner has a capacity of 25 kg/hour.

2. Grain Boiler: The capacity of the boiler is 75kg per batch (Fig. 14.54). The boiled grains can be taken out by tilting the boiler. It saves about 60% power and time in comparison to the grains boiled on the open pan type electric heaters.

3. Boiled Grain and Chalk Powder Mixer : Boiled grain is filled to the tilting type mixing chamber (Fig. 14.55). It is allowed to cool for about 5-10 minutes by running the machine. After cooling, the required quantity of chalk powder at rate of 4 percent by weight is added manually. The mixing chamber is tilted to collect the mixed grain in to the collection chamber at the bottom. The capacity is 75 kg per batch with operating time of 5-10 min for cooling and 10-15 min for mixing.

Fig. 14.53 Manually operated grain cleaner

IIHR, Bangalore

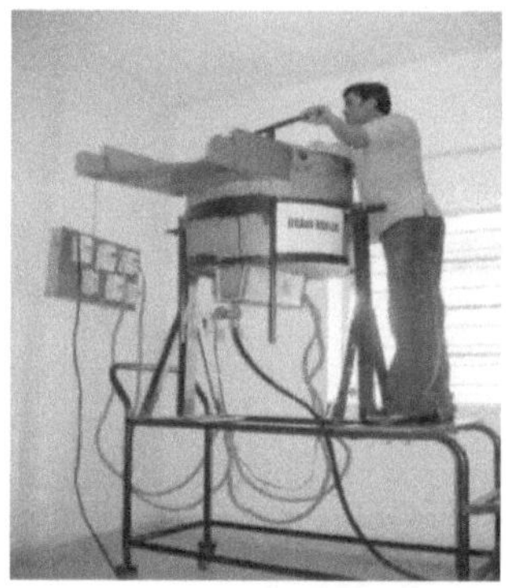

Fig. 14.54 Electrically operated grain boiler

IIHR, Bangalore

4. Grain Filler: The grain filler (Fig. 14.56) has a grain hopper, chain pulley mechanism for lifting and lowering grain hopper, grain filling mechanism with two flow control valves and frame with operator's seat. The operator opens the top valve first and required quantity (250g) of grain is filled in the transparent measuring column fixed between the two valves. Then the top valve is closed immediately. A polypropylene bag is manually held at the outlet of the bottom valve. Then the bottom valve is opened and to allow the grains to fill the polypropylene bags and then the valve is closed. The capacity of bag filler is 400 bags of 250 g each per hour per labour. Only one labour is required to operate the machine and bag filling operation.

| 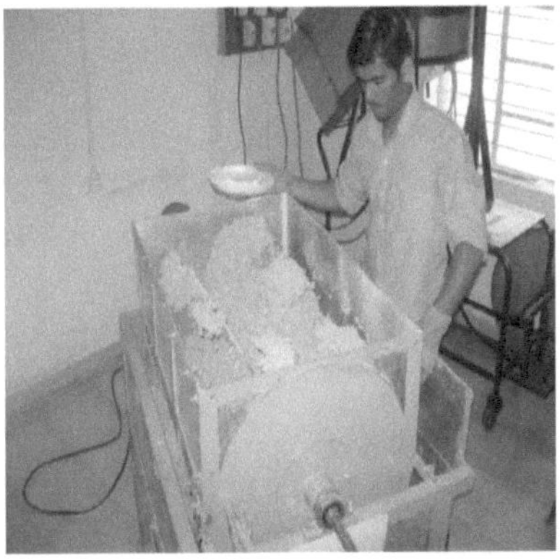 | 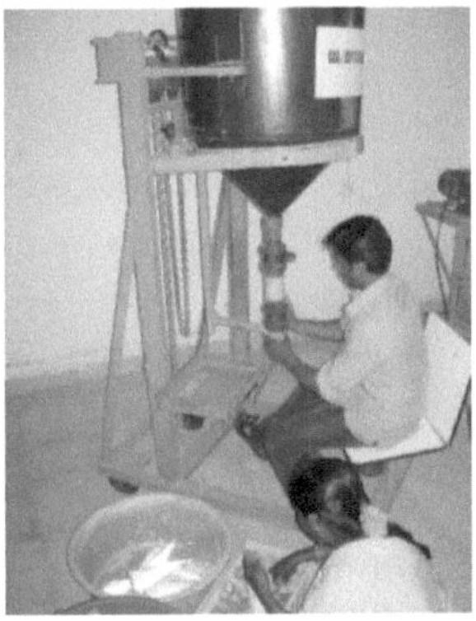 |
| --- | --- |
| Fig. 14.55 Boiled grain and chalk powder mixer<br>IIHR, Bangalore | Fig. 14.56 Grain filler<br>IIHR, Bangalore |

# *Chapter 15*

# Irrigation Systems

The system of irrigation for the crops depends upon the source of irrigation, quantity of water available and crop raised. In North India where water is available in plenty, the irrigation source is generally canal or bore well. The watering system is generally flooding the fields. In South India especially in Karnataka, Maharashtra and Andhra Pradesh where water scarcity is there, the drip irrigation system is becoming popular for horticultural crops.

The centrifugal pump sets are generally used for lifting water from lakes & open wells. The submersible pump sets are used for lifting water from bore wells. However, the water drawn from the borewell /aquifer is generally much higher than the recharge of the aquifer. The water table in South Karnataka has gone down from 50 mtr to 400 mtr during last 30 years. Almost all the open wells have dried. The existing bore wells are failing due to lowering down of water table and drilling of new bore well may not yield water at most of the places. The lakes are also drying due to irregular and reduced rain fall. The water table in North India has also gone down due to excessive exploitation using flooding system of irrigation. In view of this, the main emphasis for developing irrigation source at the farm shall be :

- Rain water harvesting and
- Bore well recharging.
- Rain Water Harvesting:

The rain water harvesting is collection of all the water falling as rain in the field or the catchment area. The field is surveyed and contour map is prepared. The lowest point in the field is selected for making the pond. The drains are made in the field to bring all the runoff water to the field pond. The water entering into the field from other areas is also collected in the pond through drains. The size of pond depends upon the catchment area of the pond, rain distribution in the area, and the water use pattern for irrigation. The pond can be used as a percolation pond for raising the water table in the bore well or open well in the field. The pond can also be lined with plastic film to store the water and the same water can be used directly for irrigation. The top of the pond can be covered by the plastic film (by providing proper structure) to reduce the evaporation and growth of the algae. The water from the green houses and buildings can also be diverted to the pond. The gutters of the greenhouses and the drainage pipes of the buildings can be connected to a pipe and water can be stored in a separate storage tank as this contains less silt and can be used directly for drip irrigation after filtering through sand/gravel filter and mesh filter.

## Recharging of bore wells

Recharging of bore wells should be done to prevent them from drying up, improve the output / discharge) and improve their water table. It is usually done by the following method. First, a pit should be dug in the region surrounding the casing pipe and cement rings should be installed in it. Cement rings matching the pit diameter should be installed till the base of the pit. The size of the pit should be one meter in diameter and 3 meter in depth. At the bottom of the pit, filter holes are drilled in the casing pipe and steel mesh of 4 mm opening is fixed tightly to the bore well casing pipe. This casing pipe will function as a filter. The pit should be filled with boulders from the base up to a height of 0.3 meter. A second layer of 0.3 meter height filled with jelly stones of 40 mm should follow this. The third layer of 0.3 meter height should be filled with jelly stones of 20 mm size. The fourth layer should be filled with charcoal of 0.6meter height. The fifth and sixth layers should be of big jelly (40mm) and small jelly (20mm) each of 0.3 meter height. A nylon curtain of 40 mesh should be spread on the layers and the remaining space of the pit should be filled with sand until 0.3meter from the ground level. A pipe should be fixed to collect the rainwater from the rain water harvesting tank / pond/ field drain and this pipe should be connected to the pit around the borewell.

By following the above method, we can recharge our borewells and keep them alive for longer time. Those having open wells too can adopt this system. By directing the filtered rainwater to the open wells, their water tables could be improved. The dried borewells (yielding water previously) can also be recharged and used for increasing the water table in the area.

## Irrigation Systems

### 1. Open channel and flood irrigation

This system is generally adopted where irrigation water is available in plenty. It is mostly followed in North India. Water from borewell or canal is taken to the field through open channels. The fields are divided into small plots for irrigation. For fruit crops the basins are formed around the trunk of the tree and the individual basin is irrigated. The channels can be lined with bricks and cement mortar and plastered to reduce the seepage. The underground pipes are also used to reduce the conveyance losses through seepage.

**2. Underground pipes**

The bore well or water from any source is put into the underground pipes through pump stand. Air vents are provided at regular intervals for escaping the entrapped air. The pipes are installed at least 0.6 meter deep, to avoid the damage from the soil working implements. The water hydrants are provided at the required locations. The irrigation of the fields is done by further conveying the water through open channels or portable aluminium or HDPE pipes.

## Sprinkler system

The sprinkler system is saving water over flood irrigation and is most suitable for undulated fields. This system is very popular in Haryana, where soil is sandy and there is water shortage. It is also very common in hilly tracks of Coorg in coffee plantations. Water from water source is filtered through mesh filter, pumped in portable aluminium or high density polyethylene (HDPE) pipes at a pressure of 3-5 kg/cm$^2$. The sprinklers are installed on pipelines at regular interval with help of risers (Fig. 15.1) . The HDPE pipes suitable for drip irrigation have 63 to 110 millimetre diameter and can withstand a pressure to 4 to 5.9 kg/cm2. The sprinkler for field crops operate at 2 to 4 kg/cm$^2$ pressure, spray diameter is 10 to 20 meter and discharge is 10 to 80 litre per minute per head. The sprayers for coffee plantations work at 4 to 7 kg/cm$^2$ pressure, spray diameter of 50 to 75 meter and discharge of 16 to 30 lpm. The sprinkler systems are easy to install and have a working life of up to 40 years.

The spray guns with spray diameter of 24 to 63 metre, working pressure of 2 to 7 kg/cm$^2$ and discharge of 130-1750 lpm are available for irrigation of vegetables and tea and coffee estates.

| | | |
|---|---|---|
|  Premier Irrigation |  http://agriorissa.blogspot.in |  Premier Irrigation |
| Sprinkler head | Sprinkler on HDPE pipe | Sprinkler on aluminium pipe |

Fig. 15.1. Sprinkler irrigation system.

**1. Drip Irrigation**

Drip irrigation delivers water directly to the root zone of the plant. This is done by using a network of mainlines, sub-mains and lateral pipe lines with emission points spaced along their lengths (Fig. 15.2). Each dripper/emitter, orifice supplies a measured, precisely controlled uniform application of water, nutrients, and other required growth substances directly into the root zone of the plant.

It increases yield up to 150%, saves water up to 70% compare to flood irrigation and more land can be irrigated with the water thus saved. Crop grows consistently, healthier and matures fast and early maturity results in higher and faster returns on investment. Fertilizer use efficiency increases up to 30%. Cost of fertilizers, inter-culturing and labour use gets reduced. Fertilizer and chemical treatment can be given through Micro Irrigation System itself. Undulating terrains, saline, water logged, sandy & hilly lands can also be brought under productive cultivation. Drip irrigation system is most suitable for fruit, vegetable, ornamental and plantation crops. It can also be combined with plastic mulch and weeds will be reduced and fertilizer efficiency will increase. The quantity of water given is generally about 0.75 times of pan evaporation reading (ET).

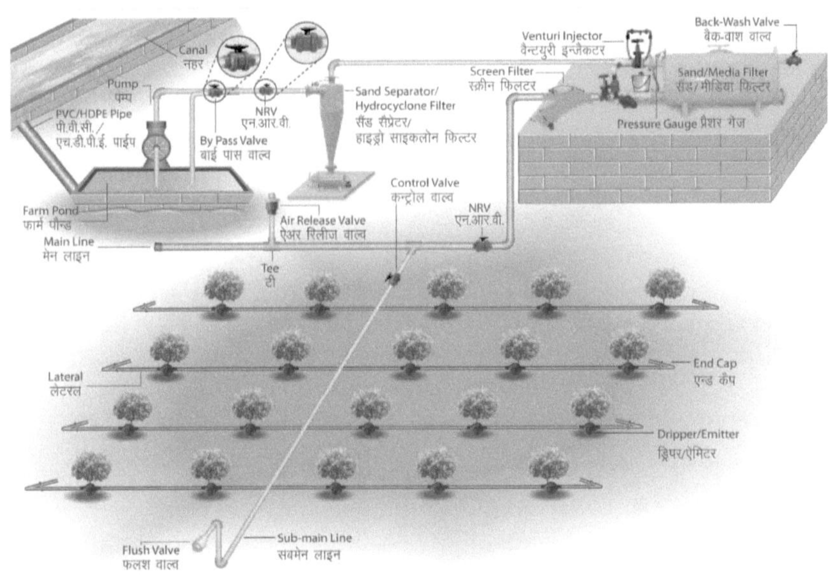


**Layout of Drip Irrigation System (ड्रिप सिंचाई पद्धति का रेखाचित्र)**

http://www.ncpahindia.com/

Fig. 15.2 Layout and components of drip irrigation system along with venturi fertigation

The main components of drip irrigation systems are (a) Water source and water pump., (b) bypass or pressure regulating valve, (c) hydro cyclone filter, (d) screen filter, (e) sand filter, (f) fertilizer tank, (g) air vent valves, (h) main pipe line, (i) sub-main pipe line, (j) lateral pipes and end plug and (k) drippers/ emitters.

**Water pump**

The working pressure of drip irrigation system is 1 kg/cm$^2$ and centrifugal pump set is generally used for pumping the water into the system. The submersible pumpset of the borewell can also be used directly for drip irrigation system.

**Pressure Regulating Valve / Bypass Valve**

This valve line controls the pressure in the irrigation pipeline. This has have spring loaded system to release the water from the irrigation system to control the pressure in the irrigation system. These are provided near the irrigation main pump and protect the irrigation pipe line from bursting, in case all the valves are closed by mistake.

**Hydro cyclone filter**

The rotation of water creates centrifugal action in hydro cyclone filter to separate the particles heavier than water (Fig. 15.3). It generally removes sand and silt particles bigger than 75 micron and heavier than 2.65 specific gravity. The height of cyclone filter is 1.0 to 2.8 meter, diameter 20 to 60 cm and flow rate is 40 to 200 lpm. The volume of sand collection chamber is 5 to 80 litre.

**Screen filter**

These screen the sand and silt particles generally bigger than 100 micron. These are available with 25 to 100 lpm for irrigation pipes of 1.5 to 4 inch diameter(Fig. 16.3).

**Sand / Gravel filter**

These filters can filter sand and silt particles of size bigger than 75 micron. The filtration media is crushed silica sand or quartz gravel of particle size 1 to 2 mm. These are available in vertical or horizontal type (Fig. 15.3). The filtration capacity is 40 to 60 lpm. These are to be flushed regularly by reversing the flow of water. The necessary pipes and valves are provided for flushing these. In automated systems the solenoid valves are operated by controllers for flushing these.

**Emitting line /Inline dripper**

These seamless extruded LLDPE pipelines have emitters fixed permanently inside. These are available generally in 16, 18 & 20mm diameter and fitted with inline drippers of capacity 0.9 to 3.8 litre per hour with spacing of 30, 40, 50, 60, 75 and 90 cm. These are suitable for banana, floriculture, vegetables and spices. The drippers have pressure compensation range of 0.5 to 4.0 kg/cm$^2$. The emitting lines with other specifications are also available.

**Later pipe**

These are manufactured in linear low density polyethylene (LLDPE). These are UV (Ultraviolet rays) stabilized and are available in outer diameter of 12, 16, 20, 25 & 32 mm however 12 and 16 mm diameter are more common in horticultural crops. These are used for fitting of drippers/emitter, foggers, spray heads and jets. For field cultivation of fruit and vegetables crops on line emitters / drippers are generally fitted. The drippers are fitted by punching holes of 2.9 mm diameter.

**Drip Tape**

The drip pipe has a wall thickness of 0.15mm to 0.63mm and inflates only when water is let inside. It is available in 12, 16, 17, 19 and 25mm diameter. The dripper/ drip hole spacing is 15, 20, 30, 40, 50, 60, 75, 90 or 100mm and the discharge rate is 0.6 to 4.0 litre per hour at pressure of 0.7 to 1.0 kg/cm$^2$.

**Emitter / Dripper**

These are installed on lateral pipelines near the root zone of the plant. The discharge of drippers is 2, 4, 8 and 12 lph. However, 2, 4, 8 lph are most popular in horticultural crops. The openable drippers (Fig. 15.5) can be cleaned manually. In pressure compensated drippers discharge does not vary much with change of pressure. The discharge rating of the emitters / drippers is specified at 1 kg/ cm$^2$. The anti bug cap can be provided in some drippers, this does not allow entry of insects inside the dripper. The sealed drippers need replacement in case of clogging. Most of the drippers have turbulent flow inside for self cleaning.

**PVC pipe lines**

Poly vinyl chloride (PVC) pipe lines are mostly used for main pipes and sub main pipes in drip irrigation. The commercially available sizes (nominal outside diameter) are 20, 25, 32, 40, 50, 63, 75, 90, 110, 125, 140, 160, 180, 200, 225, 250, 280, 315, 355, 400, 450, 500 and 560 mm. The pressure rating is 2.5, 4.0, 6.0, 8.0, 10.0 and 12.5 kg/cm$^2$. These should be as per IS-4985 standards of BIS. These are commonly available in 6 meter lengths with plain ends or socketed at one end for solvent cement weld. The mostly used sizes in drip irrigation in horticultural crops are 40 to 110 mm with pressure rating of 4 or 6 kg/cm$^2$.

**Ball Valve**

The ball valves are used to control the water flow in different main and sub-main or pipelines. There size is as per the PVC pipe line used for the irrigation (Fig. 15.4).

**Flush Valve**

These are provided at the end of sub-main pipelines area used to flush out the dirt and waste material collected at the end of the sub-main pipeline (Fig. 15.4).

## PVC Fittings / Accessories

Bends, elbow, reducer and cap etc. are the accessories required to complete the pipe layout.

## Accessories of lateral pipelines

Barbed connector, grommet and take off, plug etc. are the accessories for completing the installation of drip system.

## Sprinkler systems for land scape

For landscape the pop up sprinklers are used for irrigation of lawn (Fig. 15.6). The micro sprinklers are used for irrigating the avenue plants. The micro sprinklers (Fig. 15.7). are provided on the stakes to water the nursery grafts / plants etc.

| | | |
|---|---|---|
|  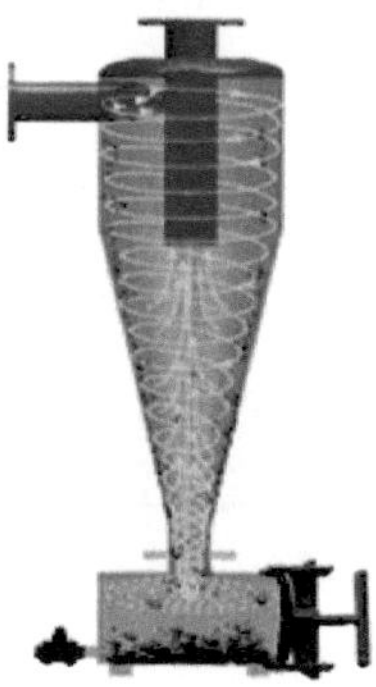 Captain Polyplast Jain Irrigation |  Captain Polyplast 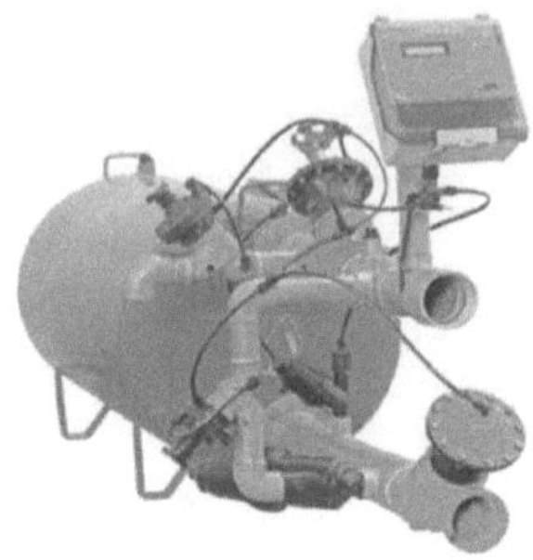 Jain Irrigation |  Datta Irrigation 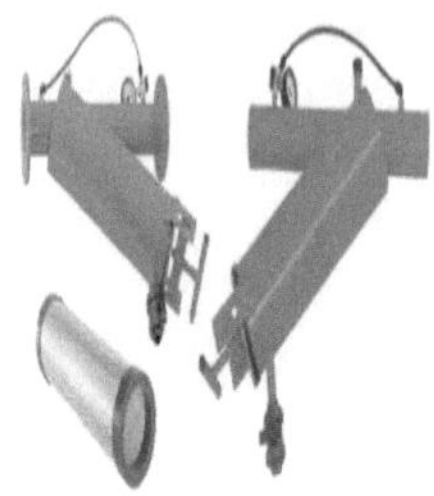 Jain Irrigation |
| Hydro cyclone filter | Sand filter & Sand filter with auto flush | Screen filter |

Fig 15.3 Filters of different types used in drip irrigation

| | | | | |
|---|---|---|---|---|
|  Sitaram Industries |  M&M Control |  Sitaram Industries |  Sitaram Industries |  Sitaram Industries |
| Ball valve | Pressure control /Bypass valve | Flush Valve | Air release valve | Foot valve |

Fig. 15.4 Valves of different types used in drip irrigation

| | | |
|---|---|---|
| 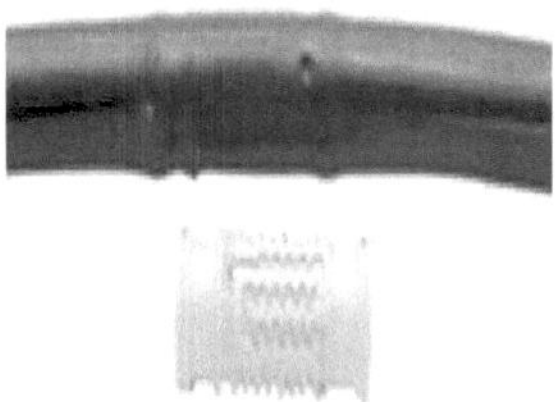 |  |  |
| Datta Irrigation | http://www.irrigationdirect.com | Premier Irrigation |
| Emitting pipe/ In line emitter | On line emitter/ dripper | |

Fig. 15.5 Emitters /drippers of different types used in drip irrigation

| | |
|---|---|
|  |  |
| Toro.com | Toro.com |

Fig. 15.6 Pop up sprinklers for lawn irrigation

| | | |
|---|---|---|
|  |  |  |
| toro.com | growell.co.uk | qingchen2010.en .made-in-china.com |

Fig. 15.7 Micro sprinlers of different types used in drip irrigation / micro irrigation systems

## Fertigation

Fertigation allows to apply the nutrients exactly and uniformly only to the wetted root volume, where the active roots are concentrated. This remarkably increases the efficiency in the application of the fertilizer, which allows reducing the amount of applied fertilizer. This not only reduces the production costs but also lessens the potential of groundwater pollution caused by the fertilizer leaching. Fertigation allows to adapt the amount and concentration of the applied nutrients in order to meet the actual nutritional requirement of the crop throughout the growing season. Other advantages of the fertigation are: (1) the saving of energy and labor, (2) the flexibility of the moment of the application (nutrients can be applied to the soil when crop or soil conditions would otherwise prohibit entry into the field with conventional equipment), (3) convenient use of compound and ready-mix nutrient solutions containing also small concentrations of micronutrients which are otherwise very difficult to apply accurately to the soil, and (4) the supply of nutrients can be more carefully regulated and monitored. An essential pre-requisite for the solid fertilizer use in fertigation is its complete dissolution in the irrigation water. To capitalize on fertigation benefits, particular care should be taken in selecting fertilizers and injection equipment as well in the management and maintenance of the system.

## Fertilizers for Fertigation

Fertigation can be achieved by using single or multiple nutrient fertilizers, in their solid or liquid form. An essential prerequisite for the use of solid fertilizer in fertigation is its complete and fast dissolution in the irrigation water. Fertilizer preparation and application of fertilizers is executed by various methods

Stock solution preparation: Solid fertilizers like ammonium sulphate, urea, potassium chloride and nitrate, and liquid phosphoric acid are mixed to prepare a "tailor made" stock solution. The stock solution is then injected into the irrigation system, at rates of 2-10 litre/m$^3$, depending on the desired concentrations of N, P and K. Clear NK, PK and NPK fertilizer solutions with at least 9-10% nutrients (N, $P_2O_5$, $K_2O$) based on cheap solid fertilizers (urea, phosphoric acid and KCl) can be easily prepared on the farm site with limited facilities under field conditions, with minimal mixing.

Compound solid fertilizer mixtures: These are manufactured for use in fertigation, with different ratios between the three major elements. The first mixture used in fertigation was 20-20-20 and was produced in the mid-sixties. The mixture 20-20-20 is generally recommended from planting to flowering stage, 14-7-21 from flowering to fruit set and 14-3-28 from fruit set to ripening and harvest. Some compositions contain microelements in the form of chelates.

Compound liquid fertilizer solutions: due to solubility, the total nutrient concentration is much lower (5-3-8; 6-6-6; 9-2-8, etc.). For specified use in greenhouses, some compositions contain microelements in the form of chelates.

**Fertilizer Compatibility**

When preparing fertilizer solutions for fertigation, some fertilizers must not be mixed together. For example, the mixture of $(NH_4)_2SO_4$ and KCl in the tank considerably reduces the solubility of the mixture due to the $K_2SO_4$ formation. Other forbidden mixtures are:

- Calcium nitrate with any phosphates or sulphates
- Magnesium sulphate with di- or mono- ammonium phosphate
- Phosphoric acid with iron, zinc, copper and manganese sulphates.

Table 15.1. Typical fertilizers used in fertigation[@]

| Nutrients | Fertilizers | Grade | Formula | Other nutrients | Solubility* (g/100g water) | Final concentration** (ppm) |
|---|---|---|---|---|---|---|
| Nitrogen (N) | Urea | 46-0-0 | $CO(NH_2)_2$ | - | 84 | N=4.6 |
| | Ammonium sulphate | 21-0-0 | $(NH_4)_2SO_4$ | 22% S | 73 | N=2.1<br>S=2.2 |
| | Urea ammonium nitrate | 32-0-0 | $CO(NH_2)_2$ $NH_4NO_3$ | - | Liquid | N=3.2 |
| | Ammonium nitrate | 34-0-0 | $NH_4NO_3$ | - | 151 | N=3.4 |
| Nitrogen (N) and phosphorus (P) | Mono ammonium phosphate | 12-61-0 | $NH_4H_2PO_4$ | - | 28 | N=1.2<br>$P_2O_5$=6.1 |
| Phosphorus (P) | Phosphoric acid | 0-52-0 | $H_3PO_4$ | - | Liquid | $P_2O_5$=5.2 |

*Contd...*

*Table 15.1 contd..*

| | | | | | | |
|---|---|---|---|---|---|---|
| Phosphorus (P) and potassium (K) | Mono potassium Phosphate | 0-52-34 | $KH_2PO4$ | - | 18 | $P_2O_5$=5.2 $K_2O$=3.4 |
| Potassium (K) | Potassium chloride | 0-0-60 | KCl | 46% Cl | 31 | $K_2O$=6 |
| | Potassium sulphate | 0-0-50 | $K_2SO_4$ | 18% S | 9 | $K_2O$=5 |
| | Potassium nitrate | 13-0-46 | $KNO_3$ | - | 21 | N=1.3 $K_2O$=4.6 |
| | Potassium thiosulphate | 0-0-25 | $K_2S_2O_3$ | 17% S | Liquid | $K_2O$=2.5 S=1.7 |
| Magnesium (Mg) and calcium (Ca) | Magnesium sulphate | 0-0-0-17 (MgO) | $MgSO_4$ | 13% S | 28 | S=1.3 MgO=1.7 |
| | Magnesium nitrate | 11-0-0-16 (MgO) | $Mg(N0_3)_2$ | - | 65 | N=1.1 MgO=1.6 |
| | Calcium nitrate | 0-0-0-27 (CaO) | $Ca(NO_3)_2$ | - | 113 | N=1.5 CaO=2.7 |

@ Fertigation Practical Aspects, International Potash Institute, Horgen, Switzerland.

* At saturation, 10°C.

** At the dripper, when dissolving 1 kg fertilizer in 100 L stock solution, injected at a dilution ratio 1:1000. For increased concentration check solubility limit of each fertilizer.

Generally two fertilizer tanks that contain the concentrated fertilizer solutions are used to separate those fertilizers that can interact. A possible combination is: a tank "A" containing calcium nitrate, potassium nitrate magnesium nitrate and microelements, whereas tank "B" contains ammonium sulphate, phosphoric acid and nitric acid; in this way P and Ca/Mg are in different tanks to avoid their precipitation. A third tank "C" contains an acid solution to control the pH of the fertilizer solution and to wash the irrigation system to avoid drippers clogging.

## Fertigation Equipment

Fertigation equipment should be able to regulate quantity applied, duration of applications, proportion of fertilizers and starting and finishing time.

## Fertilizer Injectors

Several techniques have been developed for applying fertilizers through the irrigation systems and many types of injectors are available on the market. There are tow main techniques: the ordinary closed tank; and the injector pump. Both systems are operated by the system's water pressure. The injector pumps are mainly either Venturi type or piston pumps. The closed tanks are always installed on a bypass line, while the piston pumps can be installed either in-line or on a bypass line.

**Fertilizer (closed) tank**

This is a cylindrical, epoxy coated, pressurized metal tank, resistant to the system's pressure, and connected as a bypass to the supply pipe of the head control (Fig 15.8). It operates by differential pressure created by a partially closed valve (choke valve Fig. 15.10), placed on the pipeline between the inlet and the outlet of the tank. Part of the flow is diverted to the tank entering at the bottom. It mixes with the fertilizer solution and the dilution is ejected into the system. The dilution ratio and the rate of injection are not constant. The concentration of fertilizer is high at the beginning and very low at the end of the operation. However, this apparatus is still in service on a very small scale in some countries because of its low cost and easy manufacture. The application of nutrients is quantitative and inaccurate and is adapted for perennial crops like citrus, fruit trees and/or crops grown on heavy soil. It is very simple to operate, Ideal for dry formulations, easy to change fertilizers and the stock solution need not to be pre-mixed. It is easy to install, requires very little maintenance and needs no electricity or fuel . Proportional fertigation is not possible and accuracy of application is limited are the major disadvantages. It creates pressure loss in main irrigation line, has limited capacity and not adaptable for automation.

**Venturi type**

This is based on the principle of the Venturi tube. A pressure difference is needed between the inlet and the outlet of the injector. Therefore, it is installed on a bypass arrangement placed on an open container with the fertilizer solution (Fig 15.11). The rate of injection is very sensitive to pressure variations, and small pressure regulators are sometimes needed for a constant ejection. Friction losses are approximately 1.0 bar. The injectors are made of plastic in sizes from ¾ to 2 inches and with injection rates of 40-2000 litres/h. They are relatively cheap compared to other injectors. In some cases the venturi is provided in line, which results in pressure loss in the system. A number of venturies can be used to inject a number of fertilizers.

It is very simple to operate with no moving parts, easy to install and to maintain and suitable for very low injection rates. Injection can be controlled with a metering valve and is suitable for both proportional and quantitative fertilization. Quantitative fertigation and automation is possible.

**Hydraulic / Piston pump**

This type of injector is powered by the water pressure of the system and can be installed directly on the supply line and not on a bypass line. The system's flow activates the pistons and the injector is operated, ejecting the fertilizer solution from a container, while maintaining a constant rate of injection. The rate varies from 9 to 2500 litres/h depending on the pressure of the system and it can be adjusted by small regulators. Made of durable plastic material, these injectors are available in various models and sizes. They are more expensive than the Venturi-type injectors. Electrical pumps are also in use. This system is very accurate, for proportional fertigation, with no pressure loss in the line and easily adapted for automation. It is expensive and has complicated design, including a number of moving parts, so wear and breakdown are more likely.

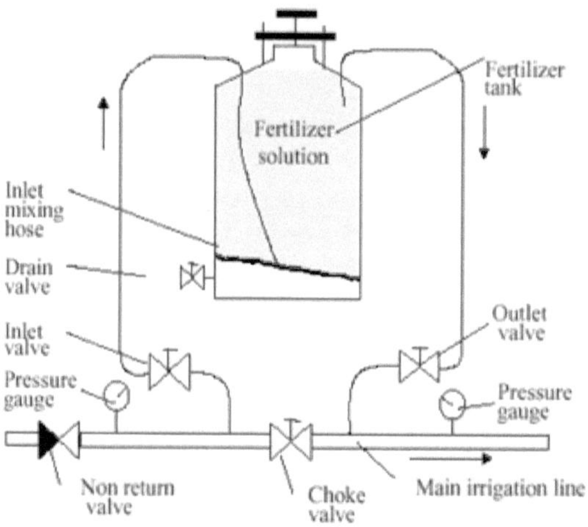

**Smart! Fertilizer**

Patricia Imas, International Potash Institute

Fig. 15.8 Pressure differential (by-pass tank) system of fertigation

Fig. 15.09 Choke valve
Jain Irrigation

http://nematology.ucdavis.edu
By-pass venturi

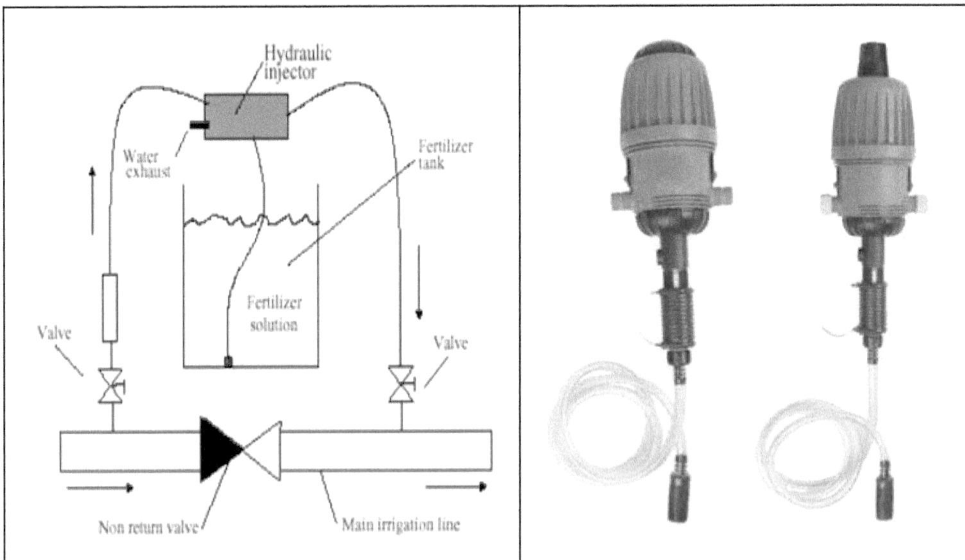


Patricia Imas, International Potash Institute | Jain Irrigation

Fig. 15.10 Hydraulic pump system of fertigation

## Fertilizer Application

The fertilizer solution in liquid form is fed into the system at low rates repeatedly, on a continuous basis, during irrigation. The flow rate of the injector should be such that the calculated amount of solution is supplied at a constant rate during the irrigation cycle, i.e., starting fertigation right after the system starts operation and finishing a few minutes before the operation ends. Regarding the choice of the fertilizers, apart from the amount and the kind, other parameters need to be considered, such as solubility, acidity, compatibility and cost.

## Solubility

The fertilizer stock solution should always be dissolved in a separate container and then poured into the suction tank . The types of fertilizer should be highly soluble and when dissolved in water must not form scums or sediments which might cause emitter clogging problems. The solution should always be agitated, well stirred and any sludge deposited in the bottom of the tank should be periodically removed. The injector suction pipe should not rest on the bottom of the tank. Hot water helps dissolve dry fertilizers. Their degree of solubility varies according to the type and the country of origin. Potassium nitrate (13-0-46) seems to have a low solubility of approximately 1:8, *i.e.*, 1 kg of dry fertilizer in 8 litres of water. The solubility of potassium chloride (0-0-62) is 1:3, while ammonium nitrate (34-0-0) and calcium nitrate (15.5-0-0) have a high solubility of approximately 1:1. Dry phosphorous fertilizers have a lower solubility than nitrates at about 1:2.5.The solubility of different fertilizers is given in Table 15.1.

## Acidity

The acidity produced by the several forms of nitrogen varies from type to type and is greatly affected by the kind of irrigation water and the type of soil. At least one check on the soil pH should be carried out at the beginning of the season and one at the end. Furthermore, a complete ionic analysis of the water is necessary.

## Quantity

A simple method for calculating the amount of fertilizer required for fertigation is to divide the annual application by the number of irrigations. Various recipes have been developed in different countries based on the conventional nutrition dosages. The total quantity of fertilizers applied is also related to the length of the growing season and the irrigation requirements.

Table 15.2 presents some of the recipes applied in for fertigation on a continuous basis, at a constant rate and feeding, during irrigation

**Table 15.2** – Net concentration of fertilizers in ppm in irrigation water recipes

| Crop | N | P | K |
|---|---|---|---|
| Citrus<br>50 | | 12 | 15 |
| Bananas | 50 | 15 | 40 |
| Tomatoes | 180 | 50 | 250 |
| Cucumbers | 200 | 50 | 200 |
| Bell peppers | 170 | 60 | 200 |
| Cabbage | 100 | 60 | 200 |
| Onion | 100 | 50 | 150 |
| Squashes | 200 | 50 | 200 |
| Potatoes | 150 | 50 | 180 |
| Groundnuts | 120 | 50 | 200 |
| Watermelons | 150 | 50 | 150 |

(mg/litre, or net fertilizer g/$m^3$ irrigation water)

The above recipes vary in accordance with the fertilizer reserves in the soil and crop stage.The above recipes are recommended for irrigation water with very low salinity. As a rule of thumb for average quality water, the maximum fertilizer concentration, which is added to the irrigation total salinity, should have an EC of about 0.5 dS/m. For higher concentrations, the salinity level in the soil root zone must be checked frequently and the application adjusted according to the soil test results.

## Automation in Irrigation and Fertigation Systems

In automation the controller and solenoid valves are the main components. Irrigation controller is an electronic device that activates the valves according to a user defined schedule. This is also called as timer or clock. The controller turns on and off the valves as programmed by the user. In other words, the controller controls the irrigation system and the user controls the controller.

The irrigation system may have to meet the requirements of different components of land scape e.g., lawns, shrubs, flowering beds and avenue trees.

The controller is able to control different solenoid valves (supplying water to different irrigation zones) for different schedules as programmed by the user. Once programmed, the controller determines when, how often and how long each valve opens.

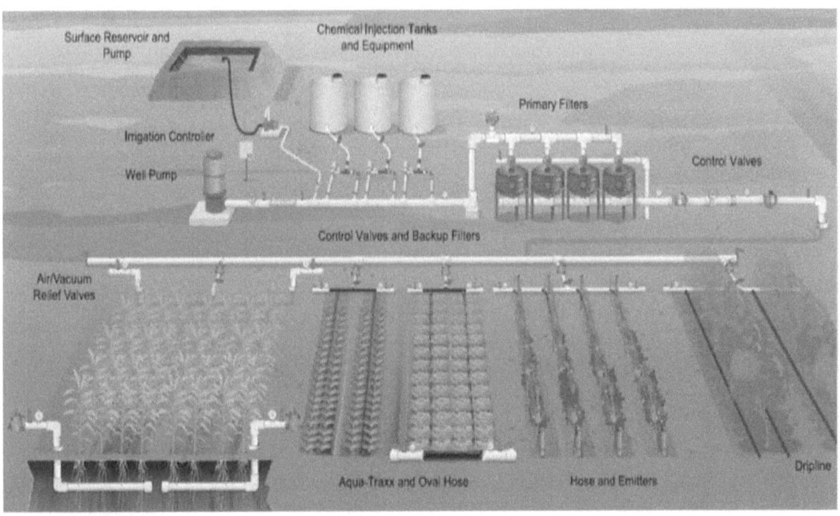


toro.com

Fig. 15.11 Fertgation system with 3 fertilizer tanks and controller for 5 zones of irrigation

The controllers get the power supply from 110/220 main or AC power supply. These are also available in battery operated models or through adaptor to the 110/220 main AC power supply.

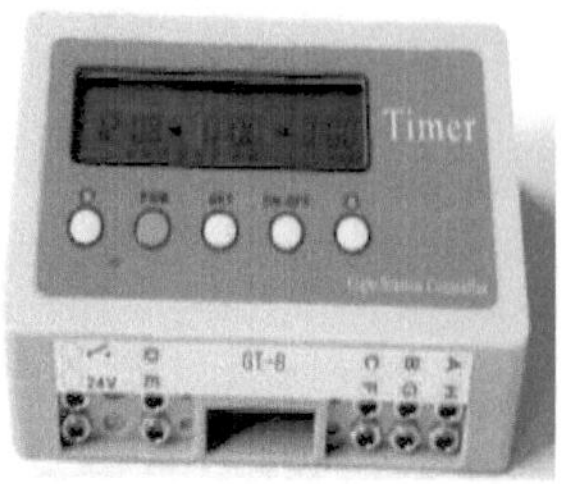


METechs

Rain Bird

Toro

Fig. 15.12 Irrigation controllers indoor and outdoor of different makes

Hunter

Agriplas

Fig. 15.13 Remote control irrigation controllers

METechs

NAANDANJAIN

toro.com

Fig. 15.14 **Solenoid valves of different sizes and makes**

Fig. 15.15 Flow sensor

The controller can also get the signal from rain sensors, weather station, soil sensors and can control the irrigation valves accordingly. The main type of sensors and their operation and utility are as follows :

1. **Rain Sensor /freeze sensor:** These sensors (Fig. 15.16) are installed on the building and sense the rain and are connected to the controller. The controller rides over the program manually fed to it and stops the irrigation system or water supply to the outdoor plants and lawn etc. These can be wired or may be transmitter type, where the rain signal is transmitted to a receiver, which inturn is connected to controller

Wired rain sensor

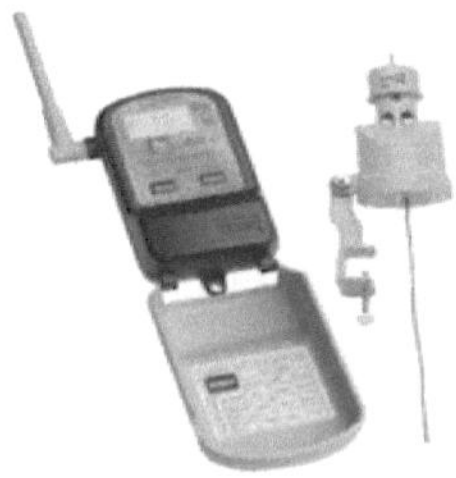

Transmitter type rain sensor with receiver

Rain sensor installation (outdoor)

Fig. 15.16 Rain sensors of different types

2. **The ET (Evapo transpiration) sensor** based controllers are connected to the weather station(Fig. 15.17). The evapo-transpiration is based on solar radiations, wind speed, rain fall and relative humidity etc. These controllers are connected to the weather station to get the requisite data and control automatically the irrigation schedule accordingly to save water.

Evepo- transpiration (ET)sensor ( transmitter type)

toro.com

ET sensor with receiver and timer/ controller

toro.com

Fig. 15.17 Evepo- transpiration (ET)sensor

3. **The soil sensors sense** (Fig. 15.18) The soil moisture based upon its electrical conductivity / resistance and send the signal to the receiver.The soil sensors are installed under the soil, near the root zone. The receiver is connected to the controller. The controller operate the irrigation valve when the electrical resistance of soil increases from the particular level or the soil moisture reduces from a particular level (usually 50% of the field capacity. Tensiometer with transducer (Fig. 15.19) are also available to transmit the data to controller. Soil sensors are also available to track the soil moisture, salinity and temperature (Fig. 15.20). The data is analysed by the controller and the irrigation is given as programmed by the user.

Flow sensors: The flow sensors (Fig. 15.15) are fitted in water supply line for the different zones. The water flow will increase in case of pipe burst or breakage. The controller will stop the water supply to the particular line and gives the alarm signal accordingly. Similar action may take place when the pipe line is blocked and pressure increases in the pipe line or water flow reduces.

Controllers are also programmed based upon the weather conditions during different weeks or months and control the irrigation water supply to the plants/ crops/lawns as per requirement during that week/month or period. These controllers automatically change water schedule based upon past or historic climate conditions for a given region.

For automation of fertigation atleast two or three stock solution tanks are used for fertilizer application. Tank 'A' may have Calcium nitrate, Tank 'B' may contain remaining fertilizers and Tank 'C' may contain acid for regulating the pH. However, the compatibility of the fertilizers have to be taken care for mixing the fertilizers in the tanks. The concentration of the solution in stock tanks is further diluted while injecting into the irrigation water by injectors. The proportion of dilution is determined by concentration in stock tank and dilution by the injector. The EC and pH are monitored in the system. The pH adjustment can be done by increasing the concentration of acid into the stock solution tank in 'C'. The pH can also be increased or decreased by varying the dilution ratio by adjusting the injector pump or venturi if possible in the system. The EC can be adjusted by varying the dilution ratio from tank 'A' and 'B' by making the necessary adjustment in the injectors. However, it is difficult to vary this in three tank system. The irrigation recipe needs change for various stages of plant like establishment of the plants to growth, flowering and fruiting etc. For taking care of this, the number of tanks has been increased upto 10 for making stock solution of each fertilizer. Eight tanks are generally used for individual fertilizer and micro-nutrients and two tanks are used for acid and base for adjustment of pH.

The schedule of irrigation and fertigation can be controlled by the controller which has additional features for monitoring the EC, pH and controlling the injectors for fertilizers. The schedule of irrigation and fertigation can be controlled as in case of irrigation controllers as already discussed in irrigation portion.

toro.com

Soil moisture sensor (transmitter type)

toro.com

Receiver for sensor

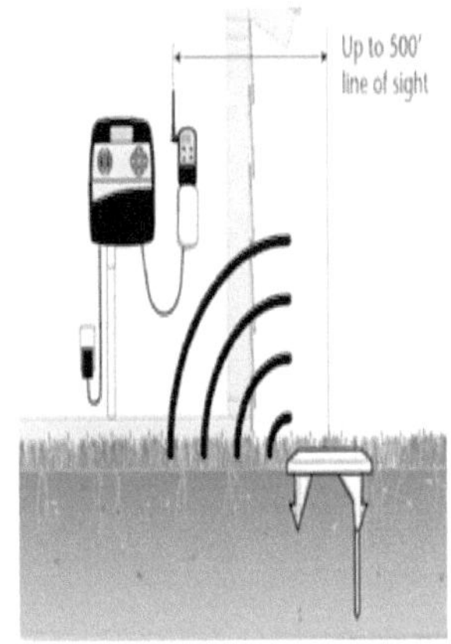


toro.com

Installation of soil moisture sensor, receiver & controller

Fig. 15.18 Soil moisture sensor for turf and its installation

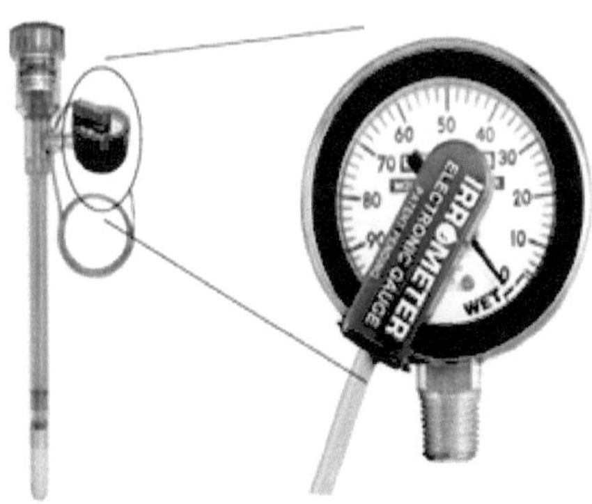


Soil moisture sensor for field tensiometer with Transducer for direct switching capabilities for Automation and Output Options

http://www.irrometer.com (WaterMark)

Soil moisture sensor for field (Electronic module)

http://www.irrometer.com (WaterMark)

Fig. 15.19 Soil moisture sensor for field

Soil moisture, salinity & temperature sensor(transmitter type) adaptable to web
toro.com

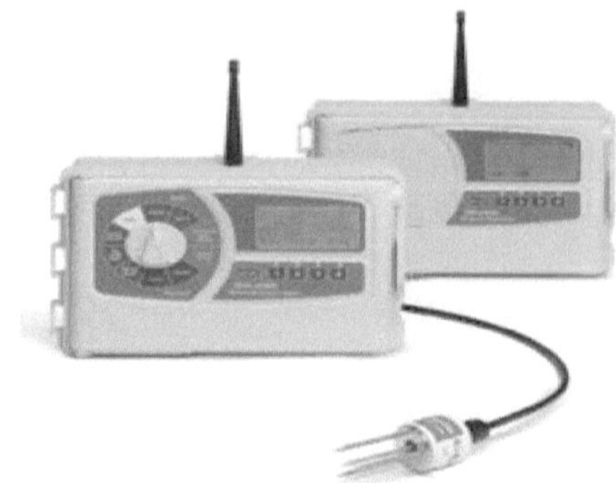

Soil moisture, salinity & temperature sensorwith transmitter
toro.com

Fig. 15.20 Soil moisture, salinity and temperature sensor

## Fertigation station

Now a days, fertigation stations are available which take care of automation of fertigation as well as irrigation. These consist of the controller and fertigation system. The irrigation water is taken from bypass for injecting the fertilizers. The injectors may be venturies or injection/hydraulic/electric pumps. Some of the components of the automated fertigation system are shown in the Fig.15.21.

Scova Impianti

a Fertizer tanks

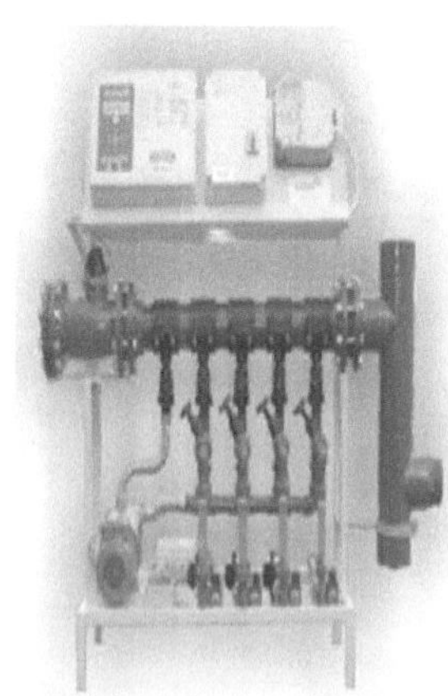

Scova Impianti
b.Dosing pumps and by pass pump

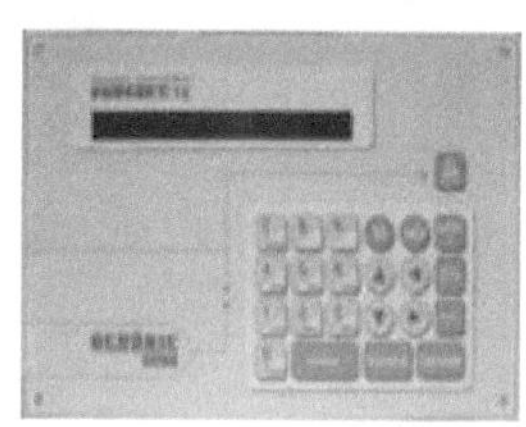

Scova Impianti

HANNA instruments
c.Electonic controller

Fig. 15.21 Components of multi tank fertigation system

The fertigation station for three tanks system is shown in Fig.15.22, it mixes the dosage of fertilizers and acid together with irrigation water. The dosage can be proportional with regulating EC and pH. The watering schedule is by clock mode or timer mode and even summing the light. The alarm is given if there is any problem in the system. This system can be upgraded for upto 27 zone walls. The injectors can be of capacity upto 5 litre/hr per channel.

The five tank system and ten tank system are shown in Fig.15.22. The five tank system consists of four fertilizer tanks/channels and one acid channel. The ten tank system consists of eight fertilizer tanks and two tanks acid/alkali solutions. The maximum capacity of fertilizer dosing channels is upto 20 litre/min. and for acid or basic solution upto 5 litre/min. These control upto 24 irrigation zones. The proportional dosing can be done to regulate the EC. The flow sensor is also provided in the main irrigation line to stop the system in case of any leakage or blockage in the system. The light sensor are also provided to adjust the irrigation based on sunshine hours.

The controllers can also be provided with ET sensors (based on ambient temperature, wind speed, ambient RH, sunshine hours/ solar radiations). The data from satellites (GIS) can also be used for site specific irrigation/fertigation. However, presently the data from GIS is available for a resolution of one square kilometre. The irrigation pumps (Fig 15.23) can be switched on by using the controller, the solar operated auto flushing system for sand filters (Fig 15.27) and screen filters (Fig 15.28) can reduce the pressure losses in the system by regularly flushing/cleaning the filters as per schedule. For disinfection of run off water in hydroponic green houses for recirculation type of irrigation water system, the ozone or UV sterilizers are used (Fig 15.25 and 15.26). This saves irrigation water and fertilizer to a large extent.

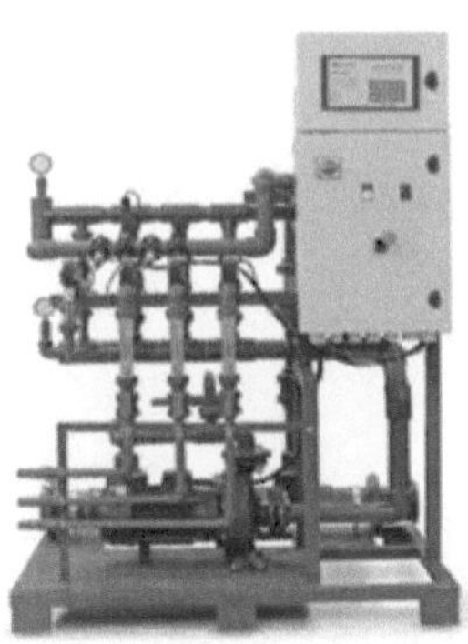

HANNA instruments Fertigation system for 3 tanks

Spagnol srl Fertigation system for 5 tanks

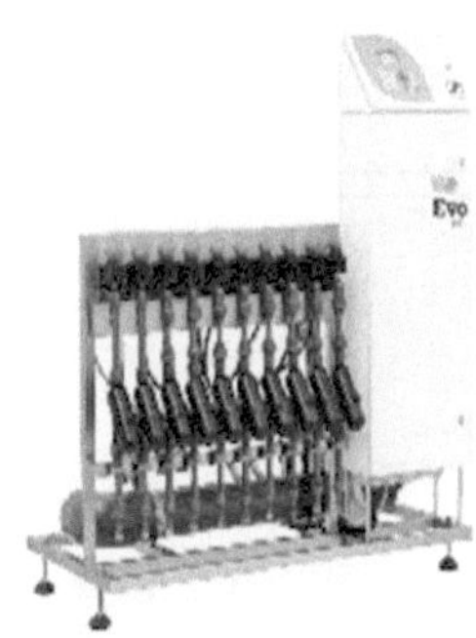

Spagnol srl Fertigation system for 10 tanks

Fig. 15.22 Fertigation stations

Scova Impianti

Fig. 15.23 Automation of irrigation pump or electric motor

HANNA instruments

Fig. 15.24 EC and pH monitor

http://advancedtreatmenttechnologies.com

http://www.purozone.co.za

Fig. 15. 25 Ozone sterilization of overflow irrigation water for hydrophonic drip irrigation system

Scova Impianti

Fig. 15.26 UV Sterilizer of runoff nutrient solution for soilless cultivation in greenhouses

Jain Irrigation

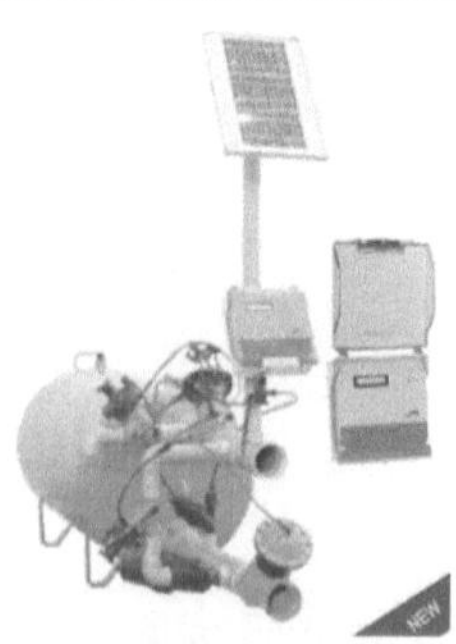

b. Jain Irrigation

Fig. 15.27 Auto flushing system for sand filters With controller & b. Solar powered controller

Scova Impianti

Fig. 15.28 Autoflushing of screen filter with controller

The controllers are also connected to computer through interface. These can be programmed and monitored through computer. These can also be programmed/ monitored through cellular/mobile phone. These can also be connected through internet and data transmission system as shown in Fig.15.29 and 15.30

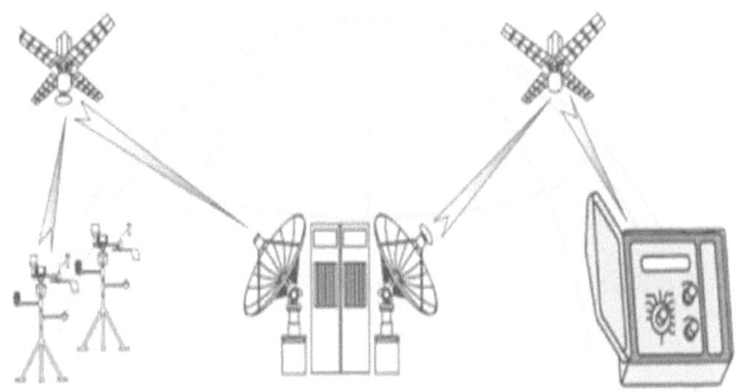

toro.com

Fig. 15.29 Controller for receiving the data from sensors via satellites

The Toro® TriComm™ System

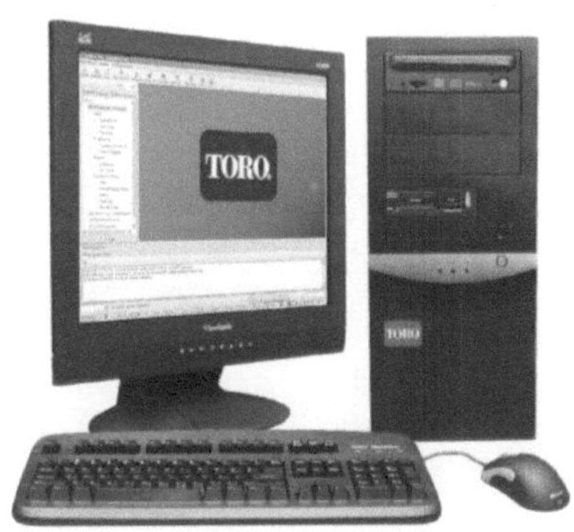


Toro, Sentinel

Fig. 15.30 computer and cellular phone with interface for control of fertigation system

To prolong the availability of natural resources of irrigation water at farm, the rain water harvesting and borewell recharging is must. To save the irrigation water and to increase the efficiency of utilisation of fertilizers, the computer/controller controlled irrigation/fertigation system is must. The cost of the irrigation controller and solenoid valves and wiring is much less in comparison to the other hardware items used in drip irrigation system of the farms or the green houses.

# *Chapter 16*
# Greenhouse Design And Construction

The northern Europe remains under snow during winter months and nothing can be cultivated. The greenhouses were developed to cultivate the vegetables under these extreme conditions, when there were no transport and storage facilities for fruit and vegetables. Initially these green houses were enclosed structures and temperature was raised inside by solar radiations or artificial heating. Initially these had glass glazing and were known as glasshouses and were used to cultivate vegetables, flowers and potted plants. The green houses can also be used to cultivate the crops in Arab countries where the ambient temperature reaches upto 52°C. These green houses are provided with evaporative cooling system or fan and pad to provide the favourable temperature and relative humidity for cultivation of vegetables and flowers. The greenhouses are also used to cultivate the grapes, papaya and mango in Japan. The crop grown under greenhouses has better yield and quality in comparison to grown in open fields.

The commercial cultivation of flowers and potted plants under greenhouse condition in India was started during 1965 by M/s. Indo-American Hybrid Seeds at Bangalore. M/s. Indo-American Hybrid Seeds arranged two seminars at national level in Bangalore during 1990 and 1993 aimed at protected cultivation. A number of industrial houses got impressed by the success of Dr. Manmohan Attavar of M/s Indo-American Hybrid Seeds and established a large number of greenhouses near Bangalore and Pune for cultivation of flowers especially roses for export. These greenhouses had the collaboration with the firms from Israel or Holland, which had practically no expertise under the Indian conditions. Most of the greenhouses failed due to faulty design.

The greenhouse is a structure covered with a transparent material for admitting natural light for plant growth. The main components of greenhouse are structure, covering / glazing and temperature control systems needed for healthy growth of plants. A large number of greenhouses around Bangalore had collapsed or overturned due to poor design of structure or foundation. Excessive temperature build up (5° to 10° C) were observed in the naturally ventilated greenhouses (NVGH) due to (a) very large size (b) wrong orientation (c) very little space between the adjacent greenhouses (d) insufficient vent area and (e) higher level of inlet opening for ambient air. Excessive tearing of glazing has also been observed due to wrong orientation of greenhouse or poor design of holding system for glazing.

The greenhouse manufacturers in India have learnt from the last 20 years experience and the greenhouses are being constructed as per requirements of Indian conditions. Alarge area has come up successfully under commercial cultivation of flowers and vegetables and vegetable nurseries under green hoses and net houses. However to improve the performance of green house these have to be designed and constructed scientifically.

## Planning of greenhouse

The greenhouse and the service facilities like structures for irrigation system, storage of materials, packaging house and cold storage has to be properly designed and located at site. The greenhouses and service facilities have to be properly connected and the necessary space has to be provided for future expansion. The soil and water should be suitable for the crop and available in the required quantity. The site should be level and properly drained for construction of green houses and buildings. Sunlight should be available for greenhouses and natural wind breaks should be present near the greenhouses. Most of the operations in the greenhouse are labour intensive and the necessary skilled and unskilled labour should be available nearby. The greenhouse has to be properly accessible to the market of the produce.

The climate is one of the most important factor for location of greenhouse. The climate decides the heating or cooling of the greenhouse and ultimately cost of production. Large number of greenhouses have come up around Bangalore and Pune due to mild climate of area, as the greenhouse could be cooled by natural ventilation and no expenditure is required on energy for cooling or heating.

## Ventilation and orientation

The orientation of the greenhouse is a compromise for wind direction, latitude of location and type of temperature control.

## Solar energy effects

Single greenhouses with latitude above 40° N should have ridge running east to west to allow low angle light to enter from side rather than ends. Below 40° N latitude (including India) the ridge of single greenhouses should be oriented from north to south, since the angle of sun is much higher. This orientation permits the movement of shadow of the gutter across the greenhouse. The location and orientation of the greenhouse should avoid falling of shadow on the adjacent greenhouses.

## Wind effects

If the greenhouse is naturally ventilated, the advantage of natural wind direction has to be taken to the maximum possible. For fan and pad greenhouse the natural wind direction should be same as the air blown by fan. The ridge of the NVGH greenhouse has to be parallel to the prominent wind direction for green house stability and to avoid the tearing of glazing. The minimum side area should be perpendicular to wind direction to reduce the cost of structure and for better stability of the greenhouse.

## Size of the greenhouse

The dimension of NV GH should not be more than 40 m x 40 m. Bigger the greenhouse, more will be the temperature build up due to poor ventilation. The length of evaporatively cooled greenhouse should not be more than 60 m.

## Spacing between greenhouses

The spacing between NVGH should be more than 6m, so that the exhaust from one greenhouse should not enter the adjacent greenhouse. For FPGH greenhouse the air from exauhst fan of one greenhouse should not enter the cooling pad of the other greenhouse, hence a spacing of 10 to 15 meter is required.

## Height of greenhouse

The bigger greenhouse needs more height and more area of the side ventilation and roof ventilation. The maximum height can be upto 5 m for 40 m x 40 m green house and this can be reduced as per the reduced size of the green house. Higher is the greenhouse more is the wind load for structure and glazing. The side ventilation can be of 2.5 m width and roof ventilation 1 m in width. The 40 or 60 mesh nylon net for side ventilation should be provided at the lowest possible level, so that plants get fresh air.

## Shading effects

To avoid the shading effect from one green house to another greenhouse these should be oriented East to West. However, the wind direction and latitude are also to be considered.

## Types of greenhouses

The greenhouses can be classified based on type of structures, type of glazing, number of spans, environmental control etc. The various types are as follows:

1. Classification as per type of structure
   a) Quonset type
   b) Curved roof type
   c) Gable roof type
2. Classification as per glazing

a) Glass glazing

b) Polycarbonate sheet

c) Fiberglass reinforced plastic glazing

i) Plain sheet

ii) Corrugated sheet

d) Plastic film

i) Ultra violet stabilized low density poly ethylene

ii) Silpaulin

e) Net house

3. Classification based on number of spans

a) Free standing or single span

b) Multispan or ridge and furrow or Gutter connected

4. Classification based on environmental control

a) Naturally ventilated

b) Evaporatively cooled or fan and pad

c) Heated green houses

Under Indian conditions, quouset type, multispan greenhouse is most suitable, because of its low cost and ease of fabrication. Ultraviolet resistant low density polyethylene (UVLDPE) single film cladding of 200 micron thickness is sufficient for Naturally Ventilated (NV) greenhouse and fan and pad (FP) greenhouses. This should be fully tightened by streching on the structure to avoid fluter and tearing. It should not be nailed or screwed to the structure as it gives the chance for tearing. The T-lock or L-lock (Fig. 16.1) should be used for fastening the sheet at structure, as this does not tear the sheet and sheet replacement is easy. The other type of fastening systems are shown in Fig. 16.2 Double inflated film of UVLDPE is required for heated greenhouses. The UVLDPE film is economical, 70 – 80 % transparent for light with 2 to 3 years usable life. The size of individual pieces of film should not be more than 200 sq meter as these are economical to replace in case of tearing. The use of silpaulin film should be avoided due to its less transparency and less life.

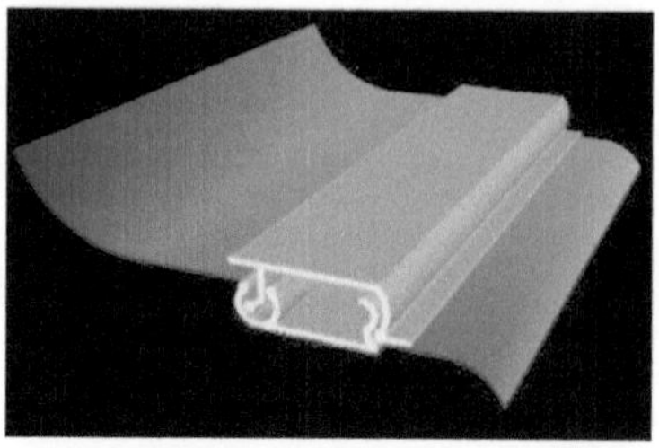

T- Lock for fitting the LDPE film
(Agra Tech, Pittsburg, CA)

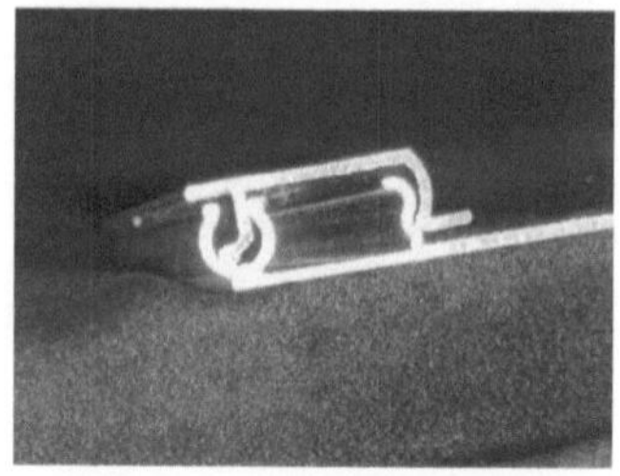

T- Lock in close-up view
(Agra Tech, Pittsburg, CA)

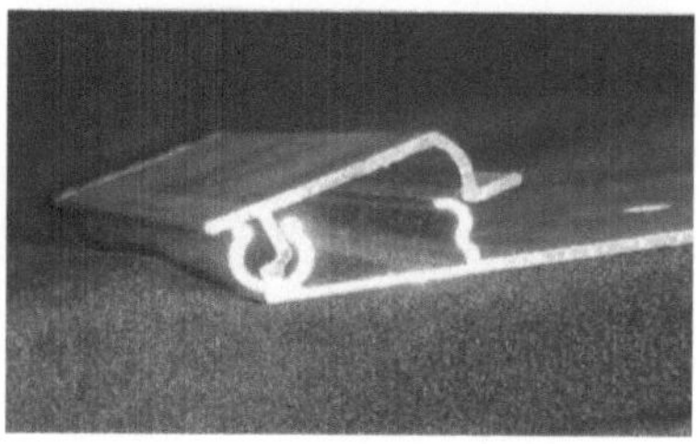

T- Lock
(Agra Tech, Pittsburg, CA)

T- Lock- Fitting view
(Agra Tech, Pittsburg, CA)

Fig. 16.1 T- Lock gripper for fitting the LDPE film on the frame of the greenhouse

NV greenhouses are most suited for mild agro-climatic conditions like Bangalore and Pune and coastal regions. FP system of cooling greenhouses can be used in dry and warm seasons of Northern plains. FP greenhouses are most suitable for nursery raising, grafting and rooting of cuttings due to high RH inside.

Greenhouses with heating system are generally required during winter in northern hilly regions and at nights in plains for better growth of plants.

The greenhouse can have more than one kind of environment control system based on the variation in the climate at location during the year.

## Structural design

The greenhouses are to be designed for necessary safety, serviceability, general structural integrity and suitability for necessary changes in future. The structure should be able to take all the necessary dead, live, wind and snow loads. The foundation, columns and trusses are to be designed accordingly. The greenhouses structures are to be designed to take up the loads as per design loads prescribed by the National Greenhouse Manufacturers Association (NGMA of USA) standards – 1994. However, wind speeds can be taken from Indian Standards and the Indian Standards can be used for designing the structures. The definition of various loads and the minimum and maximum limits as prescribed by NGMA standards are given below.

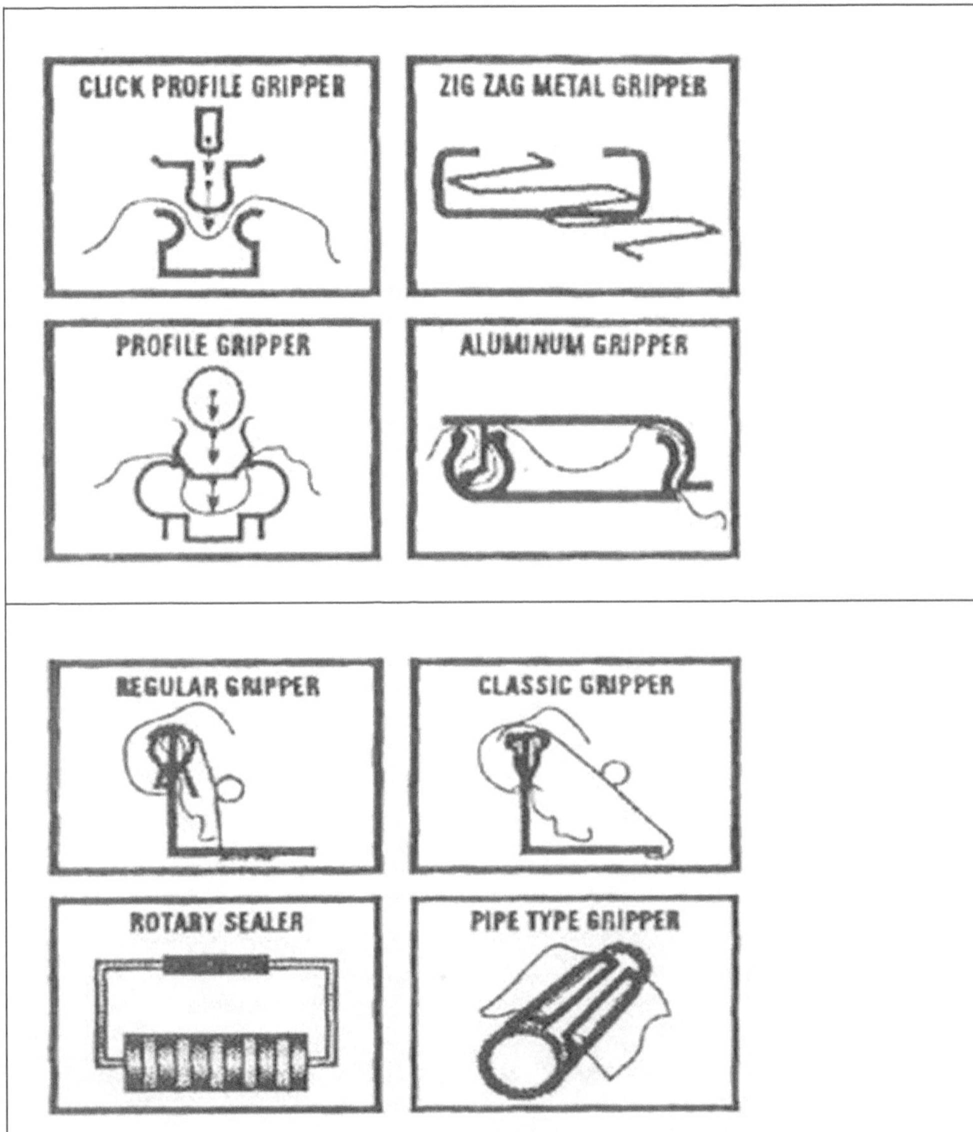


Fig. 16.2 Different type of grippers for fitting the LDPE film on the frame

## Dead load

The weight of all permanent construction including but not limited to walls, roofs, glazing materials and fixed service equipments are considered dead load.

## Live load

The live loads are temporary loads produced by the use and occupancy of the greenhouse. Live load does not include wind load, snow load, or dead load. Exterior live loads on green house roofs are the temporary loads i.e. work men and temporary equipments such as scaffolds. Interior live loads are temporary loads imposed on the structure by hanging objects. Any live load shall be considered permanent and therefore included as part of the dead load if it is imposed on structure for a continuous period of 30 days or more. The maximum roof live load is limited to 15 lb/sq. ft. (75 kg per sq. m.) of horizontal projected area.

All the roof members such as purlins, rafters, truss top members etc. shall be capable of safely supporting a minimum concentrated live load of 100 lbs (45 kg) applied downward and normal to the roof surface at their mid span, in addition, bottom chord of the roof trusses shall be capable of safely supporting a minimum concentrated live load of 100 lbs (45 kg) applied at any panel point. The structure should be able to take any partial loading and impact loads.

## Wind load

The structure and glazing should be able to take the wind loads as per the NGMA standards. The structure should be able to take wind loads during erection and construction phases. The foundation should be designed to resist overturning and sliding of the structure. The NGMA standards give the detailed procedure for calculating the wind loads and may be followed for designing. The greenhouse should be designed for minimum wind speed of 70 miles per hour ( 110 km per hour).

The wind load to be used in the design of the main wind – force resisting system for greenhouses shall be at least 10 lbs per sq. ft. (50 kg per sq. m.) as per NGMA standards – 1994.

## Material of structure

Structure should be constructed from galvanized iron pipes (square section) or angles and be able to take the necessary load as per NGMA standards. The usual size of columns and truss members for the greenhouse suitable for Bangalore conditions is 75 mm (2.5"), 65 mm (2"), 40 mm (1.5") or 30 mm (1") GI pipe B-class for 4 m span. The trusses should be able to take the necessary loads.

## Types of structures

The gutter connected Quonset multispan is most suitable for Indian conditions as per lowest construction cost. The gutters should be able to drain the rainwater falling on the roof. Vertical and horizontal drain pipes are to be provided to drain the water at least 2 m away from the foundation.

## Foundation

The foundation should be able to take necessary load as per NGMA standards. The size of foundation for greenhouse around Bangalore is 60 cm x 60 cm x 60 cm for the columns in plain cement concrete (PCC) with sand, cement, jelly ratio of

1:4:8. The column should also be covered with PCC of 15 cm dia upto a height of 60 cm to avoid the rusting of the columns. The necessary holds are welded to the column for proper anchoring in the foundation. The foundation size can also be 30 cm diameter and 90 cm depth and tractor operated digger can be used for digging the foundation pit.

## Door

The antechamber may be provided if necessary and the door of the greenhouse should be sufficiently wide for taking tractor with trailer and interculture implements inside the greenhouse. The door shall be insect proof.

## Glazing of greenhouse / net house

UV retarded LDPE film of thickness 200 micron (0.2mm ) should be used for NV or evaporatively cooled greenhouse. Double inflated LDPE film should be used for heated greenhouses. The transparency should be above 90%. The corrugated fiberglass reinforced plastic sheets become translucent and opaque after a use of 5 to 10 years and should not be used as per their high cost. The transparency of corrugated fibre glass reinforced plastic sheet becomes only around 3% after 15 years of use. The UV retarded LDPE sheet has the minimum life of 3 years and is economical.

The glazing should be tightly stretched and installed on structure on a sunny day. This will keep the glazing tight on structure and avoid water accumulation on the top.

In a shade net house the UV stabilized high density poly ethylene woven net can be used in place of glazing. These are available in shade percentage of 25%, 50%, 75% and 95%. These are available in white, green and black shades.

## Control of light intensity

The direct sun light on the plants can be reduced by providing shade net of required cut off value. These are netted out of UV stabilized HDPE and are available in 25% to 95% cut off value. LDPE films with reflective strips or bands can also be used to control the intensity of sun light in greenhouse. The use of shade with reflective strips decreases the temperature inside the FP greenhouse. For NV greenhouse the shade net can be provided above the greenhouse to avoid temperature build up.

## Environmental control

### Cooling the greenhouse

Plants need optimum temperature for maximum yield and quality. The greenhouses in plain and coastal region of India need cooling. The greenhouses in mild climates and coastal region can be naturally ventilated. The greenhouses for hot summer climates of northern plains have to be evaporatively cooled or with fan and pad (FP). The greenhouses for northern hilly areas have to be artificially heated. The greenhouses for northern plains may require both cooling and heating depending on the crop.

**Cooling load**

The solar intensity in India on bright sunny day is around 1 kW per sq. m. The transmittance of most of the glazing is around 70% during continuous use. Around 700 W energy enters the greenhouse per sq.m.

About 25% heat is used as latent heat for evaporation of water and transpiration by the plants inside the greenhouse.

About 75% heat or 500 W per sq.m. is sensible heat to be removed from greenhouse by natural ventilation or by evaporative cooling.

**Natural ventilation**

The greenhouse shown in Fig. 16.3 was 12 m x 30 m green house with 2 spans connected with a gutter. The side height was 3 m and center height was 5 m. The side ventilation area could be controlled by cranking mechanism to roll or unroll the UV resistant LDPE film. The side opening was at a height of one meter and had a width of 1 m. The total side opening was 60 square meter and the ridge opening was also 60 sq.m. It may be noticed that the temperature built up in the greenhouse is not exceeding 2° C throughout the year. Further during hot months the temperature in the greenhouse was same as the ambient temperature. The temperature and RH for the greenhouse in Fig.16.3 are shown in Fig. 16.5 to 16.7 when gerbera was grown in Bangalore.

If the ventilation is less than the required then there will be excessive temperature build up in the greenhouse. Constructing a very big greenhouse of 1 to 2 ha will result in excessive temperature build up because of poor ventilation. The inlet for ventilation should not have any obstruction or other structure nearby. The inlet for ventilation should be as close to the ground as possible so that the plants can get fresh air, however the entry of rain water by splashing is also to be avoided.

The naturally ventilated greenhouses are suitable for mild climates where maximum ambient temperature is about 35°C. These are also suitable in coastal regions having high humidity. In cold climate these are suitable during day time and the temperature inside GH can be increased upto 7 to 8°C. The temperature inside NVGH is generally above the ambient temperature. However a well designed, cropped and irrigated greenhouse can have 2 to 3°C lower temperature than ambient. This cooling is due to transpiration and irrigation of the crop.

The sensible heat of 500 watt per square meter is to be removed by the ventilation by

- Thermal buoyancy, when wind speed is low OR
- Wind effect, when wind speed is high.
- The 40 or 60 mesh nylon net is generally provided at the ventilation openings at side and at ridge.

The total ventilation area required for a naturally ventilated greenhouse fitted with nylon net with equal area for sides and ridges ventilation is given by;

$$A_T = \frac{0.0054 * A_g * S_R}{\sqrt{h} * (\Delta T)^{1.5}}$$

Where $A_T$ = Total area for ventilation- $m^2$

$A_g$ = Total floor area of greenhouse - $m^2$

$S_R$ = Maximum solar radiations received in area – $Wm^{-2}$

h = Vertical distance between center of side ventilation opening and ridge ventilation opening in m

$\Delta T$ = Difference between temperature inside greenhouse and the ambient temperature

The total area of ventilation may vary from 25 to 40 % of floor area depending upon the ambient conditions and crop.

The wall and ridge opening has to be through out length and provided with 40 gauge nylon mesh. The temperature build up inside the greenhouse can be controlled by controlling the size of side vent or roof vent by using adjustable opening by rolling or unrolling flap LDPE film over side vent or ridge vent using crank mechanism. The irrigation or misting also reduces the temperature inside the green house by 2°C to 3°C. Fig.16.4, Fig.16.5 and Fig.16.6 show the temperature and RH in NV greenhouse shown in Fig. 16.3. The average temperature build up was 1.7°C for the greenhouse when used for gerbera cultivation (Mandhar, et.al, 2001). The size of NV greenhouse has to be less than 40m x 40m to avoid excessive temperature build up.

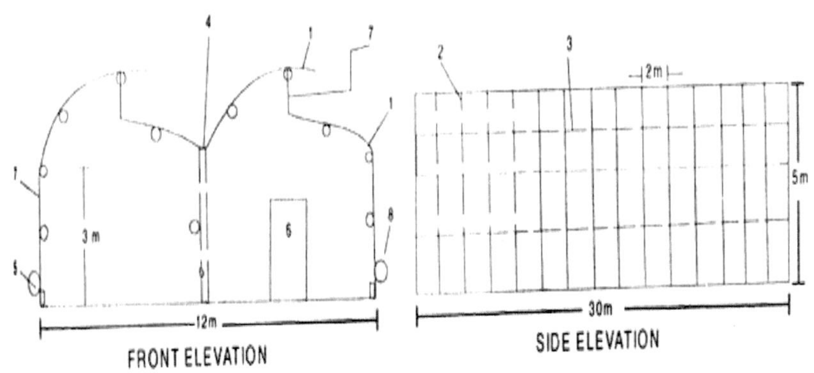


1- LDPE film 200 micron thick, 2-Truss, 3- Purlin, 4 –Gutter, 5-LDPE sheet, 6-Door, 7-Nylon mesh, 8-Crank mechanism

Fig. 16.3 Naturally ventilated greenhouse

The leaf temperature in naturally ventilated greenhouse is generally below greenhouse air temperature. The shade nets when used inside the greenhouse reduce the heat load on plants. When the wind direction is perpendicular to the ventilation opening, wind enters from upstream end and comes out from the downstream end of the greenhouse and there is increase in temperature along the path of wind. When wind direction is parallel to the opening, wind enters the greenhouse from the downstream (half) portion of the greenhouse and exits from the upstream (half) portion of the greenhouse. The wind direction inside the greenhouse is opposite to the ambient wind. There is temperature rise along the wind direction inside the greenhouse.

When wind speed is less than 1m/sec., the cooling of the greenhouse takes place mainly due to thermal buoyancy.

Maximum ventilation rate

The maximum ventilation rate taken for practical design of greenhouse is equal to the change of the complete air of the greenhouse in one minute.

$V_{max}$ = Volume of greenhouse / 60

Where,

$V_{max}$ = Maximum ventilation rate in $m^3/s$

Maximum size of naturally ventilated greenhouse

The maximum size of the NV GH should not be more than 40 m x 40 m under Bangalore condition.

## Evaporative cooled greenhouses

The evaporative cooling effect is created by sucking air through wet pads. The system is commonly known as fan and pad cooling system. The greenhouse of 12 m x 30 m size (Fig 16.4) was constructed in Bangalore and was fitted with 4 exhaust fans of 1.2 meter diameter with 1.5 hp motors. During day time 2 fans were operated the fans had 460 rpm and 509 cubic meter per minute air flow per fan. The dimension of pad was 2m X 11m. The temperature and RH for a green house (12 m x 30 m) Fig. 16.4. under Bangalore conditions with fan and pad and used for Gerbera cultivation is given in Fig. 16.5 to 16.7.

The fan and pad cooling system can give the minimum temperature nearer to wet bulb temperature. The minimum achievable temperature will be 2 to $4^0$ C above wet bulb temperature and maximum RH around 90%. No fan and pad system along with any fogging, misting, shading etc. can reduce the temperature below wet bulb temperature. Table.16.1. gives the wet bulb temperature for given dry bulb temperature and relative humidity.

**(a) Basic Air Flow Rate**

The basic air flow rate for greenhouse is 8 cfm per sq.ft. of floor area, or

$FR_{air}$ = 2.5 $m^3$ per minute per sq.m of floor area

Applicable only to

- 30 m length of greenhouse
- upto 300 m elevation
- $4^0$ C temperature rise from pad to fan
- 54000 lux light intensity

**(b) Actual Rate of Air Flow**

The total air flow rate required for greenhouse is

$TR_{air}$ = 2.5 x L x W x ($F_{house}$ or $F_{vel}$)

(the maximum value among $F_{house}$ and $F_{vel}$ is to be used)

$F_{house}$ = F elev. x F light x F temp

$TR_{air}$ = Total air requirement in cubic meter per minute for a greenhouse

L = length of house in m

W = width of house in m

$F_{elev}$. = Factor elevation = 1.00 to 1.36 for elevation 0f 300 to 2400 m

$F_{light}$ = Factor for intensity of light 0.80 to 1.60 for light intensity 43000 to 86000 lux

$F_{temp}$ = Factor for temperature rise from pad to fan 0.70 to 1.75 for temp. increase of $6^0$ C to $2^0$C

$F_{vel}$ = Factor of house length or velocity 2.24 to 1.00 for house length of 6 m to 30 m. and above.

**(c) Fan and pad**

**Selection of fan**

The fans should deliver the required air at 15 mm static pressure. The maximum centre to centre spacing between the two fans should be 7.5 m. The height of fans is to be determined based on the plant height which is proposed to be grown in the greenhouse. The greenhouse is to be provided with the necessary deflectors for directing air on plants, if necessary. The fan blades and frame are to be made of non corrosive materials like aluminium or stainless steel. above the air temperature.

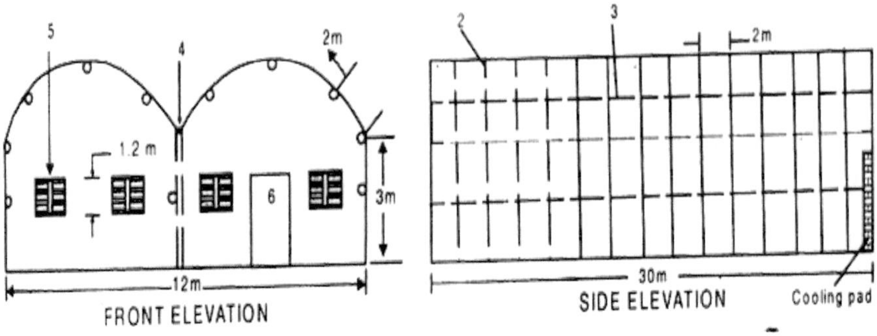


1-LDPE film 200 micron thick, 2-Truss, 3-Purlin, 4-Gutter, 5-Cooling fan, 6-Door, 7-Cooling pad

Fig. 16.4 Greenhouse with fan and pad cooling

Table.16.1 Wet bulb temperature for given dry bulb temperature and relative humidty

| **RH (%)** | **Dry bulb temperature $^0$ C** | | | |
|---|---|---|---|---|
| | 10 | 20 | 30 | 40 |
| | Wet bulb temperature $^0$ C | | | |
| 20 | 2.5 | 9.3 | 15.7 | 22.0 |
| 40 | 4.6 | 12.4 | 20.1 | 27.8 |
| 60 | 6.5 | 15.1 | 23.8 | 32.6 |
| 80 | 8.3 | 17.7 | 27.1 | 36.6 |

**Design of pad**

The cross fluted cellulose pad is preferred, These are available mostly in 100 mm thickness. One meter of pad height is given for every 20 m of pad to fan distance. However, the fan to pad distance should not exceed 60 m. The air flow rate should be of 75 cubicmeter per minute per square meter of pad. The water flow rate should be of 9 litres per minute per linear meter length of pad. The uniform distribution of water on pad is to be maintained.

## Heating of greenhouse

The heating of greenhouses in cold climates like winter in North India or Himalayan Region at high altitudes is advisable for getting better produce. Double covering of glazing with an air cushion of 2 cm to 10 cm reduces the heating load considerably. The heating load of greenhouse can be estimated as per NGMA standards

$$Q = [ A1 + (A2 * R)] * (Ti - To) * G * W * C$$

Where

Q = Heating load in Watts (W)

A1= Area of exposed cover/glazing of glass or LDPE ($m^2$)

A2= Area of exposed masonary wall ($m^2$)

R = Resistance of curtain/ exposed masonary wall to transmission of heat (expressed in

relation to transmission through glass/ LDPE film

Ti = Temperature inside the greenhouse, $^0$C

To = Temperature outside the greenhouse, $^0$C

G = Coefficient of transmission of glazing

C = Construction factor

G = 6.2 to 6.5 W $m^{-2}$ $^0C^{-1}$ (Watt per square meter per degree C) for temperature

difference of 15$^0$C to 40$^0$C between greenhouse and ambient for glass or LDPE film.

The value of G for double inflated LDPE film is only 4.0 W $m^{-2}$ $^0C^{-1}$, only 60 to

65% of G of single LDPE film.

R = 0.4 for 23 cm thick wall and 0.94 for asbestos board.

W = 1.00 to 1.26 for wind velocity of 25 to 55 from $hr^{-1}$.

C = 1.00 to 1.25 depending upon the condition and air tightness of the greenhouse.

After calculating the heating requirement of the greenhouse the heating equipment of required capacity for proper distribution of heat are to provided.

Heating systems

These can be of the following types

a.Boiler

1.With hot water tube

2.With steam pipes

b.Unit heaters

c.Infrared heaters

d.Solar heaters

## Boiler

This system is used for very big greenhouses and is a centralised system of heating. The boiler of necessary capacity is provided in the greenhouse. The fuel for boiler can be coal or fuel oil. The heating of the greenhouse is generally done using hot water at $85^0$ C or steam at $102^0$ C. Water or steam pipes are installed above the beds of crop and along the side walls of greenhouse. The steam system is cheaper than hot water system. To reduce the requirement of length of pipe to be used a number of hot water or steam pipe coils can be used and green house air circulated over them by blower for heating.

## Unit heaters

These are localised system of heating and a number of unit heaters are to be provided in the greenhouse at an height of about 3 meter to distribute heat evenly in the greenhouse. In a unit heater the fuel is combusted in the chamber at bottom. Hot fumes rise inside the heat exchanger tubes, giving heat to the walls of the tubes. Smoke exists at the top. A fan forces cool air of the greenhouse over the outside of heat exchange tubes, where it picks up heat.

## Infra –red heaters

The fuel gas (LPG) is burnt and the fumes at a temperature of about $480^0$C are passed in 10 cm diameter pipes kept overhead at a height of 1.5 m above plants. Reflectors are provided over the full length of pipe to radiate the infra red rays over the plants. The plants and soil only get heated without much heating of air. The infra red heating pipes can be provided at 6 to 10 meters interval all along the length of greenhouse. The temperature of fume gases at exist is about $65^0$ C and exhaust fan is provided for maintaining the flow of fumes.

## Solar Heating

Flat plate solar heaters are used to heat the water during day time. The hot water is stored in the insulated tanks. The hot water is circulated in pipes provided along the length of the greenhouse during night. Supplementary or emergency heating systems are provided for heating the greenhouse during cloudy or rainy days.

The greenhouses in Leh and Ladakh are trenches of about 2m width, 1m depth and about 5-10m in length. These are oriented in East-West direction. The top of the trench is covered with LDPE with suitable wooden frame. The temperature in the trench is usually higher than ambient temperature in winter. The entrapped solar radiation during day also increases the temperature inside the trench/greenhouse. At night the trench is generally covered with agro waste to reduce the heat loss to maintain the higher temperature inside the trench. These greenhouses have become very popular in Ladakh area for cultivation of vegetables during winter, when it is snow all around. These are very economical and cheap in construction.

## Environmental Control

## 1.Temperature control

In evaporative cooled / fan and pad greenhouse the thermostat can be coupled to water circulating pump and exhaust fan for controlling the temperature inside the greenhouse. The water supply to pads and exhaust fans are switched on automatically when the temperature in the greenhouse increases above the desired temperature. However, the lowest achievable temperature in fan and pad greenhouse is not below the wet bulb temperature in any case.

## 2.Relative humidity control

The humidistat coupled to water circulating pump or exhaust fan to control the relative humidity inside the fan and pad greenhouse. The maximum achievable relative humidity is 90% only in FP GH. The RH in NV GH can be increased by providing foggers.

## 3.Light intensity control

The shade net of required cut off value can be used to reduce the sunlight intensity on plants. For growing chrysanthemum the fluorescent tubelights can be used for increasing the photosysthetic period.

## Irrigation

The drippers or micro sprinklers can be used for irrigation. The irrigation and microsprinklers or foggers also reduce the temperature in greenhouse by 2 to 3 $^{0}$C. The fertigation systems can also be installed.

## Misting and fogging

The misters and foggers can also be used as per requirement to increase the relative humidity or lower the temperature in NV GH. The foggers can also be provided in heated greenhouse.

## 16.9 Maintenance of Evaporatively Cooled Greenhouses

## Maintenance of pad

The algae will grow and salts will deposit on pads if these are not maintained properly. The control of algae can be obtained without using chemicals by (a) shading the pads and sumps, (b) drying the pads daily, (c) avoiding nutrient contamination of circulating water , (c) draining and disinfecting the sump regularly, (d) replenishing 20% of circulating water each time to avoid scaling of minerals and (e) replacing old pads which have begun to settle or develop holes or blocked with mineral deposits.

## Maintenance of fan

The fans are to be maintained by (a) lubricating bearings regularly, (b) tightening 'V' belt per requirement and (c) lubricating the louvers properly.

## Protection from Birds

The damage to greenhouse clading has been noticed by the birds like peacocks. Wire mesh or grid has to be provided above greenhouse to avoid damage.

## Protection from Strong Winds

When the greenhouse is constructed in strong wind areas, the glazing/LDPE film has to be protected from tearing. At times of strong winds greenhouses, all the sides and vents including the pad are closed in FPGH. These are covered by the flaps provided for this purpose. All the fans are switched on to create a negative pressure in the greenhouse. The film clings to the structure and doesn't get blown or torn in strong winds.

## Automation / Computerization

The controller or computer and the required instrumentation can be used to control the temperature, RH, solar radiations (crop shading), irrigation, fertigation etc. The protection of the greenhouse is also controlled by the cotroller / computer as it will close all the vents and will switch on the fans for this purpose. The computer also maintains the proper records of temperature, RH, irrigation and fertigation of a greenhouse. The numbers of fans operating in a FPGH greenhouse are also controlled to save the energy. For NVGH greenhouse the vent area can be controlled by the computer to control the temperature. The vents can be closed during high speed winds and rain to protect the greenhouse and crop, respectively. For a bigger size of greenhouse the computerization is more economical than manually controlling the various operations.

## Cost of Greenhouse

The following are the approximate cost per sq.m. of floor area of different types of greenhouses and net houses at present. The bigger greenhouse will cost less per sq.m. than smaller greenhouse. The clading materials of UV stabilized LDPE film has to be replaced after 2 to 3 year of installation.

| **Type of greenhouse** | **cost Rs./sq.m.** |
|---|---|
| I. Net house | = 200 - 300 |
| 2. Naturally ventilated greenhouse with LDPE glazing | = 500 - 700 |
| 3. Greenhouse with fan and pad and LDPE glazing | = 1000 - 1500 |
| 4. Centrally heated greenhouse with LDPE glazing | = 1000 – 1200 |

The above cost do not include the irrigation, misting. fogging, computer systems and rolling benches etc.

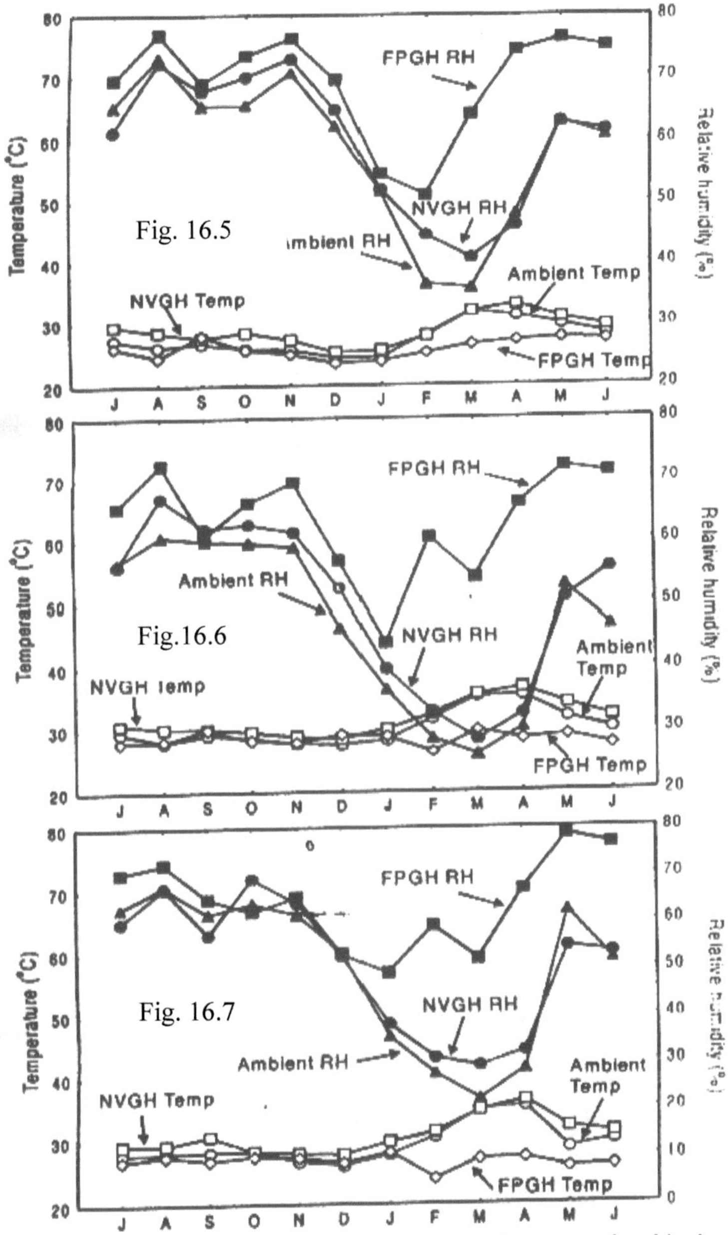


Fig.16.5. Temperature and RH in the NV and FP greenhouse and ambient at 10 a.m
Fig.16.6 Temperature and RH in the NV and FP greenhouse and ambient at 1 p.m
Fig.16.7 Temperature and RH in the NV and FP greenhouse and ambient at 4 p.m

# References


1. Abbey Rose Gardens, Nashdom Lane, Burnham,Buckinghamshire, SL1 8NJ, Tel - 01628 664 002, Fax - 01628 664 002, Email -sales@abbeyworkshop.co.uk .
2. ABC Quick Seeder, 2667 River Road 40036 Lockport, KY , U S A, (502) 845-4527 Fax (502) 845-4778 Website , http://www.abcquickseeder.com
3. ABM Tools, 79A, Pocket GG-1, Vikas Puri, New Delhi - 110018, India, Tel. + 91 98910 27608, + 91 11 2854 6891 Fax + 91 11 2854 0898, info@abmtools.com , abmtools@vsnl.net , sales@abmtools.com , abmtools.com
4. Absolute Innovations, Inc. (ABI), 1320 Third St. Osceola, IN 46561U.S.A. Phone: 877-788-7253, Fax: 574-674-5902, http://www.abiequine.com
5. Advanced System Technologies Ltd., Sentinel House, 10-12 Massetts Road, Horley, Surrey,RH6 7DE. UK. http://www.comparestoreprices.co.uk/
6. Advanced Treatment Technologies, Inc. 422 West Bailey Street, Asheboro, NC 27203, Ph: 1-855-OZONE4U (696-6348), Fax: (336) 672-5010, http://advancedtreatmenttechnologies.com
7. AG Application Equipment, Inc. , 1352 E. US 34, Earlville IL 60518, Office: 815-246-9919, Mobile: 309-824-0963, Email: info@agapplicationequipment.com http://www.agapeq.com/contactus.htm
8. Agribiz Corporation, National Highway No. 8, Near Water Canal, Vasad - 388306, Gujarat, India, Mobile , +919428902199 , 91-2692-319228,+91-2692-274507 , +91 98987-46676, http://www.agribizcorporation.com , agribizcorporation@gmail.com
9. AGRICARE , Cooting Road, Aylesham Industrial Estate, Canterbury, Kent, CT3 3EP, UK , Tel: 01227 751551 , Fax: 01227 752018 , Email: sales@agricareuk.com, http://www.agricareuk.com
10. Agricultural and Processed Food Products Export Development Authority (APEDA), Ministry of Commerce and Industry, Government of India. http://apeda.com, January 27,2012.

11. Agricultural Utilization Research Institute, (AURI), Forevergreen Chemical Free Weed Control, 19974 12 Avenue, Langley, BC, Canada V2Z 1W3, Ph: 604-534-9326, Fax: 604-530-7129 , Cell: 604-561-9325, USA Address, PMB 884, 250 H Street, Blaine WA, USA 98230-4033, Phone: 604-534-9326

12. Agriculture, Forestry and Fisheries Technical Information Society, Japan http://www.afftis.or.jp/

13. AGRIORISSA Know Your Agriculture, http://agriorissa.blogspot.in

14. Agripak International, 5 - Amber Palace, Main Shahra-e-Faisal, Adjacent to FTC Building, Karachi. Pakistan. Tel : +92-213-4556285,+92-213-4324661-4,Fax : +92-213-4556288, 6670317 WEBSITE : www.agripak.com.pk ,EMAIL : info@agripak.com.pk

15. Agriplas,13 La Belle Road, Stikland, Bellville,P O Box 696, Brackenfell, 7561, Tel: +27 21 917 7177, Fax: +27 21 917 7200, kjack@agriplas.co.za, http://www.agriplas.co.za

16. Agrupamex, S.L.,Embajadora, 3, 03190 El Pilar de la Horadada, Alicante, Spain Móvil: +34-620 928 592, Email: agrupamex@hotmail.com , http://www.agrupamex.com

17. AIT Asian Institute of Technology, P.O. Box 4, Klong Luang, Pathumthani 12120, Thailand , Telephone: (66 2) 5245000 or (66 2) 5160110-44 Fax: (66 2) 5162126

18. Alam Anwar,( 2006) Future Requirements of Agricultural Machines for Mechanizing Agriculture. Status of Farm Mechanization in India 2006 - Indian Agricultural Statistics Research Institute, (ICAR), New Delhi

19. Alard Equipment Corp. , 6483 Lake Avenue , PO Box 57, Williamson, NY 14589-0057 U.S.A. , Ph . 315-589-4511 , Fax. 315-589-3871 , http://www.alard-equipment.com/

20. Alard Equipments Corp., 2003. Z-3618, GEM stainless steel immersion washers. http://www.alard-equipment.com on 28.11.2003.

21. Albright, L..D. 1997. Specifications, Functioning and Maintenance of Equipment for Forced Cooling of Greenhouses. Proceedings of International Seminar on Protected Cultivation In India. Dec.18-19, Bangalore:l 7-24.

22. Albright, L.D. 1997. Ventilation and Shading for Greenhouse Cooling. Proceedings of International Seminar on Protected Cultivation in India. Dec.18-19, Bangalore:l 24- 30

23. All Tools Direct, Poor Peg Mill, Buckley Street, Heywood,OL10 4HZ United Kingdom. Fax: 01706 692424 , http://www.alltoolsdirect.co.uk

24. Allround Vegetable Processing , Handelsweg 5, 1619 BJ Andijk, The Netherlands Phone : +31 (0) 228 597777, Fax: +31 (0) 228 597788, E-mail: info@allroundvp.nl, http://www.allroundvp.nl/

25. Amar Agricultural Implement Works, Amar Street, Janta Nagar, Gill Road, Ludhiana, Punjab - 141 003, India ,Tel: +(91)-(161)-2491780/ 2493128/ 2814039

26. Ambica Engineering Works, N.H. No.-8, Opp. Agro Petrol Pump, Navagam, Kamrej Char Rasta City ,Surat 394185, Gujarat , India : Phone :+91-261-2344720 / Fax :+91-261-2343484 / 2621-251110 , Email ID : ambicaagromech@gmail.com, ambica_agromech@live.in ,Web Site :http://www.ambicaengineeringworks.co.in

27. American Listed.Com,Americanlisted, 200E. 5th. Ave. Ste.125A, Naperville,IL 60563, USA, 402-659-1007, http://omaha-ne.americanlisted.com/garden-house/12-row-hooded-sprayer-1500-crescent-ia_18710303.html

28. American Society of Agricultural and Biological Engineers, 2950 Niles Road, St. Joseph, MI 49085 , Phone: (269) 429-0300 Fax: (269) 429-3852 hq@asabe.org, http://www.asabe.org/

29. American Society of Agricultural Engineers, Outstanding Innovations. http://www.asae.org/resources/ae5003/model.html on 5.8.2003

30. American Society of Agricultural Engineers, Sweet cherry harvester video. http://www.afrsweb.usda.gov/donsvideos.htm on 12.8.2003

31. American Spring & Pressing Works Pvt Ltd, Malad West , P O Box No 7602, Aspee House, B J Patel Road, Near Malad Railway Station, Malad West, Mumbai, 400009 Ph 022 - 28820334, 28822331, 28822333, 28822334 , 28822336 , aspw@vsnl.com, aspee@aspee.net , http://www.aspee.com

32. ANDREAS STIHL Pvt. Ltd. Gut No.-53/01, Pune-Nashik Highway , Kuruli, Tel- Khed 410 501 Dist- PUNE ,INDIA , Tel.: +91 21 356 167 00, Fax: +91 21 356 167 01E-Mail: info@stihlindia.com, Internet: www.stihl.in

33. Andrews of Hindhead Limited, Portsmouth Road, Hindhead, Surrey, GU26 6AL, Phone 01428 605544 Fax 01428 606630, http://www.andrewsofhindhead.co.uk

34. Ankur Nurseries , 3, West View, Nappu Road, Dadar East, Mumbai - 400 014. Ph.: 6598 5564 , 9820430241 / 9820418679, info@ ankurnursery.com, http://ankurnursery.com

35. Applect Learning Systems Pvt. Ltd. A-221, Okhla Phase - I, New Delhi – 110020, Ph: (011) 40705070, Fax: (011) 47195310, http://www.meritnation.com

36. Arson Agro Industries,Mr. Rupinder Pal Singh ( Manager ), C 1, Focal Point, , G. T. Road, Moga , Punjab - 142001, India , Tel :+(91)-(1636)-222995, Email:sales@arsonagro.com , arsonagro@gmail.com , http://www.indiamart.com/arson-agro/query.html

37. AS Laser Systems Pvt Ltd, Address: 204. 205 Mayur Plaza V, Mayur Vihar Phase III,Kondli Gharoli, New Delhi 110096. MOBILE : 9711841690, EMAIL : aslasers@gmail.com, info@aslasers.com http://laserlevel.in

38. ASA-LIFT A/S Denmark , Metalvej 2, 4180 SORØ, Phone +45 5786 0700, Fax +45 5786 0707, Email info@asa-lift.com , http://www.asa-lift.com

39. Ashok Metal Products, Near S. T. Workshop,Gondal Road,Rajkot – 360004, Gujarat – INDIA, Tel.+91-281-2386106, 2369373, http://www.ashokmetal.com

40. ATCOMART Ltd. 214, Agarwal Tower-2nd Floor,Netaji Subhash Place, Pitampura, New Delhi-110 035 E-mail :contact@atcomaart.com, http://www.atcomaart.com

41. ATTRA - National Sustainable Agriculture Information Service, P.O. Box 3838, Butte, MT 59702, https://attra.ncat.org/management/contact.html

42. AUSTRAEL Machinery , P.O. Box 1513 , Mildura Via 3502, Australia , Phone 1: +61-03-5022-2888 , +61-03-5023-5989 , Fax: +61-03-5023-7888 , e. mail: afron-info@afron.com.au , Web Site: www.afron.com.au

43. Australian Grain Harvesters Association,PO Box 1435, Goondiwindi, QLD, 4390. Mob.0428422758, 0459 546 986, info@agha.org.au , http://www.agha.org.au

44. Aweta Grading And Packing, NL, Burgemeester Winkellaan 3, 2631 HG Nootdorp, The Netherlands, P.O. Box 17, 2630 AA Nootdorp, The Netherlands, Tel: +31 8 866 88 000, Fax: +31 8 866 88 002, E-mail: sales@aweta.nl , http://www.aweta.nl

45. AYTOK Drip Irrigation Equipments, 2,Organize Sanayi Yaylacik Cad. No:30,42300 - Konya ,Turkey ,Tel : 0090 332 2390603, Fax : 0090 332 2390653 ,URL : http://www.wowyeah.com/co/tokmak77 or http://www.aytokmakina.com

46. Bajaj ProcessPack Limited, Add: B-136, Sector-63, Noida, Uttar Pradesh - 201 001, India, Tel : +(91)-(120)-4639950 / 4372848, Fax : +(91)-(120)-2775137, Mobile : +(91)-9560333996, Email : sales@bajajmachines.com, office@bajajmachines.com

47. Bally, I.S.E., Johnson, P.R., Kulkarni, V.J., 1999. Mango production in Australia. Proceedings of the Sixth International Symposium on Mango. Acta Horticulturae: 509 (1): 59-67, ISHS, Pattaya, Thailand.

48. Bankura Precision Agricultural Equipments , Mr. Joginder Singh , V.p.o.-Miranpur Near Bus Stand , District-Muzaffar Nagar : Uttar Pradesh : India : 251315 Phone : +91-99977-66999, Fax : +91-01396-244147, Email : bankuraprecision@gmail.com Web Site : http://www.bankuralaserlandleveler.co.in

49. Bargain Brands, Airport House, King's Mill Lane, South Nutfield, Surrey, RH1 5JY, Tel - 01737 821266 , http://www.bargainbrands.co.uk

50. Bathla Aluminium (P).Ltd, # 9 Auto Towers, J . C. Road, Bangalore 560 002, Karnataka, India, Phone : 080-22223081 / 22223089 / 22233824 / 22235131 , Email : info@bathla.com , http://www.bathla.com

51. Beasley Power Equipment , Pam & Alan Beasley, 19 Commercial Park Avenue, (Hwy 16 across from WalMart), Taylorsville, NC 28681, Office: 828-635-0089, Fax: 828-635-0065, Mobile: 704-880-5645, Email: pam@beasleypowerequipment.com, http://www.beasleypowerequipment.com

52. Beaupre Aerial Equipment, 2400 Minnehaha Avenue South | Minneapolis, MN 55404 , USA. Ph.: (612) 338-3336 , Fax: (612) 338-5433, http://beaupre-inc.com

53. BEI International, 1375 Kalamazoo Street, South Haven, MI 49090 Ph1-800-364-7425 http://www.beiinternational.com, sales@beiintl.com

54. Beri Udyog Pvt Ltd, 100-101, HSIIDC, Industrial Area, Sector 3 Karnal Haryana - 132 001 (India), (91)-(184)-2221571, Fax (91)-(184)-2221574, http://www.fieldking.com

55. Best-B2B.com (China Best B2B Corporation), http://best-b2b.com

56. Bhadra Industries, P.O. Kanimangalam,Thrissur(Dist), Kerala,India, Pin: 680021, Ph: 0487-2448754 Mob: 9446997879 , E-mail: bhadraindustries@yahoo.co.in, http://www.bhadraindustries.com

57. Bhansali Trailors Pvt Ltd. Mr. Sanjay Bhansali, Dhantara, Gurudwara Road, Kopargaon, P.b.no.51, Ahmed Nagar State :Maharashtra, India :423601 Phone :+91-2423-223391 / 223392 Fax :+91-2423-223394 Email ID : bhansaliudyog@rediffmail.com Mobile Web Site :http://www.bhansalitrailors.in

58. Bhardwaj, K.C., M.M.Pandey, G. Singh, 2004, Horticultural Tools and Equipment,Central Institute of Aggricultural Engineering, Bhopal, India

59. Bhavani Tractor Plough ,Bhavani Estate, Near Sapna Theatre, Sarkhej , Ahmedabad, Gujarat - 382 210, India , Tel: +(91)-(79)-26820886/ 26820116, http://www.indiamart.com/bhavani-tractorplough

60. Bianchi Orchard Systems, LLC, 286 West Evans Reimer Road, Gridley, CA 95948-9536 Phone: 530-846-5625, Fax: 530-846-2931, http://kilbymfg.com

61. Bijay Kumar (2009), Indian Horticulture Data Base, 2008, National Horticulture Board, Gurgaon.

62. Binger Seilzug GmbH & Co. KG , Saarlandstr. 246 , 55411 Bingen Deutschlands, Tel.: +49 (0)6721/309648-0 , Fax: +49 (0)6721/309648-50 , info@binger-seilzug.de , http://www.binger-seilzug.de

63. Bir Singh & Sons, Chandigarh Road, Samrala - 141114, Punjab, India, Ph : 91-1628-262316, http://birsinghsons.tradeindia.com

64. Boulard T., Kittas C., Papadakis G., Mermier M.1998, Pressure field and air flow at the opening of a naturally ventilated greenhouse. Journal of Agricultural Engineering Research, 1998, 71, 93-102.

65. Boulard T., Kittas C., Papadakis G., Mermier M. Pressure field and air flow at the opening of a naturally ventilated greenhouse. Journal of Agricultural Engineering Research, 1998, 71, 93-102.

66. Boulard, T. , G. Papadakis, C. Kittas, M. Mermier, 1997, Air flow and associated sensible heat exchanges in a naturally ventilated greenhouse , Agricultural and Forest Meteorology, Volume 88, Issues 1–4, 30 December 1997, Pages 111-119

67. Brown Manufacturing Corp., 6001 East Hwy 27, Ozark, AL. 36360 Ph. 800-633-8909, Fax 334-795-3029 , www.brownmfgcorp.com , http://www.brownmanufacturing

68. Bull Machines Pvt Ltd, S.F. no. 5/1 – A, Trichy Road, L & T By-pass junction, Coimbatore – 641103.India. Phone.no. 0422-2270859 / 6586677 / 6586655, Fax.no. 0422-2270712 / 2270955, http://www.bullindia.com

69. Burkard Nurseries, 690 N. Orange Grove Blvd.Pasadena, CA 91103, 626-796-4355

70. Buxton's Ltd, Coppice House, Teddessley, Penkridge, Stafford, Staffordshire, ST19 5RP, Tel. 01785 712397 Fax. 01785 713339, enquiries@buxtons.net , http://www.buxtons.net

71. CanadianOutdoorEquipmentCompany,199LakeshoreRdEUnitA,Mississauga, Ontario, Canada 905-990-1750, http://www.canadianoutdoorequipment.com

72. Captain Polyplast Ltd. Survey No. 267, Plot No. 10-A &11, N.H. 8-B, Shapar-Veraval, Distt. Rajkot -360002, Gujarat, India, Ph. +91-2827-253006, Fax +91-2827-252056, E mail info@captainpolyplast.in, www.captainpolyplast.in

73. Captain Tractors Private Limited,Padavla Road, Taluk Kotdasangani, Veraval, Shapar, Rajkot, Gujarat - 360 002, Tel: +(91)-(2827)-252547/ 253916,Email: info@captaintractors.com , http://www.indiamart.com/captaintractors,

74. Carpi Meccanica Di Carpi Giuliano & C. Via Nova 45, 42017 Novellara (Re) - I Tel. +39 522 654144 Fax +39 522 662184 Internet www.Europa-Dma.Com , Info@Europa-Dma.Com , Tecnico@Europa-Dma.Com

75. Caterpillar Corporate Headquarters, 100 North East Adams Street , Peoria, Illinois USA 61629, 001 (309) 675-1000, http://india.cat.com/

76. Caustier France, Grading and packaging of fresh fruits, citrus and vegetables. http://www.caistier.com/Andleterre/grading/summary.

77. Central Agricultural Research Institute, PB 181, Andaman and Nicobar Group of Islands, P B 181 Port Blair-744101, (Andamans & Nicobar Islands) 744 101 Ph. 03192-250341, Fax. 03192-251068, http://cari.res.in director@cari.res.in

78. Central Institute for Subtropical Horticulture, (Indian Council of Agricultural Research) Rehmankhera, P.O.Kakori, Lucknow UP, INDIA, 227107 , Fax: 91-0522-2841025, Phone: 91-0522-2841022, 23, Email: director@cish.ernet.in, http://www.cishlko.org

79. Central Institute of Agricultural Engineering, Berasia Road, Nabi Bagh, Bhopal (Madhya Prdesh) 462 038, Ph. 0755-2737191, Fax. 0755-2734016, director@ciae.res.in, http:// www.ciae.nic.in

80. Central Institute of Post-Harvest Engineering and Technology, PAU Campus,Ludhiana (Punjab) 141 004, Ph. 0161-2308669, Fax. 0161-2308670, director.ciphet@gmail.com, http://www.ciphet.inBottom of Form

81. Central Plantation Crops Research Institute, Kasaragod (Kerala) 671 124, Ph. 04994-232333, Fax. 04994-232322, directorcpcri@gmail.com, http://www.cpcri.gov.in

82. Central Potato Research Institute, Shimla (Himachal Pradesh) 171 001, Ph. 0177-2625073, Fax. 0177-2624460, directorcpri@gmail.com, http://cpri.org.in

83. Central Tuber Crops Research Institute, PB 3502, Sreekariyam, Thiruvananthapuram , (Kerala) 695 017, Ph. 0471-2598193, Fax- 0471-2590063, ctritvm@gmail.com , http://www.ctcri.org,

84. Chadha, K.L., M.L. Choudhary (2007), Report of the Working Group on Horticulture, Plantation Crops and Organic Farming for the XI Five Year Plan (2007-12), Ministry of Agriculture, Government of India

85. Changzhou Dongfeng Agricultural Machinery Group Co., Ltd. 10 Xinye Road, Xinzha Town, Zhonglou, Changzhou, Jiangsu, China (Mainland) Zip: 213012 , China , Ph. +86-519-3256576, http://www.asia.ru/en/CompanyList/?company_id=47247

86. Charlie's Machine & Supply Inc., 406 Pearl Street, Boulder, CO 80302, Ph. 303-443-9595, Fax: 303-443-7474, email: info@charliesmachineandsupply.com , http://www.charliesmachineandsupply.com

87. Chemical Containers, Inc. , P.O. Box 1307, Lake Wales, Florida, 33859-1307, USA

88. China E-Store Wholesale Plaza , Hong Kong http://www.p-wholesale.com/subcat/19/727/agricultural-machinery-p4.html

89. Choosing Gas Powered Leaf Blowers, http://www.plant-care.com/gas-powered-leaf-blowers.html, October, 1,2012.City : Castagnaro Vr ,Tel: 0442 92025, Fax: 0442 675088, info@spedo.it , http://www.spedo.it/4/contacts 2177802.html

90. CJ Industries , Brondesbury Park , Cardigan, Ceredigion , Wales UK SA43 1NA Telephone: 01239 615300, Fax: 01239 615300 , http://www.cjindustries.co.uk/

91. Claydon Yieldometer Ltd. Gaines Hall, Wickhambrook, Newmarket , Suffolk, CB8 8YA Tel: 01440 820 327, info@claydondrill.com, http://www.claydondrill.com

92. CMI Country Manufacturing, P.O. Box 104, 333 Salem Ave, Fredericktown, OH 43019 Ph: 740 - 694 -9926 E-mail: info@countrymanufacturing.com , Fax: 740 - 694 - 5088 , http://www.countrymanufacturing.com

93. CNH America LLC, Farm Equipment Headquarters, 613 Airport Rd, Pendleton, OR 97801, 503-276-6222, E mail: na.topservice@newholland.com http://www.flexicoil.com

94. Coconut Development Board, Ministry of Agriculture, Government of India. http://coconutboard.nic.in, January 27,2012.

95. Coffee Board of India , Ministry of Commerce and Industry, Government of India. http://www.indiacoffee.org January 27,2012.

96. Compac Sorting Equipment Inc. 751 N. Elko Street, Visalia, California 93291,USA Tel : +1 559 334 0118, Fax: +1 559 334 0122, Email: usasales@compacsort.com, Website: www.compacsort.com

97. Conservation Technology & Machinery Ltd, Units 1&2 Brocca Bank, Low Row, Richmond, North Yorkshire, DL11 6PR Tel: 01748 886622, Email: enquiries@ctmltd.co.uk, Website http://www.ctm-ltd.co.uk

98. Cornell International Institute for Food, Agriculture, and Development (CIIFAD), B75 Mann Library, Ithaca, NY 14853 USA, Telephone:(607) 255-0831,ciifad.cornell.edu/ sri/2004.html , E-mail: ciifad@cornell.edu

99. Corona Clipper, 22440 Temescal Canyon Road, Corona, CA 92883, http://www.coronatoolsusa.com

100. Crendon Machinery, PO Box 95 Donnybrook 2 Ramsay Tce Donnybrook , Western Australia 6239, Australia, Tel. (Australia) 1300 761 922 ,08 97311502 (International) 618 97311502, Fax (Australia) 08 97311657 (International) 618 97311657 , E-mail info@crendon.com.au , Web Site ; www.crendon.com.au

101. Crocus Co.UK Limited, Nursery Court, London Road, Windlesham, Surrey GU20 6LQ. info@crocus.co.uk., http://www.crocus.co.uk

102. D.T. Brown & Co, Bury Road, Newmarket, CB8 7PQ . U.K., Tel : 0845 3710532, Fax: 0845 3710534, http://www.dtbrownseeds.co.uk

103. Daroli Industries, Majestic Road, Moga - 142001, Punjab, India,Phone :91-1636-223557, Fax:91-1636-223557, Mobile:+919814029811, +9198722885

104. Dasmesh Mechanical Works Pvt. Ltd., Raikot Road, Malerkotla, Distt. Sangrur, - 148 023, Punjab. INDIA. Ph: +91 88726 88726 , +91 94170 25726 , +91 94170 55726,Fax,: +91 1675 261726, Email : info@dasmesh.com Website : www.dasmesh.com

105. Dasmesh Mechanical Works, Main Road, Amargarh-148022, Sangrur. Pb. (INDIA), Phone: 91- 01675 284221, Fax No. 01675 285999,E - mail: info@dasmesh.net , Internet : www.dasmesh.net

106. Datta Irrigation Company, Industrial Estate, Savda Road, Post Faizpur - 425 503, Maharashtra , India Tel: +(91)-(2584)-222996, Mobile: +(91)-9422279300, http://www.dattadrip.in

107. Deccan Farm Equipment Pvt. Ltd., C-35, MIDC, Shiroli, Kolhapur - 416122, Maharashtra, India http://deccanfarm.tradeindia.com/ , Ph : 91-231-2468904 , Fax : 91-231-2468905, Mobile : +919545441100, +919764999860, +919923147700

108. Department of Animal Husbandry Dairying & Fisheries (DADF), the Ministry of Agriculture, Govt. of India, http://dahd.nic.in

109. Dharma Technologies ,1st cross, 60 Feet Road Saptagiri extension west, Tumkur-572102, Karnataka, India, Ph- 0816- 2292159, M- 9886737260, E-mail: shylaja_vittala@sify.com

110. Dharti Agro Engineering ,Survey No. 35, Plot No.6, Nr. Saurashtra Paper Board Mill, Behind Olympic Pipe, Gondal Road, Shapar, Rajkot - 360024, Gujarat, India,Ph : 91-2827-252258/252220, Fax:91-2827-252258, http://dhartiagro.tradeindia.com

111. Dieter Schmid, Fine Tools, Georg-Wilhelm-Str. 7 A, 10711 Berlin , Germany, Phone: +49 30 342 1757,Fax: +49 30 342 1764, http://www.fine-tools.com/garten.htm

112. Digvijay Industries, J-79, MIDC, Kupwad, Sangli. (Maharashtra)- 416 436, Phone: 0233-2644484 Cell Phone: 09372143097, Email: contact@digvijayindia.com, http://www.digvijayindia.com , http://www.digvijayindia.com/html/Anardhana.htm

113. Directorate of Onion and Garlic Research, Rajguru Nagar, Pune, (Maharashtra) 410 505, Ph. 02135-222026 Fax. 02135- 224056, director@dogr.res.in, http://nrcog.mah.nic.in

114. Directorate of Cashew Research, Darbe Post, Puttur, Dakshin Kannada-574202, Karnataka, India, Ph. 08251-231530, Fax. 08251-234350, dircajures@gmail.com, http://www.cashew.res.in

115. Directorate of Oilpalm Research, Pedavegi, West Godavari, (Andhra Pradesh) 534 450 India, Ph. 08812-259409, Fax 08812-259531, dopr2009@gmail.com, http://dopr.gov.in

116. Dobmac Agricultural Machinery, Industrial Drive, Ulverstone Tasmania PO Box 93, Ulverstone Tasmania 7315, Australia, +61 03 6425 5533 - International: +61 3 6425 5533 Fax 03 6425 5847 - International: +61 3 6425 5847 36-38 , dobmac@dobmac.com.au, http://www.dobmac.com.au

117. Dodman Ltd, Hamburg Way, North Lynn Industrial Estate, King's Lynn, Norfolk, PE30 2ND, phone +44 (0)1553 774755, fax +44 (0)1553 772227 , sales@dodmanlimited.co.uk , http://dodmanlimited.co.uk/

118. Dole Food Company, PO Box 5700 Thousand Oaks, CA 91359-5700., Ph. 800-356-3111, www.dole.com

119. Dolphin Irrigation Systems, No. 56/1, Kottigepalya, Sree Gandhad Kavalu Magadi Main Road, Bengaluru, Karnataka - 560 091, India , Tel: +(91)-(80)-23481019, http://www.indiamart.com/dolphin-irrigation/irrigation-equipments.html

120. DRANGEN Manned Planter, Harvester and Weeder , Pendragon Specialties, LLC, DBA Pendragon Fabrication, 2485 Corporate Circle, Unit 5, East Troy, WI 53120, Phone/Fax: 262.642.7793, Cell Phone: 262.844.3440, Email: info@pendragonfabrication.com, http://www.pendragonfabrication.com

121. Draper Tools Ltd, Hursley Road, Chandler's Ford, Hampshire, SO53 1YF. Tel: 023 8026 6355 , Fax: 023 8049 4201/4209, http://www.drapertools.com

122. Duijndam Machines BV , 2913 LR Nieuwerkerk a/d IJssel 2e Tochtweg 127, The Netherlands, Ph. +31 (180) 63 20 88, +31 (180) 63 20 04, Fax +31 (180) 63 30 22 http://autoline-eu.co.uk/duijndam/

123. Dumelow International Limited, Unit 7, Boundoak Industrial Estate, Eversley Road, Arborfield, Reading, RG2 9PN, UK, Ph.+44 118 9487000 .http://www.dumelow.co.uk

124. Dural Irrigation Pty Ltd is located at 270 New Line Road Dural NSW 2158, Phone 1800 244 389 , 02 9651 4111, Fax 02 9651 2638 Email customerservice@duralirrigation.com.au http://www.duralirrigation.com.au

125. DuroKon, 12210 Southwest Main Street ,Portland Oregon, USA97281, Tel: (503) 395-8616, http://www.durokon.com

126. Edp Australia Pty Ltd.,33 – 37 O'Brien Street, (P.O. Box 251), Mooroopna, VIC 3629, Australia , Ph: +61 3 5820 5337 , Fax: +61 3 5825 2758, Email: sales@edp.com.au , http://www.edp.com.au

127. Electrostatic Spraying Systems, Inc. 62 Morrison Street, Watkinsville, GA 30677, Phone/Fax (Toll Free): 1-800-213-0518, (Local): 706-769-0025, Fax: 706-769-8072 http://www.maxcharge.com

128. Ellips B.V. , Esp 300,5633 AE Eindhoven , The Netherlands, Phone: +31 (0) 40 - 245 6540,fax: +31 (0) 40 246 7183 , email: info@ellips.nl, website: www.ellips.nl

129. Emdek Sendirian Berhad , 6, Jalan Lombong Emas 6, Seremban Light Industrial Park, Off Jalan Tun Dr. Ismail,70200 Seremban, Negeri Sembilan, Malaysia. Tel : (606) 7641320, 7641216 Fax : (606) 7641310 Email : emdek@tm.net.my, emdek@streamyx.com , http://www.emdek.com.my/

130. Emerald Equipment, Emerald Seed & Supply, 6535 NE 11th Street, Redmond, OR 97756, Ph. (800) 826-8873,(541) 504-0307 Mailing address: PO Box 215, Terrebonne, or 97760 Oregon, emeraldss@ykwc.net , http://www.emeraldseedandsupply.com/

131. Envco Environmental Equipment Suppliers, Australia Ph 1800 622 343 New Zealand Ph 0800 623 336, info@envcoglobal.com, http://www.envcoglobal.com

132. Eros Metal Works (P) Ltd., G-5, MIDC Area, Hingna Wadi Road, Nagpur, Maharashtra, India : 440028,Ph : +91-7104-237775/ 236288 Fax : +91-7104-232748/ 237361, Email : emwpl_ngp@sancharnet.in, Web Site : http://www.erosmetalworks.com , Website : www.erosgroup.co.in

133. Etani Sdn. Bhd. ,No.6 Jalan Saujana Indah 7, Taman perindustrian Saujana Indah, 40150 Shah Alam, Selangor,Malaysia,Tel: 03-7845 1155, 03-7845 1166, 03-7845 1177, Fax: 03-7831 5776, Email: sales@etani.com.my , Website: www.etani.com.my

134. Exmark Mfg. Co., Inc., Industrial Park N.W.P.O. Box 808, Beatrice, NE 68310-0808, Call (402) 223-6300,Fax (402) 223-6384, http://www.exmark.com

135. F.lli Spedo , 37043 Castagnaro (Verona) - via Polesine, 1- ITALY - tel: +39-0442-92025, - P.Iva 00307540237 - info@spedo.it, http://www.spedo.it/en/

136. Falcon Garden Tools Pvt. Ltd., 2524 Lane -5, Janta Nagar, Gill Road, Ludhiana-141003(INDIA) , Tel : +91-161-2811899, 2812144 Fax : +91-161-2812144 , 2503156, E-mail : falcongardentools@rediffmail.com , falcontools@rediffmail.com, http://www.falcontools.com

137. Falcon Industries, 2524 Lane -5, Janta Nagar, Gill Road, Ludhiana-141003(India), Tel:+91-161-2494840, 2490877 Fax :+91-161-2503156 , E-mail : falcontools@rediffmail.com

138. FELCO SA (Headquarter)Mélèzes 4CH-2206 Les Geneveys-sur-Coffrane, Switzerland, Tel. +41 328 581 466 / F +41 328 571 930, info@felco.com, www.felco.com

139. Feucht-Obsttechnik, Obsttechnik-Mahtechnik-Nubtechnik, Europastr. 16, D-71576 , Erbstetten , Tel. (+49) 7191-64195, Fax. (+49) 7191-72094, FEUCHT-OBSTTECHNIK

140. Fiskars Brands, Inc. 2537 Daniels St , Madison, WI 53718, USA, Ph. 1.866.348.5661, warranty@fiskars.com , http://www2.fiskars.com

141. Food and Agriculture Organization, Harvesting and field handling, Prevention of post-harvest food losses: Fruits, vegetables and root crops. http://www.fao.org/docrep/T0073E/T0073E03.htm

142. ForConstructionPros.com, 1233 Janesville, Ave.,Fort Atkinson, WI 53538.

143. Forestry Tools, PO Box 870, Willoughby NSW 2068, Australia, Tel :02 9417 7751 , Fax:02 9417 7426, E-mail: info@forestrytools.com.au

144. Fox Coulee Aviation Inc. ,Box 1208 , Drumheller, Alberta T0J 0Y0, Phone: 403-823-3000, Toll Free: 1-877-923-3009, Fax: 403-823-3666 ,403-820-1252 , Email: brian@foxcouleeaviation.com, mark@foxcouleeaviation.com , http://www.foxcouleeaviation.com

145. Fresh Pik Produce, Inc., 7802 Sadie Road, Kenly NC 27542, Phone: (252) 237-1260, Fax: (252) 237-6133, jsharp@freshpik.com, http://www.freshpik.com

146. G K Machine, Inc., 10590 Donald Road NE, PO Box 427, Donald, Oregon 97020-0427

147. Ganesh Agro Equipments, Kalol Mehsana Highway, Village Vadpura, Tal-Kadi, Dist. Mehsana,Kadi - 382705, Gujarat, India, Phone:91-2764-273442/267446, Fax:91-2764-267446, Mobile:+919825319817, +919925031817, +919898615911, http://ganeshagro.tradeindia.com/

148. Ganesh Raj Industries, Kalol-Mehsana Highway, Vadpura – 382704, Taluka: Kadi, District: Mehsana, Phone: + 91 - 2764 - 273442 / 267164, 9898615911, Tele/Fax: + 91 - 2764 – 267446, bipin@ganeshraj.com, http://www.ganeshraj.com/contact.php

149. Garden Décor, Mont Lebanon, Mansourieh, Autostrade, Beside Besaibes Interior, Face Bouza Bachir. Lebanon . Phone : +961-4-533690, Email : info@gardendecors.co.uk , http://www.gardendecor.net

150. Garden Organic , Coventry, Warwickshire, United Kingdom, CV8 3LG, Tel: +44 (0) 24 7630 3517, Fax: +44 (0) 24 7663 9229, Email: enquiry@gardenorganic.org.uk , http://www.gardenorganic.org.uk

151. Garden Power Corporation , Box 633, 334 Church Street, Sweet Grass, Montana, 59484-9998, United States , Tel: 403-475-5177, Fax. 403-203-7559, http://www.toolworlddirect.com/

152. Gardens On Line Shopping Pty. Ltd., P.O. Box 13, Brookvale, New South Wales, Australia 2100, Tel. (02) 9453 5672, http://www.gardensonline.com.au

153. Gearmore, Inc., 13477 Benson Ave., Chino, CA 91710, Ph. 909 548 4848, 800 833 3023, Fax, 909 548 4747, Email sales@gearmore.com, http://www.gearmore.com

154. General Rental Center of Monticello, Monticello USA Ph. 763-295-2300, http://www.generalrentalmonticello.com/

155. Gillison's Variety Fabrication Inc ., 3860 W. Monroe Road Hart, MI 49420 , 3033 Benzie Hwy. Benzonia, MI 49616-9747, http://www.gillisons.com/

156. Global Agri Tech Engineers, Goldin (India)Engg.Co, No. 10/11, B. I. D. C. Estate, Gorwa Alembic Road, Gorwa, Vadodara - 390 016, Gujarat, India , Ph : +(91)-8447580361, Mobile: +(91)-9824240130, Telephone / Fax: +(91)-(265)-2291540, Email: globalagritech1@yahoo.com , goldin1971@yahoo.com , http://www.globalagritech.in/

157. Global Citrus Resources, Inc. , 302 S. Massachusetts Ave., Suite 119, Lakeland, Florida, USA 33801 , Phone (863) 606-8666 , http://www.globalcitrusresources.com/

158. Goglobal Solutions SDN BHD, 29-3, Jalan OP 1/6, Pusat Perdagangan One Puchong, 47160 Puchong, Selangor Darul Ehsan, Malaysia , 603-80606388 , 603-80610118, info@goglobalsol.com , http://www.goglobalsol.com

159. Golden Industries, H-174(A1) RIICO Industrial Area, Malviya Nagar, Jaipur - 302 017, Rajasthan, India , Tel :+(91)-(141)-2759951, Mobile: +(91)-9829214731/9414053658, http://www.indiamart.com/golden-industry

160. Good Fruit Grower, 105 South 18th Street, Suite 217, Yakima, Washington 98901, Ph 509-853-3520, 1-800-487-9946, Fax 509-853-3521, http://www.goodfruit.com

161. Goodwill Agrotech, 12-13, Wariana Industrial Complex, Block-2, Leather Complex Road, Jalandhar - 144013, Punjab, India, Phone:91-98-77023041 Mobile: +919814141031, +919877033041, http://www.goodwillagrotech.com/

162. Granite Engineering & Manufacturing Co Pty Ltd, 13 Reynolds St, Mareeba, QLD 4880, Australia

163. Green Land Engineers Chowk Civil Hospital , Circular Road Daska, Sialkot - 51010 - Pakistan Tel: +92-52-6619164 - 6611064 FAX: +92-52-6602864 E-mail: myvip0786@yahoo.com Web: http://www.glecpvt.co.cc

164. Green Planet Machines Pvt. Ltd., 35, Rani Jhansi Road, Bearing Market, Jhandewalan, New Delhi: 110055 ,Ph : +91-011-23522807, +91- 09311124407, 09868154407 Fax : +91-011-23523807, Email : info@greenplanet.in Web Site : http://www.greenplanet.in , http://www.gardenlawnmachines.com, http://www.mulcher.com/

165. Green View, No. 4, Ambika Arcade, Ambernath Road, Lokhandi Pada, Near Savalkar, Chouk, Old Panvel, Navi Mumbai - 410 206, Maharashtra, India , Mobile: +(91)-7498121590 Tel +(91)-(22)-27481208, Fax: +(91)-(022)-27455047 http://www.indiamart.com/greenview/query.html

166. Greenfingers Trading Limited, 10 Lindsay Square, Deans Industrial Estate, Livingston, West Lothian, EH54 8RL , Ph. 0845 345 0728 http://www.greenfingers.com/

167. Grimme Landmaschinenfabrik GmbH & Co. KG, Hunteburger Str. 32, 49401 Damme, Germany, Phone +49 5491 666-0, Fax +49 5491 666-2297, E-mail: grimme@grimme.de, Internet: www.grimme.de

168. GRO-MOR Inc.281 West Mountain Rd. PO Box 44 Adams, MA 01220, E-mail: gromor@verizon.net , Phone: 413-743-2064 , Fax: 413 743 2935, http://www.gro-morent.com

169. Growell Hydroponics & Plant Lighting Ltd. , http://www.growell.co.uk/

170. Gursukh Agro Works, Vill Jhakroudi, Samrala, Ludhiana, 141114, India, Ph. +91-9646000555, 9417100150

171. Guru Nanak Agriculture Implements Mfrs.G. T Road Bye Pass, Near Focal Point, Jalandhar, Punjab - 144 001, India , Tel: +(91)-(181)-2600740/ 2602670/ 2600040,Email:gurunanakagri1974@gmail.com, http://www.indiamart.com/gurunanak-agriculture-implements/query.html

172. Guru Nanak Agriculture India (P) Ltd., NH - 26 Lalitpur Road, Babina, Jhansi, Uttar Pradesh, India , 284001, , Ph : +91-510-2740649Fax : +91-510-2740649 / 2740704 Mobile : +91-9415412037Web Address : Email info@gnagro.in Site : http://www.gnagro.co

173. Guru Nanak Iron & Steel Mfg. Co., Old Railway Road, Jalandhar - 144008, Punjab, India , Mobile :+919814440963, http://www.genco.co.in

174. H & W equipment for vineyard & winery, Niagara on the Lake, ON phone: (905) 468-5016, fax: (905) 468-5676, e-mail: info@vineyardmachines.com

175. Haifa Chemicals Ltd, http://www.haifa-group.com

176. Haith Tickhill Group Companies, Vegetable grading machinery. http://www.haith.co.uk/grading .html.

177. Ha-Ko Enterprises - Manufacturing. , Plot 1 - A Cama Indl. Estate, Service Road, Off. I. B. Patel Road, Goregaon (East), Mumbai - 400 063, Maharashtra, India. , Tel: : 91-22-26852596 / 6570 4946 Fax : 91-22-26864217, response@lawncareindia.com , http://www.lawncareindia.com,

178. Hamilton Equipment, (Honey manure speader), 567 S. Reading Road, PO Box 478 ,Ephrata, PA 17522-0478, USA, Phone: 717 733 7951, Fax: 717 733 1783,Email: sales@haminc.com, http://www.honeyspreaders.com/contact.html

179. Hanan, J.J. et.al. 1978. Greenhouse Management. Springer-Verlag Berlin Heidleberg.

180. HANNA instruments, Michigan, 3820 Packard Road, Suite 120, Ann Arbor, Michigan 48108, Phone: 800 504 2662, Fax: 734 971 8155, laboratory@hannainst.com, http://www.hannainst.com/usa

181. Hare Krishna Rural Life, http://hkrl.com/gardening.html

182. HCL Group, 702 Omahu Road, Riverlands, Blenheim 7274, New Zealand, Ph + 64 3 579 2377, Fax + 64 3 579 2378, www.hclgroup.co.nz

183. Helen Yemm, How to grow: a perfect lawn, Seven basic tips for growing a magnificent lawn, The Telegraph,16 Jun 2009

184. Helper Robotech Co. 109-3 Yuha-ri, Jangyu-myeon, Gimhae-si, Gyeongsangnam-do 621-834 Korea, Phone: 82-55-3145858, Helper Robotech www.helpersys.co.kr, http://helpersys.en.ec21.com

185. Henan Fenglu Machinery Co., Ltd. , Zhengzhou City,China (Mainland) Tel:86 371 87513089, Fax:86 371 65863109, http://www.ttnet.net/ttnet/gotoprd/MN510/999/0/151303139343439313.htm

186. Hire-IT, 2 Rose Street, Melbourne, Victoria ,Melbourne, Australia, Ph 03 9848 7022, http://www.hire-it.com.au/

187. Hitachi Koki Co., Ltd., Shinagawa Intercity Tower A,15-1, Konan 2-Chome, Minatoku, Tokyo 108-6020, Japan, Phone: (81) 03-5783-0616, Fax: (81) 03-5783-0620, E-mail: Engine@Hitachi-Koki.Co.Jp, http://www.tanaka-usa.com

188. Hitachi Power Tools / Tanaka Power Equipment, 3950 Steve Reynolds Boulevard, Norcross, GA 30093, Toll Free Phone: 1-800-829-4752, Toll Free Fax: 1-800-830-7593 Email : tanakaorders@hitachi-powertools.com

189. Holland Transplanter Co. 510 E 16th St. P.O. Box 1527 Holland MI 49422-1527, USA . Ph. 616-392-3579, 800-275-4482,Fax 616-392-7996 http://www.transplanter.com/

190. Honda Appointed Dealer , John B Backhouse Airmyn, Goole, East Yorkshire DN14 8LD Tel. 01405 762876 , http://www.honda-uk.com/tillers.htm

191. Honda Motor Co., Ltd, http://world.honda.com/

192. Honda Siel Power Products Ltd, 3 & 4/48, 2nd Floor, Enkay House,Malcha Marg Shopping Complex, Chanakyapuri, New Delhi - 110 021, Plot No. 5, Sector - 41 ,Greater Noida Industrial Development Area, Distt. Gautam Budh Nagar - 201 306,Uttar Pradesh, India, Tel : 91-120-234 1050-59, Fax No. : 91-120-234 1078-79 , E mail: ho.mktg@hspp.com, http://www.hondasielpower.com, http://www.hondasuperstore.com

193. Hoover Fence Co., 5531 McClintocksburg Rd., Newton Falls, Ohio 44444-9265, Ph. (330) 358-2335, Fax: (330) 358-2624 http://www.hooverfence.com/about/

194. HORTECH S.R.L. , Viale dell'Artigianato, 20, City: Agna , Province: (PD) , Italy, Zip: 35021 , Tel.: +39 049 9515369 , Fax: +39 049 5381080, E-mail: info@hortech.it ,Web: www.hortech.it .

195. Howard M. Resh, 2002, Hydroponic Food Production: A Definitive Guidebook of Soilless Food-Growing Methods , Woodbridge Press Publishing Co., Santa Barbara, California, U.S.A.

196. http://www.feucht-obsttechnik.de/

197. Hunter Industries,1940 Diamond Street, San Marcos, CA 92078, Tel: (1) 760-744-5240, Fax: (1) 760-744-7461, http://www.hunterindustries.com

198. Hydralada Company Limited , PO Box 352, 19 Liverpool Street Hastings 4215, Hawke's Bay, New Zealand, Ph + 64 6 873 0000, Fax + 64 6 879 4172, www.hydralada.com, http://www.napierwebdesign.co.nz/hydralada

199. Hydroshop , Israel, Ph.. 972-3-7166663, http://www.420hydro.com/aboutus.asp

200. Hymatic Agro Equipment Pvt. Ltd., A-64, Sector-2, Noida - 201 301, Uttar Pradesh, India, Tel: +91-120-2520552, 2522051, Fax: +91-120-4320552, siddharth@hymatic.in , http://hymaticagro.com

201. Hymax Agro India, Mitushi Enterprises ,F-168/C, Rajouri Garden, New Delhi -110027 , (India), Phone +91-11-25194836 , Fax +91-11-25110341, Email mitushi_enterprises@hotmail.com, ktrehan@hotmail.com, http://www.hymaxagroindia.com/

202. IBS Engineering Supplies, 163 Edith Street, Innisfail, Qld 4860, Australia, Ph: 61 7 4061 3188, Fax: 61 7 4061 1214, http://www.ibseng.com.au, admin@ibseng.com.au

203. Ideal System Co., Ltd.358-14, Galsan-dong, Dalseo-gu, Daegu 704170 Korea Phone 82 - 53 - 5861800 ,Fax 82 - 53 - 5857219 Web site, www.idealsys.co.kr , http://www.tradekorea.com/product-detail/P00005210/Pin_Method_Grafting_ Robot.html

204. IMAC S.R.L., Via G.Giordani, 17 - 40054 Budrio (Bologna) – Italy, Tel: (+39) 051-8-80-37 Fax: (+39) 05- 80-80-11, E-Mail: info@imac-rondelli.it Web Site- www.imac-rondelli.it

205. Independent Lawn Mowing Contractors of Australia, PO Box 2351, Dandenong Nth 3175, Ph. 1800 501 662 , 1800 501 770, http://www.indmowing.com.au/

206. Indian Institute of Horticultural Research, Hessarghatta Lake, Bangalore-560089, India,Ph. 080-28466353, Fax 08028466291, director@iihr.ernet.in , http://www.iihr.res.in

207. Indian Institute of Spices Research, Marikunnu P.O., Calicut, Kerala, India – 673012, Ph. 0091- 495 2731410Fax: 0091- 495-2731187 ,director@spices.res.in, http://www.spices.res.in

208. Indira Gandhi Agricultural University, Krishak Nagar, 492006, Raipur, CG, India, http://igau.edu.in

209. Indo-German Agricultural Sprayers & Pressing Works , Near Chand Cinema, GT Road Ludhiana-141008( India), Tel: 0091161749507 /749508 Fax: 0091161745366 , E-Mail: gardentools@kisan76.com , www.kisan76.com

210. Institute of Japan Association for Techno-innovation in Agriculture, Forestry and Fisheries (JATAFF-Institute) , 446-1 Ippaizuka, Kamiyokoba, Tsukuba, Ibaraki 305-0854, Japan, Tel: +81-29-838-2196; Fax: +81-29-838-2337, http://www.afftis.or.jp/nouki/ki3b3h.htm

211. Intellimac, Optimal Lab Technologies, 5, I Cross Street, Thirumalai Nagar, Perungudi, Chennai - 600 096, India. Phone: 91-9710544699 John@intellimac.com , http://www.intellimac.com/

212. Interlink Pumps and Sprayers Pty Ltd.,681 Fifteenth St, Mildura, Victoria, Australia, Ph: (03) 5022 1133, Fax: (03) 5021 3496, matt@interlinkag.com.au, http://www.interlinksprayers.com.au/

213. International Crops Research Institute for the Semi-Arid Tropics (ICRISAT),Patancheru 502324 , Andhra Pradesh, India ,Tel: +91 40 30713071, Fax: +91 40 30713074,E-mail: icrisat@cgiar.org , http://www.icrisat.org/vasat/learning_resources/crops/groundnut/gnut_prodpractices/html/mod8les3/resources/2328.html

214. International Potash Institute , Coordinator India.c/o Dead Sea Works Ltd., POB 75 Beer Sheva 84100 Israel. ipi@ipipotash.org

215. International Potash Institute, IPI, Baumgärtlistrasse 17,P.O. Box 260,CH-8810 Horgen, Switzerland, Tel:+ 41 43 810 49 22, Fax:+ 41 43 810 49 25, http://www.ipipotash.org/en

216. Irrigation-Mart, Inc., 300 South Service Road East, Ruston, Louisiana 71270-3440 Ph.800-SAY-RAIN (729-7246), Email:info@irrigation-mart.com, http://www.irrigation-mart.com

217. IRROMETER Company, Inc. , P.O. Box 2424, Riverside, CA 92516-2424, or 8835 Philbin Ave.Riverside, CA 92503, Phone: 951-689-1701, Fax: 951-689-3706, sales@IRROMETER.com, techsupport@IRROMETER.com , http://www.irrometer.com

218. ISEKI & Co. Ltd.,https://www.iseki.co.jp/english/index.html

219. Iseki Noki, Co. Ltd., Kazuno Akita http://ss.gircas.affrc.gu.jp/engpage/jarq/34-1

220. Ishii Products, Round shape fruit grading system. http://www.ishii-ind.co.jp/english_pages/product.htm .

221. J.S.P. Lawn Mowers and Garden Tools, Guru Kirpa colony, Village Jaspal Banger Ludhiana,Punjab, India., Phone: +91 161 3245 896, Mobile: +91 98151 66720, http://www.jspgardentools.com

222. Jai Hind Agriculture Industries, Roorkee Road, Opp Pal Dharmshala, Muzaffarnagar Uttar Pradesh, India , Phone:91-131-2442344, +919412842344, http://jaihindagricultureindustries.tradeindia.com

223. Jain Irrigation Systems Ltd. Jain Plastic Park, Jalgaon - 425 001. Maharashtra , India. Tel : +91 257 225 8011 ,Fax : +91 257 225 8111 , http://www.jains.com

224. Jameson, 1451 Old North Main Street, PO Box 1030, Clover, SC 29710, Ph. 803-222-6400 800-346-1956, http://www.jamesonllc.com

225. Jaulent Industrie, 1475, chemin des poulidets, 82000 Montauban, Tel : +33 (0)5 63 67 81 84 ,Fax : +33 (0)5 63 67 80 30 ,http://www.jaulent.com/en/

226. Jayshreedies, No. 22/B, Revabhai Estate, CTM, Ahmedabad - 380 026, Gujarat, India, Tel. +91-79-25857821, Fax. +91-79-25857832, Website: www.indiamart.com/jayshreedies, www.indiamart.com/jayshreedies

227. JBT FoodTech , 70 West Madison, Suite 4400, Chicago, IL 60602, USA , Phone: +1 312 861 5900, Fax: +1 312 861 5897 , Email: jbtcorporation@jbtc.com , http://www.jbtcorporation.com

228. John Deere Landscapes, Deere & Company World Headquarters, One John Deere Place Moline, Illinois 61265, http://www.deere.com , https://www.johndeerelandscapes.com/

229. John, P Jacob, A Handbook on Post Harvest Management of Fruits and Vegetables, Daya Publishing House, Delhi.2008

230. Jyotirling Agro Pvt. Ltd. 16/26, Opposite Kolhapur Steel, Nagaon Road, Kolhapur - 416122, Maharashtra, India, Phone :91-231-2469642, Fax :91-230-2468937, http://jtorlingpvt.tradeindia.com

231. K.S.Worldwide Exports Pvt. Ltd. Chandigarh Road, Near Octroi, Nichi Mangli, Focal Point. City :Ludhiana, Punjab,India :141010 Phone +91-161-5000412 / 3243003 Fax :+91-161-2672299, Email : kulbir2649@gmail.com , Web Site:http://www.ksexport.co.in

232. Kalbag Nandan, Horticultural Consultant, 201/4, Cherish Homes, Swami Vivekanand Nagar, Virar (West) - 401303., Tel. : 99757 82285, 0250-2503082, Email : ntkalbag@hotmail.com & info@gardentia.net, http://www.gardentia.net

233. Kalsi Agros, Samrala Road, Opp Markfed Godown, Khanna - 141401, Punjab, India, Phone:91-1628-227790, http://kalsiagros.tradeindia.com

234. Kamdhenu Engineering Company, H - 40, Additional Industrial Area, M.I.D.C., Ambad, Nashik - 422 010, Maharashtra, India, Tel. No. : +91 - 253 - 2380085, Telefax : +91 - 253 - 3053762, Email : response@kamdhenuengineering.com, http://www.kamdhenuengineering.com

235. Kamsun Engineering Industries , 1/16, 2nd Street,Venus Colony Alwarpeth , Chennai Tamil Nadu , India - 600018 , Ph. +91-44-24328479 Email : kamsunind@yahoo.co.in ,Web Site : http://www.kamsun.com

236. Kanhaiyalal Ramratan Krishi Yantra Laghu Udyog, N.H. 11, Rampura Dabari, Jaipur-Chomu Road, Jaipur (Rajasthan) INDIA , Phone: +91-1423-224485 , 224685, E-mail: info@krkylu.com Website: http://www.krkylu.com

237. Karshni Traders, Shanti Complex, Shop No. 2, Shingaar Cinema Road City, Ludhiana, Punjab, India Phone +91-181-6530001+91-9780309656 , Email sidh_aggarwal_20@hotmail.com ,Website http://www.indiapowersprayer.in

238. Kennco Manufacturing, Inc. 1105 3rd Street NE, Ruskin, FL 33570, Phone: (813) 645-2591, Toll Free: (800) 645-2591, Fax: (813) 645-7801, http://www.kenncomfg.com

239. Kent Tractors Limited, Mincing Court Farm, Duck Lane, Shadoxhurst, Nr Ashford, Kent TN26 1LT,Telephone +44 (0) 1233-732439 Fax +44 (0) 1233-733944, Mobile +44 (0) 78606-42295 E mail andrew@kenttractors.com, http://www.kenttractors.com

240. Kerian Machines Inc., 1709 Hwy 81 S, PO Box 311 , Grafton, ND 58237 USA , Phl:(701) 352-0480, Fax: (701) 352-3776, Email: sales@kerian.com, http://www.kerianmachines.com/contact.html

241. Kevin McNamara & Son Pty. Ltd,Tennis Court Designers , Fact 30/513-515 Maroondah Highway, Ringwood VIC 3134 , Australia, Ph: 03 9876 8474 ,Fax: 03 9876 8082, Email: info@asltennis.com, http://www.tenniscourts.com.au

242. Khedut Agro Engineering, Plot No.6, Survey No.191, Shantidham Society Road, Near Orke Farm, 8-B, National Highway, Veraval (Shapar).Ta. Kotada Sangani, Dist. Rajkot. (Gujarat). Phone No. : +91-02827-253312. +91 94262 06420 Email : khedut.agro@ymail.com dinesh_patel63@yahoo.com , http://khedutagro.com

243. Kisan Engineering Works, C-57, M.I.D.C., Awadhan, Dhule, Phone no. – 02562 239812 / 9422289812, kisanengg239812@yahoo.co.in

244. Kishan Agro Industries, Gandhi Chamber, Gondal Road,Opp. Bombay Petrol Pump,Rajkot – 360002, Gujarat - India. +91-9825079989, +91-281-2388131, 2388135, 2241016, info@kishanagro.com , www.kishanagro.com

245. Kittas C., Boulard T., Papadakis G. 1997 , Natural ventilation of a greenhouse with ridge and side penings. Sensitivity to temperature and wind effects. Transactions of the ASAE, 1997, Vol 40, No 2, 415-425.

246. Kittas C., Papadakis G., Giaglaras P. 1998, Heat and mass transfer processes in the greenhouse ecosystem. International Conference on Agricultural Engineering, Oslo, August 24-27 1998. Papers published in CD-ROM.

247. Kittas C., Papadakis G., Giaglaras P. 1998, Heat and mass transfer processes in the greenhouse ecosystem. International Conference on Agricultural Engineering, Oslo, August 24-27 1998. Papers published in CD-ROM.

248. Kongskilde Juko Ltd Oy , Opintie 4 23100 MYN, Country : Finland , Phone : 2.4393200 Fax No : 2.4393210 , E Mail : mail@ksf.kongskilde.com

249. Korvan Industries Inc., Oxbo International Corporation, 270 Birch Bay Lynden Road, Lynden, WA 98264, USA.Tel: + 1 360 354 1500, Fax: + 1 360 354 1300, E.mail address: korvan@korvan.com ,Website: http://www.korvan.com

250. Kovai Classic Industries, No. 1, Iyer Hospital Road, Singanallur Post, Tamil Nadu, Ph. 0422-2270012

251. Kubota, Co., Ltd., Takeda, Kumamoto, http://ss.jircas.affrc.go.jp/engpage/jarq/34-1

252. Kverneland Group, N-4355 Kvernaland ,Norway ,Phone:+ 47 51 42 94 00 , + 47 51 42 94 21, Fax: + 47 51 42 94 01, E-mail: contact@kvernelandgroup.com, http://ien.kverneland.com

253. KW Automation, P.O. Box 121, Old Bowling Green Road, Palmwoods, Queensland, Australia, Phone: +61 (07) 5445 9549, Fax: +61 (07) 5445 0499, http://www.kwautomation.com.au/

254. Lall Behary Dey & Company, No. 44, Hidaram Banerjee, Lane, Ground Floor, Kolkata, West Bengal - 700 012, India , Tel. +(91)-(33)-22374123, +(91)-(33)-22251233, M +(91)-9830797831, Website : http://www.mullicktools.com

255. Land Power Agrotech Private Limited, G. T. Road, Near Barnala Chowk, District, Moga, Punjab - 140 001, India, Tel: +(91)-(1636)-226633, http://www.indiamart.com/land-power-agrotech/potato-digger-harvester.html, Website : http://www.landpowerindia.com

256. Land- und Kommunaltechnik, Am Fuchsgraben 5b, D-77880 Sasbach,Tel: +49(0)7841/6068-0, Fax: +49(0)7841/6068-10, eMail: mail@lipco.com, http://www.lipco.com/english/products/TUNNEL_Sprayers/tunnel_sprayers.htm

257. Lee Shuknecht & Sons, 4458 Ford Road,Elba, NY 14058, USA , Ph- 585-757-6628 Fax- 585-757-9395, http://www.shuknecht.com/

258. Lidselmash, 70, Sovetskaya str., 231300, Lida, Belarus, l. + 375 154 52-76-82 , b. + 375 29 608-75-66, http://en.lidselmash.by

259. Littau Harvester, 855 Rogue Avenue, Stayton, OR 97383 , Tel:(503)769-5953, 1-866-262-2495, Parts:(503)767-5953, jared@littauharvester.com, http://www.littauharvester.com,

260. Lloyd Engineering, The Highland Works, Stansbatch, Leominster, Herefordshire, UK, HR6 9LL, http://www.screeningbucket.co.uk

261. Lona Lawn Mowers Pvt. Ltd., New LonaMowers, Kallukadav Road, Pulinchodu, lwaye, Kerala, India 683101, Phone: (91)-484-2626574 , Email: sales@lonamowers.com ,http://www.lonamowers.com/about

262. Ludhra Overseas, Babu Jagjivan Ram Chowk, 120 Feet Road, Basti Guzan, Jalandhar - 144 002, Punjab , India, Tel:+(91)-(181)-2208671 Mobile:+(91)-9779053671, Website : http://www.ludhratools.com

263. M&M Control Service, Inc., Ph. 800-876-0036, 847-356-0566 Tel 847-356 - 0747 sales@mmcontrol.com , http://www.mmcontrol.com

264. Macarthur Mowers & Marine , 16 Argyle Street, Camden NSW 2570, Tel: 02 4655 5575, Fax : 02 4655 5578 , Email: sales@macmm.com.au

265. Machinery & Engineering Ltd. , P.O. Box 1342 , Pardess-Hanna, 37110 , Israel , Tel.: +972-6-6371203 , Fax.: +972-6-6370246 , e. mail: contact@afron.com , Web Site: http://www.afron.com

266. Machinery Group Co., Ltd. No. 328, Xinye Road, Changzhou, Jiangsu, China. 213012Tel.:+86-519-83256666,Fax:+86-519-83260303,+86-519-83261722,Website: http://www.df-tractor.com, sales@df-tractor.com , http://www.dfamgc.com

267. MAFELL AG, Beffendorfer Strasse 4, D-78727 Oberndorf a.N. , Germany, Tel. +49 7423 / 812-0 , Fax +49 7423 / 812-218 ,E-Mail: export(at)mafell.de, http://www.mafell.de

268. Mahendra Agricultural Industries, Delhi Road, Baraut - 250611, Uttar Pradesh, India, Mobile: +919917547789, +919456035890, +919411066559 http://mahendraagri.tradeindia.com/grass-digging-khurpa-1034803.html

269. Mahesh Patel, Bungalow No. 1, Satellite Complex, Prem Chand Nagar Road, Opposite Mansi Tower, Ahmedabad - 380 015, Gujarat , India , Tel : +91-79-26732095, +91-79-26753501, Website : http://www.nifindia.org/

270. Malhotra, Vinay 2007, Implements Industrial Cluster at Moga, Proceedings Tractor & Farm Machinery Manufacturers ' Meet, Nov. 16-17,2007, Central Institute of Agricultural Engineering, Bhopal.

271. Malviya Agrico,18, Industrial Esate, Rani (Rajasthan) – 306115,India, Phone - +91 2934 222133, Telefax - +91 2934 222215, http://www.malviyaagrico.com

272. Mamta Iron Works, Nursery Road, Mahadev Nagar, P.O. Devsar, Billimora - 396380, Gujarat, India, Ph :91-2634-283549 Fax: 91-2634-284445, Mobile:+919825119779 http://mamtaironworks.tradeindia.com/

273. Mandhar S.C. et.al (1999) Design and fabrication of greenhouse for commercial production of high value crops, Floriculture Today 5(5) : 11-20.

274. Mandhar, S.C. et.al.1997, Engineering Factors for Design of Greenhouses, Proceedings of International Seminar on Protected Cultivation in India. Dec. 18-19, Bangalore: 182.

275. Market Farm Implement - 257 Fawn Hollow Road - Friedens, Pa. 15541Ph. (814 )-443-1931, http://www.marketfarm.com/

276. Martignani Ing. C. di Martignani Stefano & C. S.R.L., Via E. Fermi, 63, Zona Industriale Lugo 1,48020 S.Agata sul Santerno (Ra),Italy, Tel. +39 0545 23077 - Fax +39 0545 30664, e-mail: infoweb@martignani.com , http://www.martignani.com

277. McConnel Limited, Temeside Works, Ludlow, SY8 1JL, England Phone: +44 (0)1584 873131 Fax: +44 (0)1584 876463, sales@mcconnel.com, http://www.mcconnel.com

278. Mears, D.R. 1997. Multispan Low to Medium Cost Greenhouse Facilities Design - Fundamentals. Proceedings of International Seminar on Protected Cultivation in India. Dec.18-19, Bangalore:5-16

279. Mechanical Transplanter Co. 1150 Central Ave. Holland, MI 49423, USA, Ph: (616) 396-8738 Fax: (616) 396-3619 Toll free: 1-800-757-5268 email: mtc@mechanicaltransplanter.com , http://www.mechanicaltransplanter.com

280. Mehar & Mukand Tools.K-96-97, Phase-VII, Focal Point, Ludhiana - 141 010, Punjab , India , Tel :+91-161-2676351, Website http://www.mmtsgardentools.com

281. Metal Weld Engineering Pvt. Ltd. No. 58, F. B. Scheme, No. 78, Vijay Nagar, Indore, Madhya Pradesh - 452 010, India, Tel +91-731-2802888, Fax +91-731-2802888Email: metalweld111@gmail.com, http://www.indiamart.com/metalweldengineering

282. METechs Corp.,Mapleton Plaza at Essa Road, 555 Essa Rd, Barrie L4N 6A9,Cana da;Ph. 1-800-705-6269,Fax: 1-705-999-9715, http://www.metechs.com

283. Meyer's Equipment Manufacturing Corporation, 601 West Center Avenue, Dorchester WI 54425 ,Ph: 715-654-5200, Fax: 715-654-5558, E-mail: emcdorchester@tds.net , http://www.emcspreaders.com

284. Michael Raviv and J. Heinrich Lieth, 2008 , Soilless Culture: Theory and Practice,Elesevier, Amesterdam, The Netherlands.

285. Michael. A.M. and T.P.Ojha, 1987, Principles Of Agricultural Engineering Vol. I , Jain Brothers, East Park Road, Karol Bagh,New Delhi, India

286. Midwest Vineyard Supply, Inc, Contact: Mike Switzer, 2300 S Twin Bridge Road, Decatur, IL 62521 Ph: (217) 864-9896, Fax: (217) 422-7793, sales@midwestvineyardsupply.com , http://www.midwestvineyardsupply.com

287. Millcreek Manufacturing Company, 34 Zimmerman Road, Leola, PA 17540, U.S.A. Ph: 1-800-311-1323 / Fax: 717-656-0952, http://www.millcreekmfg.com

288. Millennium Landscape and Lawncare Inc., 2583 W. Plains Rd, Mason, MI. 48854, millenniumland@acd.net, http://www.millenniumland.com/

289. Ministry of Commerce and Industry, Government of India. http://commerce.nic.in.

290. Miraj Steels Pvt. Ltd, Relience Estate, Zalod Road, Dahod, 389151 Gujarat , India, Ph. 02673 222410, http://rajb2b.com/yp/48696-miraj_steels_pvt_ltd.htm

291. Mississippi State University, Caution with hooded sprayers, http://msucares.com/crops/cotton/sprayer.html

292. Mittal Engineering Works, 28, Bagpat Road, Mittal Nagar, Meerut. (U.P.)- 250 002 INDIA, Ph: +91-121-2513256 Fax: +91-121-2513289, E-mail: info@mew.co.in, http://mew.co.in

293. Modern Industries, 52, Industrial Estate, Hoshiarpur-146 001, Punjab, IndiaPh : 91-01882-251636, Fax : 91-01882-250236, E-mail : moderninds@gmail.com , Website : www.moderninds.com

294. Monosem Inc. , 1001 Blake St.,Edwardsville, KS- 66111,Ph. 913-438-1700, http://www.monosem-inc.com/

295. Mori Team Machinery Co., Ltd.No.209, Shen An Road , Chang Hua, Taiwan , Tel: +886-4-7373668, Fax: +886-4-7371618 , http://www.garden-machine.com

296. Motobloki, ChP, R-N Zhulyany, St. Sadovaya, 58Kiev, Ukraine, Tel.: +380 (97) 1278441, +380 (93) 9525395,http://www.ua.all.biz/en/g941917

297. Mudhar Allied Traders, C 126, Maya Puri Ind. Area, Phase II,New Delhi – 110064 Tel. : 28115482, Cell. : 9810026399, E-mail : mudharnp@yahoo.co.in , http://www.mudharallied.com

298. My World: The Fastest Leaf-Blower on the Plateau, http://seniorhiker.wordpress.com/2009/11/10/my-world-the-fastest-leaf-blower-on-the-plateau/, October, 1,2012.

299. NaanDanJain Irrigation (C.S.) Ltd.,Post Naan 76829, Israel. T:+972-8-9442180, F:+972-8-9442190,E: mkt@naandanjain.com www.naandanjain.com

300. National Agro Industries, Link Road, Industrial Area-A, Opp. Transport Nagar, Ludhiana - 141003. India ,Ph. +91 - 161 - 2222041, 5087853, Fax +91 - 161 – 4621299, E-mail : sales@nationalagro.com, Website : www.nationalagroinds.com, www.agriculturalimplements.com

301. National Committee on Plasticulture Applications in Horticulture (NCPAH),10th Floor, Internationa,l Trade Tower, Nehru Place, New Delhi - 110 019,Tel. : 011 – 46511275, Fax : 011 - 46508501, Email : ncpah@ncpahindia.com , ncpahindia@yahoo.co.in ,web : http://ncpahindia.com ,

302. National Greenhouse Manufacturers Association of USA. Standards - Design Loads in Greenhouse Structures, Ventilating and Cooling Greenhouses, 1994

303. National Innovation Foundation, India, PO Box 15051, Vastrapur, Ahmedabad 380 015, Gujarat , Telephone: +91-79-2673 2456, Fax: +91-79-2673 1903, info @nifindia.org, http://www.nif.org.in

304. National Research Centre for Grapes, PB No. 3, Manjri Farm Post, Pune (Maharashtra) 412 307, Ph. 020-26956002 Fax. 020-26956099, dirnrcg@gmail.com, http ://nrcgrapes.nic.in

305. National Research Centre for Seed Spices, Tabiji, Ajmer (Rajasthan) 305 206, Ph. 0145-2443238, Fax. 0145-2681001, nrcss.director@gmail.com, http://www.nrcss.org

306. Nawab Ali (2007), Farm Mechanization: Status, Policy and Issues,Proceedings of Tractor and Farm Machinery Manufacturer's Meet- Nov. 16-17, 2007 Central Institute of Agricultural Engineering, Bhopal.

307. Neptune Sprayers (Agro Division) , Neptune Packaging Pvt. Ltd. ,78, Janki Nagar Main, Near Jain Mandir, Indore -452001, Madhya Pradesh, India , Phone: +(91) -(731) - 4278662 Fax: +(91) -(731) -4215662 , Email: info@neptunesprayers.com , URL: www.neptunesprayers.com

308. Newhouse Manufacturing Co., Inc. , 1048 North Sixth Street, Redmond, OR 97756 , (541) 548-1055, (800) 587-1055 , newhouse-mfg.com , http://www.newhouse-mfg.com

309. Ningbo Beilun Daqi Hengxing Machinery Co., Ltd. Company , Room 1602, The Pacific Mansion, No. 500 Mingzhou Road, Beilun, Ningbo, Zhejiang, China City/Province: Ningbo/Zhejiang Country/Region: China Tel. : 86-574-86890832 Fax : 86-574-86960836 Mobile: 86-13567889720 http://sunwell.en.made-in-china.com

310. Northern Tool + Equipment, 2800 Southcross Drive West, Burnsville, Minnesota 55306, USA, Ph: 952-894-9510, Fax: 952-894-1020 , http://www.northerntool.com

311. Northwest Rentals, 8026 Jacksboro Highway , Fort Worth, Texas 76135, Ph. 817-237-6767 , Fax 817-237-1121, Email: northwestrentals@sbcglobal.net, http://www.northwestrentals.net/mb

312. Oesco, Inc , P.O. Box 540, Route 116 ,Conway, MA 01341 , USA, Ph. 800-634-5557, 413-369-4335 , Fax 413-369-4431 , info@oescoinc.com , http://www.oescoinc.com

313. Ohio State University, 2003. Yield, berry quality, and economics of mechanical berry harvest in Ohio. http://www.ohioline.osu.edu/ rc299/rc299_8.html on 25.7.2003.

314. Olimpias S.A. , Machinery and Packaging Lines Classification of Fruits and Vegetables, Lappa Achaias - 35 km New National Road Patras - Pyrgos, 27 052 , Ph: +30 26930 31828 , Fax: +30 26930 31768 , Email: info@olimpias.gr , http://olimpias.gr

315. Omson House, Model House Road, Basti Sheikh, Jalandhar - 144 002 India, Tel: +91-181-243 0440, 5083280,Fax: +91-181-243 2318, E-mail : info@omsonsgroup.com , http://www.montstar.com/contact.html

316. Omson House, Model House Road, Basti Sheikh, Jalandhar - 144 002 India, Tel: +91-181-243 0440, 5083280,Fax: +91-181-243 2318, E-mail : info@omsonsgroup.com, http://www.montstar.com/contact.html

317. One John Deere Place, Moline, Illinois 61265, http://www.deere.com

318. Onions Potatoes Mechanization, info@onions-potatoes.com , http://www.onions-potatoes.com/ harvest/ harvest-mechanized.php,

319. OPICO Limited, Cherry Holt Road, Bourne, Lincolnshire, PE10 9LA ,United Kingdom, Tel: 01778 421111 , Fax : 01778 425080, 01778 391663 , Email: ask@opico.co.uk, Web: www.opico.co.uk

320. Orchard Machinery Corporation , 2700 Colusa Highway, Yuba City, CA 95993, Ph. (530) 673-2822, (530) 673-2822, Fax (530) 673-0296 , http://www.shakermaker.com

321. Orchard Valley Supply, Inc., 1521 Mountain View Drive, Quakertown, PA 18951, Ph: (888) 755-0098, Email: info@orchardvalleysupply.com , www.orchardvalleysupply.com,

322. Oregon, Lakeview Vineyard Equipment Inc. 40 Lakeshore Rd. RR#5, Niagara-on-the-Lake, Ontario Canada L0S 1J0 , Ph.: 905-646-8085 , Fax: 905-646-3289 , Toll Free: 1-866-677-4717, http://www.lakeviewvineyardequipment.com, www.gregoireharvesters.com/

323. Outsidepride.com, Inc., 915 N. Main, Independence, OR 97351, Ph: 800-670-4192 ,Fax: 503-606-0659, support@outsidepride.com

324. Oxbo International Corp., 7275 Byron Rd, Byron NY 14422. http://www.oxbocorp.com

325. P. Tuckwell Ltd, UK,Shop Street,Worlingworth,Suffolk, IP13 7HU , ,http://www.tuckwell.co.uk

326. Pacific Northwest National Laboratory, 902 Battelle Boulevard,Richland, WA, Ph. 1-888-375-PNNL (7665), http://picturethis.pnl.gov, http://picturethis.pnl.gov/picturet.nsf , http://www.pnl.gov

327. Padgilwar Agro Industries, No. 192, Wardhaman Nagar, Behind Inox Cinema, Nagpur - 440 008, Maharashtra, India , +(91)-(727)-6777770, Mobile, +(91)-9823052421,Fax, +(91)-(712)-2763478, padaagri@rediffmail.com, www.padagro.com

328. Pandey M. M.,1997, Farm Machinery Research Digest, All India Coordinated Research Project on Farm Implements and Machinery, Central Institute of Agricultural Engineering, Bhopal.

329. Panesar Agriculture Industries, Bathinda Road, Near Handiaya Chowk, Opposite Power House, Barnala, Punjab - 148 101, India, Telephone: +(91)-(1679)-284084/ 284484, E mail. panesarcombine@sify.com http://www.panesarindia.com/,

330. Papadakis G., Briassoulis D., Scarascia G., Vox G., Feuilloley P. Stoffers J.A., 2000, Radiometric and thermal properties and testing methods for greenhouse covering materials. Journal of Agricultural Engineering Research, 2000, 77(1), 7-38.

331. Papadakis G., Manolakos D., Kyritsis S., 1998, Solar radiation transmissivity of a single span greenhouse through measurements on scale models. Journal of Agricultural Engineering Research, 1998, 71, 331-338.

332. Papadakis, G.,A. Frangoudakis, S. Kyritsis Mixed, 1992, Forced and free convection heat transfer at the greenhouse cover, Journal of Agricultural Engineering Research, Volume 51, January–April 1992, Pages 191-205

333. Paras Engg. Works, New G.H 6/112, White Apartment, Paschim Vihar. Delhi, India, 110087, Ph. +91-011-25270107/28363083, +91-9810472529 , Fax +91-011-28363083, Email. paras.lawnmover@gmail.com ,Web Site : http://www.paraslawnmower.com

334. Patricia Imas, Recent Techniques in Fertigation of Horticultural Crops in Israel, Presented at the IPI-PRII-KKV Workshop on: Recent Trends in Nutrition Management in Horticultural crops, 11-12 February 1999. Dapoli, Maharashtra, INDIA

335. Paul A. Lauria Enterprises Inc., Garden City Park, NY 11040, Email: NoPests@MosquitoZapper.com , www.leafmulcher.com

336. Paul V. Nelson, et. al.1978. Greenhouse operation and management. Reston publishing company, One, Reston, Virginia, USA.

337. Peach Ridge Orchard Supply, Inc., 8405 Fruit Ridge Avenue, Sparta, MI 49345 Phone: 1-800-452- 6748, Fax: 616-887-8239, E-mail: sales@peachridge.com, http://www.peachridge.com/pruning/pruning/

338. Peal Engineering, Vegetable handling systems. http://www.pealeng.com/graders

339. Pellenc Australia Pty Ltd, 14 Opala St,Regency Park SA 5010, Phone : (08) 8244 7700, Fax : (08) 8244 7788,Email : admin@pellenc.com.au , http://www.pellenc.com.au

340. Pellenc SA, Route de Cavaillon BP 47, 84122 Pertuis cedex, France , Tél : 33 (0)4 90 09 47 00, Fax : 33 (0)4 90 09 64 09 , www.pellenc.com, mail : pellenc.sa@pellenc.com

341. Perfect Agro Industries, B-11. Focal Point Extension, Jalandhar City. Punjab India, Tel : +91-181-2602652, 2601147.Fax: +91-181-5001222. Email : perfect@perfectagro.com

342. Peterson Donald Dr. , Working toward mechanical harvest for tree fruits, Good Fruit Grower / January-15th 2006, http://www.goodfruit.com/Good-Fruit-Grower/January-15th-2006/Working-toward-mechanical-harvest-for-tree-fruits/

343. Peterson, D.L. 1998, Mechanical harvester for process oranges. Applied Engineering in Agriculture, 19(5) L 455-458.

344. Philip George, Philipji.com, http://www.philipji.com/item/2012-01-16/coconut-climbing-machine Phone: (503) 678-5525 , (877) 678-5525 , (503) 678-5693 , http://gkmachine.com ,http://vimeo.com/gkmachine, http://vimeo.com/8459694

345. Piedmont Tobacco Equipment Co.Inc., 5870 Hwy. 268, Dobson, NC, USA 27017 , (336) 368-4600 , http://www.piedmonttobaccoequipment.com

346. Pillon Srl, Via della meccanica, 2/4 Viale Dell'Artigianato, 36042 Breganze (Vi), Tel. +39 0445 308159, Fax. +39 0445 300480, E-mail: info@pillonsrl.it, sales@pillonsrl.it, http://www.pillonsrl.it

347. Popular Steel Works,1325/1, Shivaji Udyamnagar, Kolhapur - 416008, Maharashtra , INDIA, Phone 91+231+2657376 , 2657424 Fax 91+231+2657988, Email:- psw@psw.co.in , kenyapsw@psw.co.in , popularpsw@sancharnet.in , kpr_popular@sancharnet.in, Website:http://popularpsw.in

348. Porwal Coconut Machines, Sy. No.47,48 (1), 9th K.M., B.H.Road, Mallasandra, Tumkur – 572107, Karnataka, India . +91-97415-76054 Fax: +91-816-2206555/2206505, http://porwalcoconutmachines.com/, info@porwalcoconutmachines.com ,

349. Power Tools Direct, PO Box 3361, Littlehampton, BN16 9GQ, Tel: 0845 8380 661, Email: sales@powertoolsdirect.com , http://www.powertoolsdirect.com

350. Premier Irrigation Adritec Pvt Ltd. 17-1C, Alipore Road ,Kolkata–700027,West Bengal, India Phone: (033) 40121111,Fax:(033) 40121155, E-mail: sales@pial.in , http://www.pial.in/

351. Primco Industries, M-79, Mohan Garden, Nr. Pipal Wala Chowk,New Delhi - 110059, India, Phone:91-11-25267081, Fax:91-11-25283202/25251242, Mobile:+919810681614, +919250031500

352. ProGreen Plus, 7130 Oak Ridge Highway, Knoxville, TN 37931, (865) 235 1799, coveryourgrass@ progreenplus, http://progreengrass.com/leaf-blower-maintenance-routines/

353. Progress, Prairie Farmer, 812 Kelley Dr. O'Fallon, IL 62269, Tel: 618/726-7528, jflint@farmprogress.com, http://farmprogress.com/story-going-under-hood-to-treat-pigweed-0-53490

354. PUROZONE cc.,Unit 1d, River Park, 77 De Waal Road, Diep River, Cape Town, Ph. (021) 705 3079 Fax (021) 706 3078, http://www.purozone.co.za

355. Quality Deer Management Association, PO Box 160, Bogart, GA 30622, 1-800-209-DEER (3337) , http://www.qdma.com

356. R - Tech Agri Equipments, Old No. 2 / 95, New No. 2 / 232, Bombay Nagar, Narasimhanaicken Palayam, Coimbatore - 641 031, Tamil Nadu , India , Tel: +(91)-(422)-2460296, Mobile : +(91)-9944284440, http://www.rtechagri.com

357. R P Overseas,Khingra Gate, Adda Hoshiarpur, Jalandhar, 144001 (PB), India,Tel: 0091 181 2442449 E-mail: sales@rpoverseas.com , http://www.rpoverseas.com

358. Rain Bird Corporation, 970 West Sierra Madre Ave., Azusa, CA 91702, Ph. (626) 812-3400, http://www.rainbird.com

359. Rajasthan Mechanical Works Ltd. A-188(B) Road No. 6-D V.K.I. Area, Jaipur - 302 013, (Rajasthan, India).Phone : +(91)-(141)-2331549/2332239/2231654, Fax : +(91)-(141)-2332631, E - Mail :brahmpuri@hotmail.com, brahmpuri@sancharnet.in, http://www.brahmpuri.com

360. Ramco International, Dhogri Road, Near Pathankot Bye Pass, Jalandhar - 144 004, Ph. +91-181-2611707, 2611708, Fax:+91-181-2611511, 2290554, E-mail : ramco@jla.vsnl.net.in, ramco@gardentoolsindia.com, http://www.gardentoolsindia.com, http://www.indiamart.com/ramco-int/

361. Ramsay Highlander , 45 Gonzales River Rd., Gonzales, CA 93926, Phone: (831) 675-3453, Toll-Free: (800) 777-2314, Fax: (831) 675-2127 E-mail: info@ramsayhighlander.com , http://www.ramsayhighlander.com

362. Rapid Tools Direct, 1 Leeds Road, Huddersfield, West Yorkshire, United Kingdom, HD1 6DG, Tel: 0800 1583551 ,Fax: 01484 480265, Email: info@rapidtoolsdirect.co.uk , http://www.rapidtoolsdirect.co.uk

363. Rapid Tools, 3218 Rainbow Court, Burlington, Ontario, Canada L7M2T8, Ph: 905-336-0553 , Fax: 905-336-9303 , Web: www.rapidtools.net , Email: rapidtools@cogeco.ca

364. Ray's Supply Company, 5120 Old Haywood Road, Mills River, NC 28759, Ph: (828) 891-9200, Fax: (828) 891-9248, Rayssupplycompany.com

365. REMU Ltd , Inhantie 7, 63700 Ähtäri, Finland , Tel: +358 20 7431 160, Fax: +358 20 7431 161

366. REMU USA Inc.8 Hemlock St Old Orchard Beach ME 04064, Maine, USA, Tel: +1 888 600 0018, Fax: +1 207 934 4879, usa@remu.fi, http://www.remu.fi/ , info@remu.fi

367. Renaldo Sales & Service - Niagara Energy - Pioneer Propane, 1770 Mile Strip Road, North Collins NY, 14111 USA, Ph. 800-424-5564, Outside US: 716-337-3760, Fax: 716-337-2756, http://www.renaldo.org/

368. Richard.C.F., Thomas.E.W. and Joseph. C.S., 1998. Yield, berry quality and economics of mechanical berry harvest in Ohio. Bulletin Research, The Ohio State University, U.S.A: 1-22.

369. RK Sprayer Equipments P Ltd, No.4-5, JDA Complex, Fruit Mandi, Narwal, Jammu, Jammu-180001 , 01912476429

370. Royer Industries Inc. 6856 Howlett Road, PO Box 3001 , Oshkosh, WI 54902 , Corporate Office (920) 688-2999 , Service (888) ROYER02 , Facsimile (920) 688-3087 , www.royerind.com , royer@royerind.com

371. R-Tech Agri Equipments, #2/95, Bombay Nagar, Narasimhanaicken Palayam,Coimbatore-641031., Tamil Nadu,INDIA Mobile : +91-9944284440, +91-7845574423 Email : rtechagri@gmail.com , sreevardhan@treeclimber.in , info@treeclimber.in, web : www.treeclimber.in

372. S R Agrotech, 1st Main Road, Vinayaka Nagar Tumkur, Karnataka - 572 101, India, Telephone : +91-816-2277084, Fax : 91-816-4021049, Mobile :+91-09448866543

373. Sagar Engineering, I. G. Road, Chikmagalur, India, 577101, Phone : +91-8262-230883, Fax : +91-8262-230883, Mobile : +91-9900341841, Email : sagar.coffeemachinery@gmail.com , Web Site : http://www.sagarengineering.com

374. Sandbrook MacRae Ltd, Pohangina Valley West, RD 14, Ashhurst, New Zealand , Tel: +64 6 329 4794, Fax. +64 6 329 4784 , info@seedspider.com , http://www.seedspider.com/

375. Sanei Industry Co.Ltd . , Hokkaido,Japan, ymiura@sanei-ind.co.jp , http:// www.sanei-ind.co.jp /english/index eng.html,

376. SAPLANTCO SDN BHD , (wholly owned by Sawit Kinabalu) , Lot 104, Kian Yap Industrial Estate, Phase 3A, Mile 5 1/2,Jalan Tuaran, Lorong Durian 5, 88450 , Kolombong, Kota Kinabalu,Sabah , Malaysia

377. Sareko Tools & Forgings, D-20, Focal Point, G. T. Road, Bye Pass, Jalandhar - 144 004, India , Tel: 0091-181-2602266, 5053166, Fax: 0091-181-2602166, E-mail : info@sarekotools.com , www.sarekotools.com

378. Saubhagya Agro Equipments Pvt. Ltd. Plot No. 2 & 3, Survey No.549, Dhandha, Idar Road, Ahmedabad - 383001, Gujarat , Phone : 91-79-222562/233563 Mobile : +919825070662 Fax : 91-79-222562 ,

379. Scova Impianti s.r.l. via Emilia Ovest, 83/d, 41013 Castelfranco Emilia (Modena) ,Tel: +39 059 923202 Fax: +39 059 920358 , Email: info@scovaimpianti.com , http://www.scovaimpianti.com

380. SFM Technology, 9 Bancombe Court, Martock, Somerset, TA12 6HB, Tel: 01935 822285, Fax: 01935 826199, http://fruitharvesting.com/contact.php

381. Shakermaker.com, 2700 Colusa Highway, Yuba City, CA 95993, (530) 673-2822 (Office), (530) 673-0296 (Fax) http://www.shakermaker.com

382. Shanghai Huawei Water Saving Irrigation Co., Ltd., Shanghai, China

383. Shanghai Sheng Qian Industry Co., Ltd , No. 888, North Sichuan Road, Shanghai, P. R. China, 200080, Phone:86-21-61558216 , Fax: 86-21-23025819, Mobile:18221501250

384. Shindaiwa, 400 Oakwood Road, Lake Zurich, IL 60047-1564, Toll Free: (877) 986-7783, Phone: (847) 540-8400, Fax: (847) 540-9670, http://www.shindaiwa-usa.com

385. Shiva Engineers, Patel Avenue, Plot No.165, Flat No.1, Right Bhusari Colony, Nr. Kothrud Bus Depot, Paud Road, Pune - 411038, Maharashtra, India , Phone:91-20-27129610/27127104/25284631, Fax:91-20-27127104/25284631, http://www.tradeindia.com/selloffer/3107428/Vegetable-Washer.html

386. Shiva Industries, Mandipet Tumkur, Karnataka,India,572101

387. ShopWiki Corp . 17 West 18th Street, 8th Floor, New York, NY 10011, privacy@ShopWiki.com , http://www.shopwiki.co.uk/ , http://www.shopwiki.com/

388. Shriram Temp-X-Changers (India), 991/2A G.I.D.C. Makarpura Industrial Estate, Vadodara, Gujarat, India 390010, Tel : 91-265-2645-129, Fax : 91-265-2638-504 , Web Site : http://www.fruitprocess.com

389. Sigma Agricultural Machines, A-2,Industrial Estate , Ludhiana -141003(India ) , Ph: 91-161-2531545 Fax No: 91-161-2531843

390. Silkwood Wines, Pemberton Winery , Channybearup Road, Pemberton WA 6260, Western Australia , www.silkwoodwines.com.au

391. Singh Gyanendra ,( 2006) Agricultural machinery industry in India (Manufacturing, marketing and mechanization promotion),Status of Farm Mechanization in India 2006 -Indian Agricultural Statistics Research Institute, (ICAR), New Delhi

392. Sitaram Industries, 132, Karnavati Park,Opp. New Baroda Express,Near CTM Cross Road, Amraiwadi, Ahmedabad, Gujarat - 380026, India. Mobile: +(91)-90990 20145 / 98250 20145 parag.sitaram@gmail.com, http://www.sitaramindustries.net

393. Skybird Agro Industries , 72, Azad Nagar, Harjit palace, 100 feet road, Amritsar - 143006 (India), Phone: +91-183-2589114 ,Tele/Fax: +91-183-2583614, Email:info@skybirdagro.com , skybirdagro@hotmail.com , www.skybirdagro.com ,

394. Smart! Fertilizer Management, 15 Eshkol St. Hod Hasharon 45343, Israel, Tel: +972-523-597964 ,Fax: +972-777041348, http://www.smart-fertilizer.com

395. Smolsky ,M. (2012). How to Plant a Lawn, http://www.homeandgardenideas.com/ outdoor -living/lawn-care/seeding/how-plant-lawn, Oct., 5,2012.

396. Snehal Plastic, Ravi Randal Krupa, Opposite Lathiya Motor, Near Amardip Welding, Panchshil Society, Rajkot, Gujarat - 360 003, India , Tel: +(91)-(281)-6454134, http://www.indiamart.com/snehal-plastic/

397. Sormac B.V., Huiskensstraat 68 , 5916 PN Venlo , The Netherlands , P.O. Box 419, 5900 AK Venlo , The Netherlands , T: +31 (0)77 351 84 44 , F: +31 (0)77 320 48 01, E: info@sormac.nl , http://www.sormac.co.uk/en/contact

398. South Texas Ag Research, Rio Grande Valley, http://www.star-texas.com

399. Southern Agro Engine Private Limited , 24A, III Phase, Industrial Estate, Guindy, Chennai - 600 032, India. , Ph : 91 - 44 - 22500378 / 22501325 / 22500133 / 22501134 , Fax : 91 - 44 – 2250039, Email : marketing@southernagro.com , exim@southernagro.com , Website : www.southernagro.com

400. Spagnol SRL, Greenhouse Technologies, Via Paludotti, 5/7,31020 Vidor (TV) Italy Tel. +39 0423 986083, Fax. +39 0423 987984, Email: info@spagnolaut.com, http://en.spagnolaut.com/contacts.php

401. Spapperi S.R.L. , Zona Artigianale , San Secondo (PG) , Italy, Tel. (+39) 075.857.81.56 Fax (+39) 075.857.88.48, http://www.spapperi.it

402. Spectrum Electrostatic Sprayers Inc., 7500 San Felipe, Suite 860, Houston, Texas 77063 Phone: (713) 783-5771 , Fax: (713) 783-5772 , email:spectrumsprayers@comcast.net , http://spectrumsprayer.com

403. Spedo F.Lli Di Spedo Pietro Danilo & C S.A.S., Via Polesine, 119 - 37043- Tel: +39-0442-92025 - P.Iva 00307540237 - info@spedo.it, http://www.spedo.it/en/

404. SpeeCo, Inc., 15000 W. 44th Ave, Golden, CO 80403, Ph.1-800-525-8322, speeco@speeco.com , http://www.speeco.com

405. Speed Link, Tel: +92 321 2390 002, http://www.tractors.pk/potato-digger-spinner.html

406. Spices Board of India (2009), Ministry of Commerce and Industry, Government of India. http://www.indianspices.com, January 27,2012.

407. SPL Technologies (P) Ltd. Lower Ground: 77-80, Shiv Vihar (West), Vikas Nagar,New Delhi-110059 Ph. +91-11-25352643, 25352644, 25352645,Fax_011-25352649 E-mail info@spltech.in Website http://www.spltech.in

408. Spraying Systems Co., PlotNo 303&304, 4th Phase, 5th Main, 2nd Cross, Peenya Industrial Area, Karnataka, Bangalore 560 058, India, Phone: 080 39853200, Fax: 080 39853388 Email: ssipl@sprayindia.com http://www.spray.com

409. Sri Narayana Engineering Industrisies (P)Ltd, P B No 2, Shoranur, Palghat District, Pin- 679122 ,Kerala, India, Tel : 0466 – 2220439, Tel : 0466 – 2220343 E-mail :info@snei.in , sneisrr@dataone.in , http://www.snei.in/gardening-tools.html

410. Stavley Machine & Tools Company, Near Unison Engineering Industries, Pathankot, Bye Pass Chowk, Jalandhar, Punjab - 144 012, India Tel: +(91)-(181)-2610873, Mobile: +(91)-9814100078/ 9876300254 Fax: +(91)-(181)-2610873 Email: ksunison@hotmail.com, stavley1@hotmail.com, stavley@gmail.com, http://www.stavley.in,

411. STIHL Incorporated,Oceana West Industrial Park, 536 Viking Drive ,Virginia Beach VA 23452 , USA , Tel.: +1 757 4869100 , Fax: +1 757 4869324 , Internet: www.stihlusa.com

412. STIHL Pty. Ltd., 5 Kingston Park Court , 3180 Knoxfield Victoria ,Australia , Tel.: +61 39 2156666 , Fax: +61 39 7634423, E-Mail: csc@stihl.com.au , Internet: www.stihl.com.au

413. STIHL, http://www.stihl.co.nz/STIHL-Products/Olive-Harvester/Olive-Harvester/21610-1595/SP-481.aspx, info@stihl.co.nz.

414. Stock Photo - A grape harvester in a vineyard on the southern highlands of New South Wales, Australia, http://www.123rf.com/photo 4631215 a-grape-harvester-in-a-vineyard-on-the-southern-highlands-of-new-south-wales-australia.html

415. Subhash Agro & Allied Industries , 18, Waryana Indl Complex, Kapurthala Road, Jalandhar - 144021, Punjab, IndiaPhone: +91-181-2227970/2226078, Fax:91-181-2253653,Mobile:+919814063078, +919814265455, http://grassslasher.tradeindia.com/bill-hook-with-handle-357821.html

416. Sucha Singh & Bros. Chandigarh Road, Samrala, Punjab - 141 114, India , Tel +(91)-(1628)-262361, Mobile +(91)-9815488361Email: suchasinghvirdiagritech@gmail.com , http://www.indiamart.com/sucha-singh-bros/query.html

417. Sue Courtney , 2001, Riding the Harvest - Vintage 2001 at Kerr Farm, Wine of the Week, 25 April 2001 http://www.wineoftheweek.com/stories/0104kerrharvest.html

418. Sunkist Research and Technical Services, Fontana Sunkist Research and Technical Services Center , 10509 Business Dr. Unit B , Fontana, California 92337, U.S.A. (800) 383-7141 , http://www.sunkist.com/ , http://www.sunkistresearch.com

419. Supreme Name Co., Ltd., 11F, No. 208, Min Sheng 1st Rd., 800, Kaohsiung , Phone 886-7-2292798 , Fax 886-7-2294783 Websitewww.progo-sports.com/

420. Sutton Agricultural Enterprises, Inc. ,746 Vertin Avenue, Salinas, Calif. USA 93901, Ph: 866/280-6229, or 831/422-9693, Fax: 800/482-4240, or 831/422-4201, info@suttonag.com , http://www.suttonag.com/HarvestStar.html

421. Sweere / Sweere Food Processing B.V. Postbox 4, 4730 AA Oudenbosch, The Netherlands, Ph : (+31) (0) 165 315 536, Fax (+31) (0) 165 310 246, Email: : info@sweere.net, t.sweere@sweere.net , .http://www.sweere.net/

422. Tainan District Agricultural Research & Extension Station, COA, 70 Muchang Hsinhua, 712 Tainan, Taiwan, ROC, TEL:(06)5912901 FAX:(06)5912928, E-mail: tndais@mail.tndais.gov.tw, http://www.tndais.gov.tw/briefing/briefing93_english_16.htm

423. Tamil Nadu Agricultural University (2010) http://www.tnau.ac.in, on March,05,2010, Agricultural Engineering Technologies Catalogue, Agricultural Engineering College & Research Institute, Coimbatore.

424. Tamil Nadu Agricultural University (TNAU) , Coimbatore, http://www.tnau.ac.in/

425. Tanzania Automotive Technology Centre (Tatc), Private Bag, Nyumbu, Kibaha, Tanzania, Ph : +255 732 923390, Fax: +255 73-292-1230, E mail: dg@tatcnyumbu.org, http://www.tatcnyumbu.org

426. Taskpower, PO Box 2025, 118 Vanguard Street, Stoke, Nelson 7041, New Zealand , Ph: 0800 200 744, 1800 820 150 , +64 3 5483544,Fax: +64 3 5483547, Email: sales@taskpower.co.nz, http://www.taskpower.co.nz/

427. Tata Steel Ltd (Agrico Business & Retail Initiatives ), 43, J.N Road Tata Centre 3rd Floor Kolkata-700071 , Email :tvshenoy@tatasteel.com, http://www.tataagrico.com

428. Tea Board India(2009), Ministry of Commerce and Industry, Government of India. http://teaboard.gov.in, January 27,2012

429. The 2011 Manure Expo in Nebraska, http://www.cornnation.com/2011/5/10/2163094/the-2011-manure-expo-and-why-you-dont-want-to-miss-it ,

430. The Blackmore Company, 10800 Blackmore Avenue, Belleville, MI 48111 U.S.A.Ph. U.S. & Canada (800) 874-8660, International 1+(734)-483-8661, Fax (734) 483-2387 TrayInfo@blackmoreco.com , http://www.blackmoreco.com/

431. The Canadian Outdoor Equipment Co., 199 Lakeshore Rd E, Unit A, Mississauga, OntCANADA, Ph. (905) 990-1750, http://www.canadianoutdoorequipment.com

432. The Ferguson Manufacturing Co., P. O. Box 1098, Suffolk VA 23439, 590 Madison Ave., Suffolk VA 23434, Ph. (757)539-3409 Fax (757)934-3612 , fermfgsu@nettek.net , http://www.fergusonmfgco.com/feedback.htm

433. The Reseach Magazine for the Oregon State Univesrsity Experiment Station, Winter 2001, Oregon State University, http://oregonprogress.oregonstate.edu

434. The Toro Company, 8111 Lyndale Avenue South, Bloomington, MN 55420, USA, USA 1-888-384-9939, Canada 1-800-544-5364, http://www.toro.com

435. The Toro Company, Irrigation Division, 5825 Jasmine Street, Riverside, CA 92504 Tel 1-877-345-8676, http://www.toro.com

436. The University of Arizona, (2012), Vegetable Grafting, http://cals.arizona.edu/grafting /grafting-robots on Feb. 25,2012

437. The University Of Georgia, Onion Production Guide, http://www.caes.uga.edu/applications/publications/files/pdf/B%201198-2 2.PDF

438. The Woodsmiths Store, Tel: + 44 (0) 191 252 4064 , Mobile: 07802 571 641 , enquiries@woodsmithstore.co.uk ,http://www.woodsmithstore.co.uk

439. Thstore.com, 1700 Embassy Dr, West Palm Beach, Florida 33401, USA, Ph: (561) 686-5289, thstore.com

440. Tickhill Engineering Co Ltd, Cow House Lane, Armthorpe, Doncaster, South Yorkshire, DN3 3EE, U.K., Tel: +44 (0) 1302 831911, Fax: +44 (0) 1302 300173, http://www.haith.co.uk/potato-grading-equipment.html

441. Tirth Agro Technology Pvt. Ltd., National Highway- 8B Nr. Bharudi Toll Plaza, At Bhuvana 360311, Ta Gondal , Distt Rajkot, Gujarat, 09227927962, 09227927992, http://www.shaktimanagro.com

442. Tol Inc. , 955 E. Levin / P.O. Box 1498 , Tulare CA 93274 / 93275 , USA , Phone 1: +1-559-6862844 , Phone 2: +1-800-732-2142 , Fax.: +1-559-685-1006, e. mail: contact@tol-inc.com , Web Site: http://www.tol-inc.com

443. Tong Engineering Ltd, Ashby Rd, Spilsby, Lincolnshire, PE23 5DW, Telephone: 01790 752771, Fax: 01790 753611, Sales: sales@tongpeal.com , http://www.tongpeal.co.uk

444. Tools Agricultural Production and Trade Ltd , Eker-Mak. Sti. , 1.Km. Haymana Way 06 900 Polatli / Ankara , Tel: 0 (312) 625 52 90, Fax: 0 (312) 625 52 94 , eker-mak@eker-mak.com.tr , http://www.eker-mak.com.tr/

445. Toolstop Ltd, UK, http://www.Toolstop.Co.Uk/Tulip-Hand-Shovel-With-Small-Wooden-Handle-P2134

446. Top-Power-Tools.com, http://www.top-power-tools.com

447. Transmax Engineering Industries, 9, Sahkar Vijay Industrial Estate, Navghar, Vasai East, Mumbai - 401 202, Maharashtra, India , Ph. +(91)-8373905361+(91)-(250)-2392379 +(91)-(250)-3200114, +(91)-(250)-2391571, Mobile: +(91)-9892183692, http://www.transmaxindia.com/enquiry.html

448. Transplant Systems Pty Ltd, 63 Enterprise Avenue, Berwick, Victoria,3806, Tel: 03 97699733, Fax: 03 97699722, Email: info@transplantsystems.com.au, http://www.transplantsystems.com.au

449. Trials Equipment UK Ltd. Hudson's Hill, Hedingham Road, Wethersfield, Braintree, Essex CM7 4EH, UK Tel: +44(0)1371 850793 , Fax: +44(0)1371 850905 , E mail: Sales@Trialseq.Co.Uk , http://www.Trialseq.Co.Uk/ Equipment_Engineering.Php

450. Truax Company, 4300 Quebec Avenue North,New Hope, MN 55428, Ph. (763) 537-6639,Fax (763) 537-8353 , Email: truax3@qwestoffice.net , http://www.truaxcomp.com

451. Tuthill Temperley, Wardington House, Banbury, Oxon, OX17 1SD, England, Tel: +44 (0) 1295 750513, Fax: +44 (0) 1295 750036, http://www.tuthilltemperley.com/

452. Uma Agro Industries, C-4, Rajendra Industrial Estate, Paldi Three Cross, Kheralu Road, Visnagar - 384315, Gujarat, India Phone:91-2765-232225 ,Mobile : +919825047440 , http://www.tradeindia.com/fp485398/Reversible-Plough.html

453. Universiti Putra Malaysia, Serdang, Selangor, Malaysia

454. University of Florida, Gainesville, Institute of Food and Agricultural Sciences, Citrus Research & Education Center, 700 Experiment Station Road, Lake Alferd, FL 32611; Tel. (352) 392-3261, (863) 956-1151, Fax. (863)956-4631, http://www.ufl.edu/, www.lal.ufl.edu/history

455. URBINATI, zona ind. Via della Liberazione, 18 - I-47030 San Mauro Pascoli (FC) - Italy , Tel. (+39) 0541/932018 - Fax (+39) 0541/932679 http://www.urbinati.com

456. Valley Fabrication Inc., 1056 Pellet Ave., Salinas, CA. 93901 or P.O. Box 3618 Salinas, CA 93912, Phone: 831-757-5151, Fax: 831-757-0701, http://www.valleyfabrication.com/

457. Vegcraft, McDonald Contractors Ltd. ,Crafts Farm, Carmyllie, By Arbroath, Angus, DD11 2RQ, Scotland, Ph, +44 (0) 7779 664034, 1241 860204, 1241 860318, Fax 1241 860204, Email malcolm@ vegcraft.mcdonaldcontractors.co.uk, http://www.vegcraft.mcdonaldcontractors.co.uk,

458. Vegetable Harvesting Systems (VHS), Town Drove, Quadring, Spalding, Lincolnshire, UK, PE11 4PU, Tel: +44 (0) 1775 821222, Fax: +44(0) 1775 820914, Email: info@vhsharvesting.co.uk , http://www.vhsharvesting.co.uk/contact_us.php

459. VescoPlastics, PO Box 47 Virginia 9430 South Africa , Ph. +27 57 212 00 00 Fax +27 57 212 66 66, http://www.vesconite.com

460. Vine Tech Equipment , 335 North Gap Road, Prosser WA 99350, USA,Office: (509) 788-0900, Fax:(509) 788-0901,jackm@vinetechequipment.com, http://www.vinetechequipment.com

461. Vishnu Sankar, (cocnut climber)Tirunelveli, Tamilnadu, India

462. Vishvkarma Machine Tools , 63, Mahesh Nagar (Near Panch Dev Mandir & Gurudwara), Ambala Cantt - 133 001. Haryana. India, Phone : +91-171-4000763,2661763(O), Fax : 91-171-2662122, Email : vmtvices@yahoo.com , http://fruitgrading.net/

463. Visser International Trade & Engineering B.V. Postoffice box 5103 ,3295 ZG 's-Gravendeel, The Netherlands ,Main office,Mijlweg 18,3295 KH 's-Gravendeel,The Netherlands , Ph: +31 (0) 78 673 9800, Fax: +31 (0) 78 673 3434, E-mail: sales@visserite.com http://www.visserite.com

464. VST Tillers Tractors Ltd., Post Box No. 4801, Whitefield Road,Mahadevapura , Bangalore - 560 048. India, Ph: +91 - 80 - 28510805 / 06/ 07 | +91 - 80 – 67141111, Fax: + 91 - 80 – 28510221, Email: vstgen@vsttillers.com, Website : www.vsttillers.com

465. Wave Impex, 7 Indra Market, Gill Road, Ludhiana-141003(Pb.) India, Ph. + 91-161-2531000 Fax +91-161-2537777 , Email mail@waveimpex.com Website www.waveimpex.com

466. Wayne's Farm Machinery,4804 Taylorsville Road, Taylorsville, KY 40071, 502-477-2561 , E- mail wcox@shelbybb.net , http://www.waynesfarmmachinery.com

467. Weed Control Products and Tools, Weed Pullers, Weed Poppers and Consumer Shopping Resources, http://www.ergonica.net

468. Weening Brothers Mfg Inc., 3475 Hwy 88, RR # 1, Bradford, Ontario , Canada, L3Z 2A4 Ph: (905) 775-3839 , Fax: (905) 775-8769 , Email: sales@weeningbrothers.com , http://www.weeningbrothers.com

469. Weifang Langpak Machinery Co.,Ltd Company, No.222 Dongfeng East Road, Kuiwen District, Weifang City, China, Phone:86-0536-6103062 Fax:86-0536-6103062 Mobile:15965062380 Website:http://www.walking-tractor.com

470. Weifang Xiaer Imp.& Exp.Co., Fushou West Street, City weifang , State Shandong Country, China, 261021 , Website http://dannylee007.en.busytrade.com

471. Welsh Onion Transplanter , 1988, Journal of The Japanese Society Of Agricultural Machinery: 60(4) http://www.afftis.or.jp/nouki/ki3b3i.htm

472. WesSpur Tree Equipment , 1680 Baker Creek Place, Bellingham, WA 98225 , Toll Free (800) 268-2141 Regular (360) 734-5242 Fax (360) 733-6311 www.wesspur.com

473. Whitney, J.D.1999, Field last results with mechanical harvesting equipment in Florida orange. Applied Engineering in Agriculture 15(3): 205-210

474. Wikco.Com Inc., 1467 N Grant Ave., Casa Grande, AZ 85122, 520-316-0446 Email : sales@wikco.com, 520-316-0446, http://wikco.com/

475. Williames Pty Ltd, Wills Street, Warragul Victoria 3820 Australia, Tel: +613 5623 5755, Fax: +613 5623 5165, Email: sales@whti.com.au , www.whti.com.au

476. WOLF-Garten Tools, E.P. Barrus Ltd, Launton Road, Bicester, Oxon OX26 4UR, http://www.wolfgarten-tools.co.uk/multi-change-adjustable-fruit-picker

477. WOLF-Garten, MTD Products Aktiengesellschaft, Geschäftsbereich , Industriestr. 23, 66129 Saarbrücken, Ph: +49 (6805) / 790, Fax: +49 (6805) / 79442 E-Mail: info@WOLF-Garten.com, http://www.wolf-garten.org

478. Worklift (UK) Ltd. , Elizabeth Farm, Thornford Road, Berkshire RG19 8EL England Phone: +44-1635-269333, Fax: +44-1635-269444 , e. mail: worklift.uk@virgin.net

479. Yammer Noki, Co., Ltd , https://www.yammer.com, http://www.whti.com.au/, http://brain.naro.affrc.go.jp/

480. Yogesh Industries,No. 34, Chawari Bazar, Delhi, Delhi, India - 110 006,India, +(91)-(11)-23274447 Mobile : +(91)-9968651268, http://trade.indiamart.com/details.mp?offer=2498663673

481. ZETA Agricultural Machines and Equipments, Tel: +420 566623086 , Fax: +420 566628913 , Mob: +420 602503168,Mob: +420 773638693 Email: zeta@zeta-zr.cz, http://www.zeta-zr.cz/

482. Zhejiang Zhongli Tools Manufacture Co., Ltd., Huchuan Industry Area, Jinyun, Zhejiang,China, http://www.made-in-china.com/showroom/grace955/companyinfo/Zhejiang-Zhongli-Tools-Manufacture-Co-Ltd-.html

483. Zimidara Agro & Allied Industries, Sarhind Road, Village-Baran, Patiala - 147001, Punjab, India , Mobile : +919872482548 , http://www.tradeindia.com/fp584291/Roto-Seed-Drill.html

484. Zwicker Land Design, Box 583,Claysville, PA 15323, Ph.(724) 663-7795, jeri@zwickerlanddesign.com, zwickerlanddesign.com

www.ingramcontent.com/pod-product-compliance
Ingram Content Group UK Ltd.
Pitfield, Milton Keynes, MK11 3LW, UK
UKHW041133071026
15587UKWH00009B/453

9 789351 302650